Hi-Fi Stereo Handbook

by

William F. Boyce

HOWARD W. SAMS & CO., INC.
THE BOBBS-MERRILL CO., INC.
INDIANAPOLIS • KANSAS CITY • NEW YORK

FIRST EDITION

FIRST PRINTING—1972

International Standard Book Number: 0-672-20918-7
Library of Congress Catalog Card Number: 72-84430

Preface

Since the first edition of this book in 1956, constant change and improvement in sound communication, storage, and reproduction techniques, and the development of low-cost, high-quality stereo compacts have made hi-fi stereo an economic reality for everyone. These remarkable developments are highlighted by such effects as precision dynamic coupling of pickup devices to much-improved mediums; transistorization; integrated circuits; miniaturization; combination of separate chassis into single integral units; acoustic-suspension, compact speakers; and four-channel systems, encoders, decoders, and synthesizers—all giving dramatically improved performance with more quality and realism at less cost than was hitherto available.

It has been more than a century since geniuses of music creation brought forth the need for mediums, techniques, and equipment to capture and store their great works. The world will forever lament that it was not able to record and store the work of these highly disciplined and talented artists of yesterday. At least from this point on, we can achieve this for our own age and for future generations.

Modern concepts and standards of high fidelity, or hi-fi, provide for the ultimate in the endeavor toward reproduction of sounds exactly as they were originally created. True fidelity, as you will learn in the first chapter, is only occasionally achieved, but modern recording techniques provide results with maximum control of source, bearing, vocal, instrumental, and other elements arranged with emphasis, presence, balance, and three-dimensional effects, to create a final mix that can be dramatically superior to live programs. The art of recording, storing, and reproducing sound electronically has progressed to the point where you, with eyes closed, can almost imagine yourself sitting fifth row, center at Carnegie Hall.

But once your appetite for high-fidelity surround-sound has been whetted, what then? What equipment should you buy? How much will it cost? And these considerations are only the beginning. This book was prepared as a reference and guide for all persons interested in high-quality sound reproduction. In addition to supplying information which will help you in planning, selecting, and installing appropriate systems, it also describes the various system components, plus what they do and how they operate. Thus, no matter whether the subject is new to you or you are a technician or an experienced hi-fi enthusiast, *Hi-Fi Stereo Handbook* has been written for you. I trust that you will find it informative and highly useful.

William F. Boyce

Contents

CHAPTER 1

FIDELITY, SOUND, AND DISTORTION 7

What *Is* High Fidelity? – Sound – Distortion of Sounds in High-Fidelity Systems – Performance Goals

CHAPTER 2

MONOPHONIC AND STEREOPHONIC SOUND 17

Monophonic Sound – Stereophonic Sound – Stereophonic Systems – Compatibility

CHAPTER 3

STEREO TECHNIQUES 32

Basic Components of Stereo Systems – Stereo Recording – Disc Recording – Tape Recording – Stereo Broadcasting and Reception – Stereo *Versus* High Fidelity?

CHAPTER 4

PROGRAM SOURCE EQUIPMENT 55

Record Players – Application of Electron Tubes, Semiconductors, and Integrated Circuits – Tuners – Tape Recorders – Microphones – Television

CHAPTER 5

AMPLIFICATION AND CONTROL 161

Power Amplifiers – Power Amplifier Circuits – Preamplifiers – Preamplifier Circuits – Stereo Amplification

CHAPTER 6

SPEAKERS 259

Speaker Drivers – Cone-Type Radiators – Speaker Impedance – Acoustic Impedance and Resonance – Horns and Horn Drivers – Directivity of Basic Units – Dual and Multiple Speakers and Systems – Construction Features of Speakers – Electrical Divider (Cross-over) Networks – Center Speaker for Stereo

CHAPTER 7

SPEAKER BAFFLES AND ENCLOSURES 305

Baffles – Simple Enclosures – Bass-Reflex Enclosures – Acoustic Labyrinth – Horn-Type Enclosures – Compact Hi-Fi Speaker Systems – Electrical Speaker Equalizers – Stereo Speaker Systems – Choosing a Speaker System – Speaker Listening Tests

CHAPTER 8

SYSTEMS DESIGN, SELECTION, AND INSTALLATION 345

Requirements – Audio Power – Fidelity – Frequency Requirements – Distortion – Gain – Building up a High-Fidelity System – Four-Channel Systems – Elaborate Arrangements – Systems Layout – Systems Installation – Feedback

INDEX 393

1

Fidelity, Sound, and Distortion

It is universally agreed among experts on high fidelity that there is as yet no exact scientific operational definition for a high-fidelity system. Standards and specified measurements of performance of a system have not been possible to establish because of limitations of the human ear and because of variations in human taste, room acoustics, system distortions, noise, and comparative volume levels.

WHAT IS HIGH FIDELITY?

A commonly accepted concept of high-fidelity sound is that it is reproduced sound with a high degree of similarity to the original or live sound. High fidelity is felt to be achieved when the sound that is reproduced has negligible distortion from the original, when it has little extraneous noise, and when the volume levels and room acoustical effects are pleasing to hear. This reproduced sound might even be more pleasing to the listener at the output of the system than the original live sound would have been if heard at its source.

A reproduction of sound is something like a photograph. The picture cannot carry the original scene to the viewer in every detail. Some features of the picture may be de-emphasized, whereas other features may be emphasized intentionally, or distortion may be introduced for purely aesthetic reasons. Distortions of this sort can greatly improve the illusion that the photographer is trying to create. In the same way, the picture can be spoiled by undesirable distortions and effects, such as poor focus, poor film, or improper lighting.

Like photography, modern high-fidelity techniques encompass controls for modification of the original (live) sound to compensate for certain defects and make provision to actually improve the effects according to an individual listener's tastes. Undesirable distortions, differences in comparative sound levels, and injection of extraneous noise are also held to a minimum so that the pleasing qualities of the original sound will not be reduced.

In addition, modern concepts of high fidelity take into consideration the listener, his ear mechanism, and his nervous response, plus his listening experience and training.

Psychophysical reactions and imagination contribute to the realism of high-fidelity reproduction. The word "presence" is used to describe the degree of realism of the reproduced sound. This term suggests that the reproduction is so real that the listener can feel the presence of the source that is causing the live sound, even though that source is many miles away or even extinct. Furthermore, psychologists have shown that the trained human mind will fill in missing sounds that should appear in a musical rendition, even though these sounds are not present in the reproduction.

The application of the term "high fidelity," then, is largely a personal matter. Everyone can be a hi-fi expert—at least as far as his own tastes in equipment and quality of sound reproduction are concerned.

SOUND

The word "sound" is used in different ways. In the psychophysical sense, "sound" means to the listener the sensation of hearing audible vibrations conveyed from any medium (such as air, usually) through the ear to the brain. As used in physics, however, "sound" means the external cause of the sensation. In hi-fi, we are concerned with both meanings.

Sources of sound are bodies in vibration. Vibrations of a low-note, bass-viol string can actually be seen. The sounds caused by such a source have only a few vibrations per second. They are therefore called *low-frequency sounds, low notes,* or *lows.* On the other hand, the tinkles of a glass or of a musical instrument such as a triangle have comparatively many vibrations per second and are said to vibrate at *high frequencies.* Such sounds are known as *high notes* or *highs* in the range of human hearing sensations.

Frequencies of sounds audible to the average person range between 20 and 20,000 hertz (cycles per second). The average human hearing system has certain characteristics of receiving, converting, and interpreting sound that are important considerations in the production of high-fidelity impressions.

Loud sounds are heard with good fidelity over a comparatively wide frequency spectrum. In other words, highs, lows, and in-between-frequency sounds are heard in proper relation to each other when all these sounds are loud; however, when the volume is reduced, the ear tends to attenuate (or cut down) the highs and lows but leaves the in-betweens proportionately louder. This is demonstrated by the curves shown in Fig. 1-1.

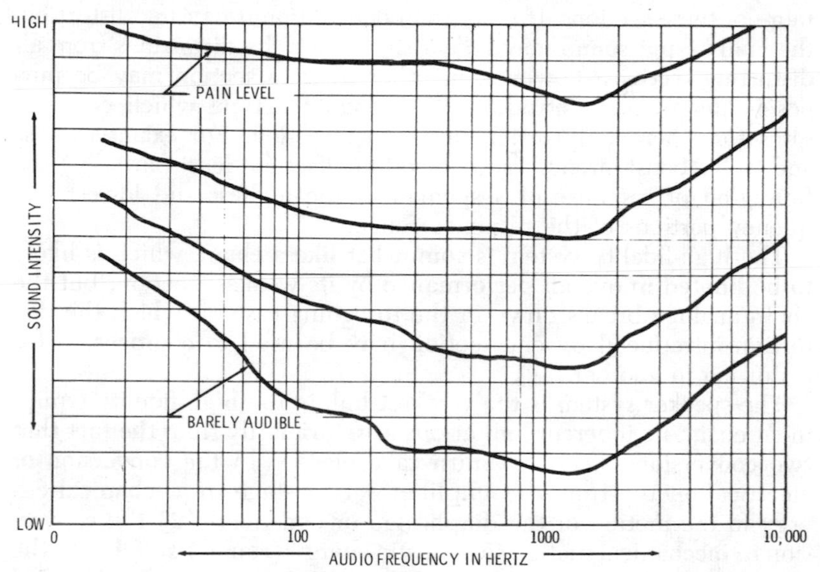

Fig. 1-1. Sound intensity required at different frequencies to produce uniform response in the sensory system of the ear.

The normal human hearing system has directional characteristics, receiving all sounds best from the forward position (source in front of observer) and with the ability to distinguish the direction of the source or of a reflection. The ear can distinguish several sounds of different frequencies at the same time to a high degree. It is possible for the ear to distinguish between sounds with frequency differences as low as 3 hertz and volume differences of 1 decibel. The *decibel* is the standard unit of measurement of the loudness of sound.

The ear can to some degree detect relative phase changes of sound. This means that a small increase or decrease in the frequency, or pitch, of one note in relation to the frequencies of other notes simultaneously played can be detected.

Requirements for the components of a high-fidelity system that can please many individual ears of various listening tastes are quite exacting because of the fineness of the mechanism of the human hearing system.

DISTORTION OF SOUNDS IN HIGH-FIDELITY SYSTEMS

The best high-fidelity systems are substantially less than perfect. The ways in which the audio-frequency output sound differs from an input or a desired ideal output sound are classified as *distortion*. A complete high-fidelity system may be divided into functional sections, as illustrated in Fig. 1-2. Distortion may be created in one or more of these sections. If more than one section is causing distortion, the final output sound may reflect the sum of the distortions from all distorting sections; however, in other cases, a section may be purposely designed to introduce distortion of a type which compensates for inherent distortion in another section. For example, bass and treble boost circuits can be used to offset (at least partially) the falling-off of response of a speaker at the highest and lowest frequency portions of the response range.

The high-fidelity system is somewhat like a chain, which is likely to be limited in overall performance by its weakest section; but the chain analogy breaks down in the foregoing case, in which the distortion introduced by one section may be used to compensate for distortion in another.

The speaker system is the weakest link in the high-fidelity equipment chain. Its inherent limitations arise primarily from the fact that two conversions of energy must take place: (1) the conversion of electrical energy from the amplifier output stage to mechanical energy in the motion of the diaphragm or cone, and (2) the conversion of mechanical motion to acoustic energy (sound) suitable to the listener's ear. Such energy conversion is known as *transduction*, and the devices which effect it are known as *transducers*. Input devices

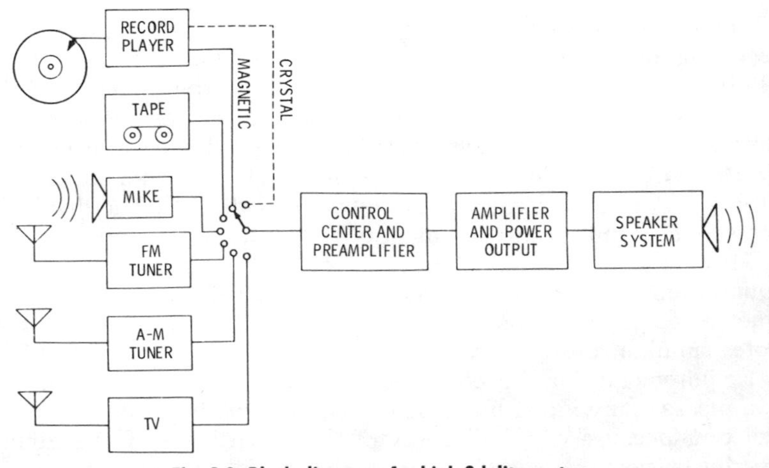

Fig. 1-2. Block diagram of a high-fidelity system.

such as phono pickups and microphones are also transducers and have many of the same weaknesses as speakers, though to a lesser degree because of the relatively low power levels at which they operate. Input devices provide transduction between sound input (or physical motion of a phono-pickup needle) and electrical output, just the reverse of the action in speakers.

The amplifier portion of the system can also contribute distortion if not properly designed or properly used, or not in proper working order. The voltage-amplifier stages are basically the least troublesome. Being of the resistance-coupled type, they usually have good response over the required frequency range with very little distortion. The power-amplifier stage and the output transformer which couples it to the speaker system are ordinarily important contributors to the overall distortion in the system.

Let us consider what we expect from an ideal system and how such a system would perform. The specifications of an ideal system are not easy to state, because, even in the actual attendance of a listener at a concert, the location of his seat, the arrangement of the orchestra, and the acoustics of the hall can greatly influence just exactly how the music sounds to the listener. Most of the tastes and reactions of the listener are conditioned by experience and are too complex to be classified in any complete manner.

To keep our discussion concrete and practical, therefore, we must concentrate on those electrical and physical features which distinguish a given system from other systems and which, in the most direct way, provide the information needed by the prospective purchaser of such a system.

The most generally accepted concept of perfection in a high-fidelity system is that which envisions reproduction sounding to the listener exactly as though he were present at the location of the original source of the music at the time this music was being recorded or transmitted. Seldom, if ever, will a system approach such a condition, but this is the earnest objective of the high-fidelity enthusiast.

Imperfections are generally classified according to their effects on performance. These effects are as follows:

1. Frequency distortion
2. Amplitude distortion
3. Spatial distortion
4. Phase distortion
5. Transient distortion

Frequency Distortion

Frequency distortion is the variation of sound output intensity with frequency, for constant input intensity. This ordinarily has the effect of limiting the range of sound frequencies which can be usefully reproduced, or at least of reducing the relative amplitude of certain frequency components so much that the sound loses its naturalness. Frequency distortion may arise electrically in the amplifier, in the transformer which couples it to the speaker system, and in the speaker voice coil. Frequency distortion may arise mechanically in the diaphragm or cone, in its mounting and orientation, and acoustically in the transfer from the diaphragm to the space into which the sound is radiated. Input and storage devices such as tapes, records, phono pickups, tuners, and microphones can also introduce frequency distortion.

Amplitude Distortion

Amplitude distortion is the failure of the instantaneous amplitude (intensity) of the sound output of any or all frequencies to be directly proportional to the instantaneous amplitude of the electrical signal input. An ideal system would have an output-versus-input amplitude relation which is plotted as a straight line and is thus *linear*. For example, in a linear system, doubling the voltage (or current) of the electrical input would double the intensity of the sound output. Tripling the electrical input would triple the output. Any variation (an increase or decrease) of the input will cause a corresponding proportional change in the output. Since in this ideal system the output is always directly proportional to the input, no amplitude distortion is introduced, and the waveform of the sound-pressure (intensity) output is an exact replica of the electrical voltage or current input waveform.

Practically, however, some amplitude distortion is introduced at some point or points in every system, so that the input-output relation cannot be plotted as a straight line; the relation is thus nonlinear to some degree. For this reason, this type of distortion is often referred to as *nonlinear distortion*.

The effects of linearity on a sine-wave input are shown in Fig. 1-3. The characteristic of the linear system is illustrated in Fig. 1-3A. With the straight-line characteristic, the ratio between input voltages before and after any change is the same as the ratio between resulting sound intensities. For example, the variation *a-b-c* on one side of zero is the same as variation *c-d-e* on the other side, for both input and output. On the other hand, this is not true for the nonlinear system illustrated in Fig. 1-3B. There, the curvature of the characteristic is such that portion *a-b-c* of the input signal produces a much smaller variation of output sound intensity than does portion *c-d-e*. The output waveform is therefore distorted. Its nonsinusoidal characteristic indicates that harmonic distortion (a form of amplitude distortion) has been introduced.

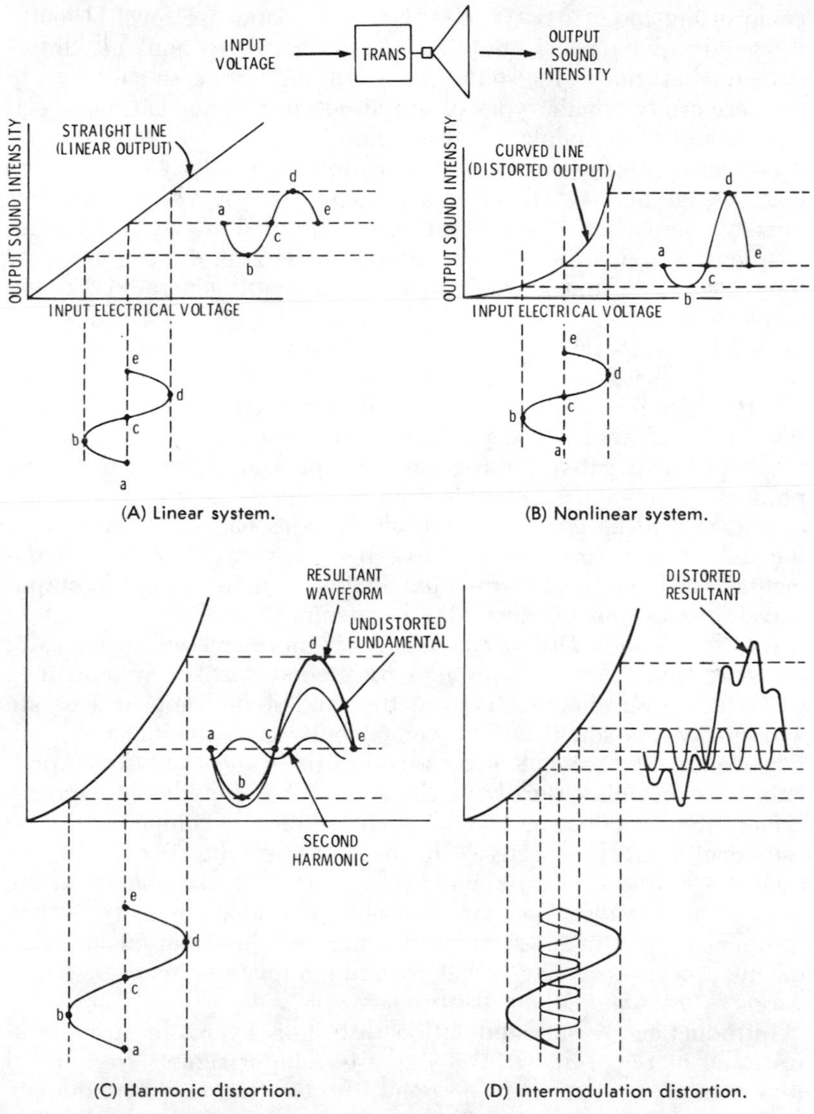

(A) Linear system.

(B) Nonlinear system.

(C) Harmonic distortion.

(D) Intermodulation distortion.

Fig. 1-3. Effects of linear and nonlinear systems on sine-wave inputs.

Nonlinearity in the output section of audio systems can result from poor or limited design in the output transformer, from changes in the radiation efficiency of the diaphragm or cone due to flexing with amplitude, from a change of effective magnetic-flux density with a change in the motion of the voice coil or diaphragm, and from nonlinearity of the air. (The ratio between compressed volume and compressing force is not constant with variation of level.) Poorly designed input devices, amplifiers, and control sections can introduce nonlinearity as the sound passes through the system.

There are two main types of amplitude distortion: harmonic distortion and intermodulation distortion.

Harmonic Distortion—Harmonic distortion results from the fact that passage of a signal through a nonlinear system generates frequency components not present in the original signal and having frequencies which are integral multiples (1, 2, 3, 4, etc. times) of the frequency of the signal from which they are generated. For example, a nonlinear system to which a pure sine-wave electrical signal of 400 Hz is applied would generate and radiate sound energy at such frequencies as 800, 1200, and 1600 Hz in addition to that at 400 Hz. Fig. 1-3C illustrates harmonic distortion due to a nonlinear system. Notice that the output waveform, which is "flattened" somewhat on the negative alternation and "peaked" somewhat on the positive alternation, is a combination of the undistorted fundamental and its second harmonic. If both alternations had been "flattened," the output waveform would have been composed of the fundamental and the third harmonic. Similarly, more complex output waveforms contain higher-order harmonics.

Intermodulation Distortion—When two pure sine-wave signals of different frequencies are applied to a good speaker system, they should have no effect on one another and should appear separate and distinct as sound output components. In a nonlinear system, however, the two signals heterodyne in the same way as the oscillator and incoming signals in the mixer of a superheterodyne receiver; they produce new undesired frequency components with frequencies equal respectively to the sum and difference of the frequencies of the original sine-wave signals. The harmonics arising from harmonic distortion are also obtained, along with frequency components of the sums and differences of these harmonics. This mixing process is similar to that used in modulation; hence its designation is *intermodulation distortion.*

Introduction of intermodulation distortion by nonlinearity is illustrated in Fig. 1-3D. In this figure two input signals are applied to a nonlinear system. At this point, the two signals intermodulate and produce a waveform with both harmonic and intermodulation distortion. Although the original frequency components are still present, the output signal also contains new distortion components with frequencies which are respectively equal to the sums and differences of the frequencies of the applied input signals. Harmonic-distortion components are also present. The same factors in audio-system design and construction which cause nonlinearity and produce harmonic distortion also cause intermodulation distortion.

Spatial Distortion

Spatial distortion is distortion which manifests itself to the listener as a real or apparent wrong location of the source of sound. It arises either from a narrow directivity characteristic of the speaker or from the failure of the system to simulate the true spatial distribution of the sources of sounds being reproduced. The directivity characteristic is a feature of the speaker system alone; on the other hand, spatial location of sound sources to simulate the original can be obtained in the speaker system only when the original sound is transmitted or recorded with this in mind. Binaural and stereophonic techniques are examples of the latter.

Phase Distortion

Phase distortion is distortion resulting when the different frequency components are reproduced in improper time relation to each other. The causes of phase distortion are generally the same as the causes of frequency distortion, and substantial frequency distortion is practically always accompanied by phase distortion.

Transient Distortion

Transient distortion is failure of a system to follow exactly sudden large changes in sound level. If the speaker system is not properly designed, pulses of sound energy tend to shock the system into oscillation at its natural frequency. The flywheel effect of these oscillating circuits causes the oscillation to continue after the true pulse which excited it has ceased. This effect is often referred to as *hangover*.

PERFORMANCE GOALS

To ensure our selection of the best type of high-fidelity system for our needs and desires, we should be thoroughly familiar with the ultimate in performance. In other words, if we have performance goals to aim at, we will best know what to look for in practical but less-than-perfect equipment. Keeping in mind the types of imperfections reviewed in the foregoing and the kinds of distortion they can introduce, we may now summarize the features of a theoretically ideal system. Such an ideal system would do the following:

1. Interpret, amplify, compensate, and reproduce sound components of any and all frequencies in the audible range with good efficiency.
2. Add negligible frequency components not in the original sound.
3. Distribute the sound in such a way that its sources would appear to be located nearly the same as they were in the original

and so that the quality of the sound would be independent of the location of the listener with respect to the speaker system.

4. Allow negligible unnatural delay of some frequency components relative to others.
5. Reproduce, without resonance effects or hangover, sudden large changes in sound volume level.

2

Monophonic and Stereophonic Sound

The word "monophonic" is derived from two Greek roots: *mono,* meaning "one," and *phone,* meaning "sound." Thus, the combined roots mean "one sound," or as we use it in hi-fi parlance, "one-source sound" (one amplifier channel, one speaker system).

The word "stereophonic" is also derived from two Greek roots: *stereos,* meaning "solid," and *phone,* meaning "sound." Thus "stereophonic" denotes "solid" or three-dimensional sound–sound coming from different sources, at different locations, with different volume levels (two or more separate channels).

Other terms used in describing monophonic and stereophonic sound are "monaural" and "binaural." Since *mono* means "one" and *auris* means "ear," the root of "monaural" is "one-ear." The term "monaural" was used for some time to describe what we now call "monophonic." However, due to the limited scope of the word "monaural," the word "monophonic," meaning one channel from start to finish, is now used instead.

The other term often associated with hi-fi systems is "binaural." This word is derived from two roots: *bi,* meaning "two," and *auris,* meaning "ear." Thus, "binaural" might be literally translated as "two-ear sound."

The association of binaural sound with stereo sound arises from the fact that the binaural system was an early method of obtaining stereo effects. The basic idea behind binaural techniques is the fact that we, as human beings, have a sensation of direction in the sound we hear because our two ears work separately. The sound sensation to the brain from one ear is kept separate from the sensation from the other ear, and both are transmitted to the brain through separate auditory nerves. The brain compares the two auditory signals received by the ears and, from the differences between them, determines the direction from which the sound came. The theory behind binaural systems is that if two sound signals, which would be heard by a pair of human ears at the source, are transmitted, reproduced in the same relationship, and applied to the corresponding ears at a remote location, all the directional effects of direct live listening will be preserved. Binaural systems are discussed later in this chapter.

MONOPHONIC SOUND

In a monophonic sound system, the sound usually emanates from only one location when being reproduced. Dual speakers, large horns, and the location of the speaker system in a corner of a room can be used to spread the sound so that it is difficult to place the sound source at one point. However, there is no stereo effect until two separate channels utilizing separate microphones, amplifiers, and speaker systems are used.

While monophonic sound may be very pleasing to listen to, stereophonic sound has advantages for the music lover that cannot be equaled by monophonic systems.

STEREOPHONIC SOUND

Modern stereophonic sound, with its directivity and depth properties, adds the third dimension to the sound. It makes clear distinction between foreground, middle, and background, as well as between right, middle, and left sound sources. Stereo has thereby been able to produce a greater amount of clarity and instrumental sound color than monophonic sound. The reception of complex sound sources is also made possible. The directional effects and small time delays from echoing and reverberation of the elemental sounds of music are separately channeled from source to ear, thus providing high-quality simulation of live music.

History of Stereo

The idea of stereophonic sound is not new. Ever since electrical sound systems were first devised in the form of telephone circuits, engineers realized the spatial characteristics of the reproduced sound are important. It is known that as far back as 1881 experiments with binaural sound were being made. In that year, performances from the Paris Opera were transmitted via a pair of telephone lines to the Paris Exposition. Each telephone line constituted a channel, with the two channels corresponding to the two ears of the listener.

In the early 1920's, shortly after standard a-m broadcasting began in the United States, experiments were made with dual radio broadcasts. However, the fact that few listeners had two receivers led to the discontinuance of these broadcasts.

On April 27, 1933, engineers of the Bell Telephone Laboratories transmitted the music of the Philadelphia Symphony from the Academy of Music in Philadelphia, to Constitution Hall in Washington, D.C. A three-channel system was used in this experiment.

Perhaps one of the most spectacular steps forward in the field of stereo sound reproduction was the introduction in 1940 of Walt Disney's *Fantasia.* This was a movie for which the accompanying sound approached true stereo characteristics. The sound was recorded using a large number of microphones, each feeding a separate recording channel. In the theater in which the picture was shown, each of the many speakers was placed in the same relative position as a microphone in the recording setup. Speakers were mounted in positions all around the perimeter of the theater, even in the back. To the audience, the sound could come from any direction, including the sides and the back. This production was a huge success, due in large degree to the realism and unusual nature of the sound. Today, practically all movies use wide-screen projection along with some form of stereo sound, but none approach the number of sound channels used in *Fantasia.*

All the foregoing progress was confined to the commercial theater or communications business. For a long time, people in the sound business realized that, until a simple and convenient method of recording stereo sound on discs was perfected, stereo could not be introduced into the majority of homes. In 1931, A. D. Blumlein obtained a British patent on a system for cutting and reproducing two-channel recording discs. Later, in 1936, Bell Laboratories engineers A. C. Keller and I. S. Rafuse obtained United States patents for two-channel disc recording. However, these early ideas did not blossom into commercial reality because materials, methods, and techniques had not advanced to the point at which production and distribution were feasible.

In 1952, another pioneer, Emory Cook, developed binaural disc records using two normal pickups spaced about two inches apart. The outer portion of the record surface was used for recording one channel, and the inner portion for the other channel. Because two pickups were needed, there were tracking problems with this arrangement and it never gained wide acceptance. It was not until single-groove systems were developed that stereo discs became practical.

Single-groove stereo records became a reality to the public in 1956, when the London (British Decca) Company developed its system. The next year the Westrex stereo-disc system was introduced in the United States. (The Westrex system has been standardized in this country, and it is the subject of detailed discussions in later chapters.) Since the advent of stereo discs in large quantities, other components of stereo systems have followed rapidly. Today there is available to the public a wide variety of stereo tuners, amplifiers, speaker systems, and other accessories, all of which are discussed in the following chapters.

STEREOPHONIC SYSTEMS

In Chapter 6, it is shown that speakers have directivity which varies with frequency. This means that the frequencies heard from a speaker vary with the position of the listener with respect to the speaker. To put it more simply: If you're in front of the speaker, you hear highs, and if you're off to the side, you don't. This is a kind of spatial distortion. But there are several other kinds of spatial distortion which engineers strive to overcome with stereo systems.

If the foregoing type of distortion is eliminated, and all the frequency components of the sound are radiated equally in all directions, all of the reproduced sound will still emanate from the point at which the speaker is located. The fact that the instruments or voices which originally produced the sound were widely separated spatially means that the single-point speaker source is not realistic. In other words, there is distortion; one might quite accurately call this "apparent-source direction distortion."

For example, if we attend a concert and don't sit too far back from the stage (or orchestra pit), we will be clearly aware that the piano is, say, to our left, the violins are to the right, and the drums possibly in the middle portion of the stage. Of course, our eyes tell us these things, but, when the orchestra begins to play, our ears will also tell us. As explained more fully later, the human auditory system is keenly directional, with the slight difference between the sound components which enter the two ears indicating to the brain the direction of the source. If, in our example, we don't hear the piano, the violins, the drums, and the other instruments as though they are coming from their respective directions, we are not hearing an exact simulation of the source material. In this case, we cannot experience the realism of high fidelity. True stereophonic sound overcomes this lack and restores to the listener a sense of the direction of the original source of the instruments creating the sound.

Recording Techniques

Realism is preserved in stereophonic sound by picking up and reproducing sounds at different depths (distances) as well as from different directions in proper or exaggerated relationships. Delays to make it seem that one sound is deeper (farther away) in relation to another may be introduced artificially within the recording equipment. This method is known as *delay stereophony.*

An old trick in the recording business is to add an artificial second channel having the same material as the first, but with a time delay introduced in it. This provides an artificial stereophonically enhanced sound. The most effective time delay is believed to be between 8 and 12 milliseconds.

Variations in intensity of the same sounds will provide a stereophonic effect if reproduced at properly spaced positions within the room. This is easily demonstrated by the ping-pong effect, where the sound of a ping-pong ball bouncing back and forth is reproduced.

Recording music from an orchestra is a refined process of organization, placement, separation, emphasis, combination, equalization, reverberation, limiting, rolling off, takes, inserts, overdubs, and editing. Usually the orchestra is arranged in a circular manner around the conductor, spaced in a manner suited to recording—which is different from arrangements made for performing for an audience. There may be an isolation booth for a singer who hears (and keeps time with) the orchestra through a low-level speaker or headphones. Each musical instrument or group of the same instrument may have one or more microphones at strategic locations to take best advantage of the output of each instrument or instrument group. The object is to separate sounds and to reduce "leakage," which is any sound picked up by a microphone but not intended to be picked up. Separation is essential to achieve clarity, or emphasis, and balance control for the production of a good record. Low-volume instruments such as flutes must be separated from louder instruments, or their presence will be obscured.

For these reasons, the musicians are sometimes separated by space or by sound-reflecting walls. Microphones are placed closer to softer musical instruments to increase their relative outputs. On occasion an instrument such as a bass viol is placed behind a V-shaped wall to prevent its high-level sound waves from spreading around the room. Dozens of microphones may be used with artful placement to develop the desired separation, presence, or intimacy. Hard music, such as rock-and-roll, requires closer "miking" and more microphones. For violins, several microphones may be placed over a group and the heights varied to change the effect during the program.

All of the audio signals thus obtained are combined in the recording room on tape. The output of each microphone is adjustable on the control engineer's panel so that the separate sounds can be narrowed, broadened, increased, decreased, and altered in frequency response, according to the art of creating a beautiful and dynamic combination of the input elements of sound. It can be seen that the engineer must be a talented artist, as well as a good technical man, to be able to produce a creative "mix." There is a striking difference between the sounds in the orchestra room and the sound output of the final recording.

The control room is essentially concerned with mixing, emphasis, reverberation, and limiting. Since each mike has its own control, the mix is accomplished in the control room, not in the orchestra room nor by the conductor. *Emphasis* or *de-emphasis* is the increase or decrease of the relative amplitude of selected portions of the af spectrum used to enhance the sound of an instrument or to delete its lesser outputs by means of electronic equipment. High frequencies are boosted to bring out the overtones of violins; this makes them "brighter." The midrange frequencies are often boosted in the recording of guitars, drums, and percussion instruments to create the hard sounds of rock-and-roll records.

Reverberation is applied by the electronic equipment to sustain musical sounds. This has the same effect as enlarging the room containing the orchestra or the output of the specific instrument so treated. It is said that this adds excitement but it is often used to "cover up" poor musical performance. Reverberation of sounds can also be created in specially designed, small echo chambers or with electromechanical devices.

Limiting is provided by automatic electronic equipment designed to instantaneously reduce volume peaks that exceed the capabilities of the electronic systems used. This protects the recording against damage and distortion due to overloading.

The *take* is the actual recording of the live music, mixed, emphasized, balanced, etc.

The *insert* is a take of that part of the program that is desired to be improved. This is spliced in place on the original master tape. Good inserts are hard to achieve because it is difficult to repeat the same texture of performance from any group at a separate time from the first or "master" play. Tempo, mix, balance, and emphasis must come together to sound the same.

The *overdub* is a technique which is used to apply the vocal performance onto the master tape after the orchestra is recorded. The master tape of the orchestra performance is played back while the vocalist accompanies this playback. The orchestra playback is wired directly from the recorder to another tape deck to produce a copy of the orchestration while the new vocal performance is added to the vocal track of the copy. This type of operation is used to improve an original poor vocalization and/or to reduce the cost of recording. However, there must be available a master tape of the orchestra without vocal. Two tapes are usually made: one with the vocal and one without it. In order to achieve this, an isolation booth is required for the vocalist. Overdubs are also used to create special effects, such as doubling the number of strings in an orchestra.

Editing is the elimination of lesser parts of the performance or reduction of length to fit standard record or tape-cartridge sizes. The best parts are pieced together to make a finer performance, or a selection may be shortened, or "clams" and studio noises eliminated. Following the editing process, master records or master processing tapes are made from the edited master. The master records and tapes are used to reproduce records, tapes, and tape cartridges at a small fraction of the cost of making the original master.

Channels

In the course of our discussions of stereo, we often use the word "channel." As applied to this subject, "channel" means a separate and distinct path for an electrical or acoustic signal. For example, an ordinary monophonic audio-frequency amplifier is part of a single channel, because it can carry only one signal at a time. If we attempt to use the amplifier for more than one signal at the same time, the two signals will interfere with each other and cannot be separated at the output. In nonstereo systems, only one channel is used. One microphone picks up the sound, which is amplified in one amplifier of conventional design, and fed to one speaker system. In a two-channel system, two microphones are used; the sound signal from each microphone is passed through a separate amplifier system, so that one sound signal has no effect on the other. As we shall see later, three amplifiers are not always necessary for three channels, because there are techniques for providing three or four electrical channels in two amplifiers, although the effect is not exactly the same as though three or four complete amplifiers were used. One or two of three or four channels can be simulated in a two-channel amplifier by additive or subtractive methods to gain the practical effect of a third or fourth channel.

Multichannel Systems

As was said, stereo sound is "solid sound," that is, sound which, even though it is artificially reproduced, seems to the listener to come from the same directions that it would if he were present at the source. One theoretical way to reproduce stereo sound would be to mount speakers all around the listener. Then we would design the system so that the speaker or speakers in the respective directions are activated at the right times with the right portions of the sound program.

Let us imagine that one listener is listening to a live performance of an orchestra in a concert hall, and a second listener is in a similar position in a similar room in a remote location to which the sound is being transmitted by a stereo system. If the piano is playing directly in front of the concert-hall listener, then the stereo system should reproduce the piano primarily from a speaker directly in front of the remotely located listener. If another instrument is located to the left side of the local listener, the listener in the remote location should hear it from the speaker or speakers to his left, and so on.

This kind of arrangement is employed in elaborate systems such as were used for *Fantasia,* and, to a lesser degree, is being used today in movie theaters and in home systems with four discrete or matrixed channels. The method is depicted in the diagram of Fig. 2-1. Theoretically, for "perfect" stereo effect, we would need an infinite number of speakers, so that there will be no "holes" in our reproduction. However, as we shall see, this is no problem because the same effect as additional speakers can be obtained by "sharing" between adjacent units. As can be seen from this illustration, such an elaborate system is very expensive, and there are many electrical problems that are not immediately apparent.

Binaural System

The foregoing method operates on the principle of providing actual sound from all the directions involved. Obviously, this method is far too complex to be practical for home use. Other methods, including the stereo methods now in general use, provide a sense of direction and depth even though as few as two channels are used.

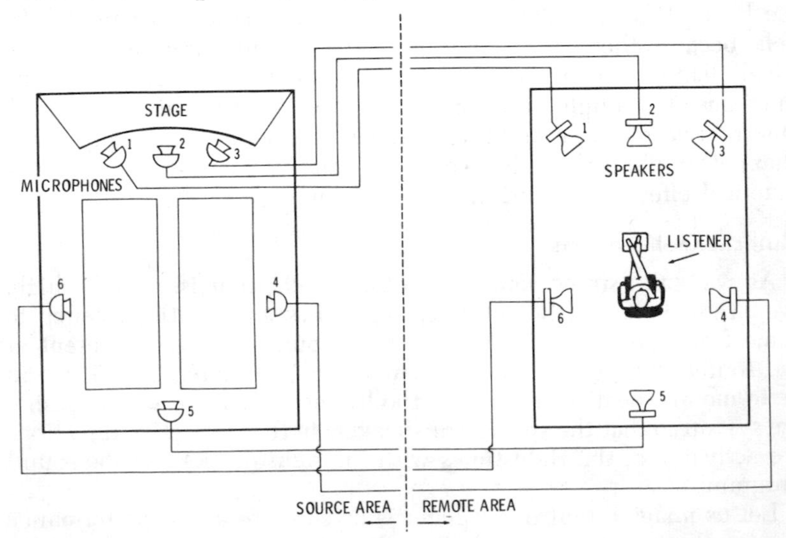

Fig. 2-1. An elaborate stereo system.

The binaural system illustrated in Fig. 2-2 is an example of a two-channel system. Rigidly correct binaural systems use an actual "dummy" head at the source location as shown. A microphone is mounted at each ear of the dummy. Thus each microphone should "hear" exactly what each ear would hear if the dummy were a human being. The whole head is used, instead of just placing microphones at the two ear locations, to simulate the effect of the human head on the sound waves. Each microphone is connected to a separate amplifier, and two audio signals are transmitted to the listener's location. There, each channel terminates in an earphone on the same side of the head as the source microphone for that channel. If we

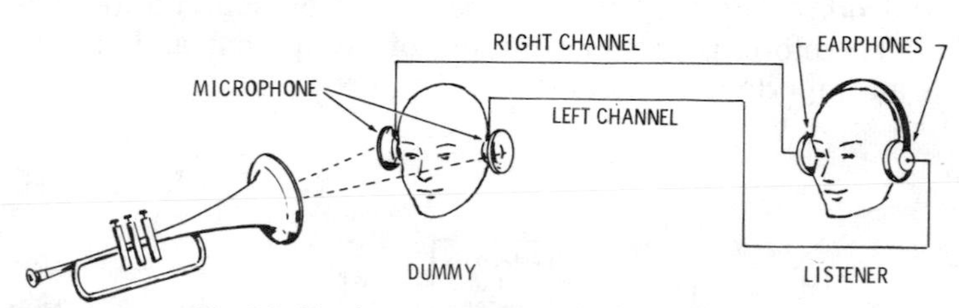

Fig. 2-2. Simple representation of a binaural system.

assume that the microphones, amplifiers, transmission systems, and earphones have a high degree of fidelity, each ear of the listener should receive the same sound as the corresponding ear of the dummy at the source. The listener's auditory system puts the two sounds together to provide indication of the directions and depth from which the sound components are reaching the dummy at the source.

Multiple-Channel Stereo

Binaural systems give a good degree of realism, but most people don't want the inconvenience of wearing headphones. That is why the stereo system of today uses speakers. However, the two channels of a stereo system operate on the principle that if the two speakers are located the same distance apart as two pickup microphones at the source, the outputs of the speakers will combine in such a way that the listener is given the sensation of sound direction and depth comparable to that of a listener at the source. A simple diagram of this kind of stereo system is shown in Fig. 2-3. The system is designed to have each speaker reproduce the sound which is present at its corresponding position at the source.

Speaker and Microphone Placement—The obvious advantage of stereo reproduction is that, added to other high-fidelity characteristics, it gives the listener greater enjoyment through the sensation of the relative direction, depth, and intensity of the various parts of the program. It affords a greater satisfaction because it gives an added dimension of realism.

However, there is difficulty in locating the listener in the same relative position as the "assumed listener" at the source. The listener (at the source) is considered to be sitting along the center line of the hall, and about equidistant from the two microphones. But suppose the remote listener's setup is such that he can't place himself at the same distance from the speakers as the assumed listener is from the microphones at the source. It has been found that small differences have little negative effect, but if the listener is sitting directly in front of one of the speakers, he hears practically nothing from the other speaker, and his sensation of sound direction is spoiled. Therefore, proper positioning of equipment and the audience is an important consideration.

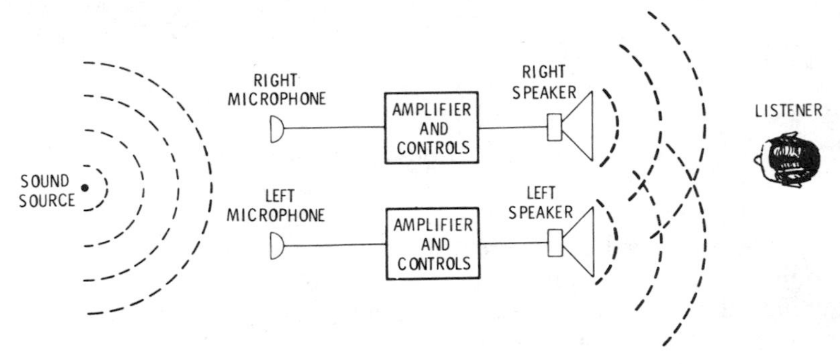

Fig. 2-3. Simple representation of the stereophonic principle.

Another important consideration is how far apart the microphones should be from each other when recording, and how far the speakers should be from each other when reproducing the recording. It is logical, and generally accepted, that the speakers should have the same spacing as the microphones. To understand better the problem of deciding what this spacing should be, let us consider the extreme cases. If the microphones and speakers were as close as they could be to each other, they would appear as one, and we would have a monophonic system. But let's assume that they're just a few inches apart. Then if the listener stands twenty feet away they will still effectively appear as one source. If he moves closer, until his distance from the speakers is comparable with the distance between them, the stereophonic effect is reestablished.

Now suppose that we spread the microphones and speakers very far apart. In the extreme, they would be so distant that they would not pick up sound or reproduce it so it could be heard. When they are closer together, but still widely spaced, the reproduced sound would be heard from two separate and distinct sources. This effect is known as the "hole in the middle," because nothing seems to be coming from the area between the two speakers, whereas the center of the orchestra would be at this point at the source.

To overcome the "hole-in-the-middle" effect, stereo may employ any number of channels. An arrangement may provide for pickup of an orchestra with six or even twelve microphones proportionally spaced around the orchestra. A corresponding number of amplifiers are used to reproduce the program. The amplifiers feed the speakers which are placed at the points where microphones were located in the original recording. Certain motion-picture extravaganzas have used these techniques.

The complete stereo system just described produces a very realistic stereophonic effect, imparting breadth, depth, and even height to the sound output. In addition, many detailed directional effects are possible. However, three or four speakers with two channels can provide nearly the same effect. Improved techniques of recording—microphone placement, channel intensity control, and speaker placement—have made it possible to approach the quality of discrete four-channel stereo with two channels (and a third and fourth speaker). This is accomplished by combining certain portions of the right- and left-channel signals and feeding the resulting signals to the other speakers.

To provide signals for the other two speakers, two additional microphones, properly positioned at right rear and left rear, may be employed at the sound source. The signals from the added microphones are mixed with the signals from the left and right front microphones, and the resultant signals are fed to the left and right channels, respectively. Thus, the left and right channels contain a certain proportion of the effects of the outputs of the other microphones.

The signals at the outputs of the preamplifiers may be divided so that the majority of the output from either channel is fed to its respective left or right speaker, but a small portion (less than a third) of the output of each channel is fed to the rear speakers. Therefore, the rear speakers have a lower input level than do the other two.

Only a small difference in the intensity of the sound from the two front speakers is sufficient to give the desired directional effect. However, the stereo effect is produced over a greater area by the other speakers. The listener can sit nearer or farther from the speakers, or move more to the right or left without losing the stereo effect. The illusion of a "curtain of sound" spreading across the room is developed.

It can be seen from the foregoing that adding the third and fourth speakers involves more than parallel connections. Technical instructions for setting up three- and four-speaker stereo are given later in this book.

Four-Channel Stereo

Stereo systems with four discrete channels are now available. Such a system has four completely separate channels, with four separate input signals passed through four preamps, four amplifiers, and at least four separate speaker systems. To many music lovers, the dramatic effect that can be produced with well engineered and matched equipment and expert installation is worth the additional cost of such a system. In addition, excellent low-cost four-channel systems are becoming available, with the result that four-discrete-channel sound (sometimes called surround sound) can be enjoyed by almost every family.

These systems should not be confused with four-speaker systems that reproduce derived or matrixed four-channel sound. Similar methods of using derived sound have been suggested for removing the "hole-in-the-middle" effect, discussed elsewhere in this chapter.

Four-channel matrixing presently is accomplished by several methods. The simplest is the Dynaco system that permits three- or four-directional (simulated channel) information to be carried on a presently compatible two-channel disc, tape, or fm broadcast. The effect of four-directional sound playback can be enjoyed on a conventional stereo system with the addition of only one or two loudspeakers and simple interconnecting components and cables.

The advantages of this system outweigh its disadvantages. Present stereo recordings can be used in this system. Some present records inadvertently have extra channel information because of peculiarities of the recording system, which actually produces left and right signals which can be vectorially added or subtracted to create a matrix of signals. This process provides several additional distinguishable signals, two of which should be discrete signals and one or two of which may be ambient signals or subsignals—or you may call them derived.

This kind of matrixing can be accomplished purposely on any two-channel stereo medium by recording the signals from the rear two microphones 180° out of phase with the recording of the signals from the two front (main-channel) microphones. This operation would be carried out while recording four discrete channels of information in preparation for encoding the four channels to provide two channel matrixes for fm stereo transmission or recording of a stereo disc or tape.

By decoding whatever portions of these additional signals ($L - R$ and $R - L$) that exist in any two-channel stereo medium, one or more additional channels may be derived from the unaltered original program material containing recorded information 180° out of phase.

When the signals from rear microphones are fed to the encoder 180° out of phase, these signals are subtracted rather than added, thereby creating additional signal components. They are supposed to be added for mono, and, therefore, under this form of processing, these rear signals will not show up in a mono playback, but will reduce whatever portions of the rear signal components are in the sum of all the sound signal components in the mono playback. This reduces the compatibility of any medium so recorded. For mono, in some cases, the program quality will be reduced to an undesirable degree.

Regardless of which kind of processing of the original program signals is used, the effect of three- or four-channel sound can be derived to a sufficient extent that acceptable and enjoyable simulated three- or four-channel stereo can be produced from an existing two-channel system. A description of the Dynaco equipment and installation information is given elsewhere in this book.

The Electro-Voice system for reproducing four-channel stereo sound utilizes all phases of any four-channel recording by encoding four separate input signals into two complex matrixed signals which can be stored on tape, processed onto discs, or broadcast over fm stereo.

A portion of the signal components of each of the four original discrete channels is encoded in each of two matrixes in different proportions according to the amplitude and frequency of the components in relation to phase (or bearing) of the original discrete but composite signals. Therefore, each matrix contains signal components from each of the original four discrete channels, and the two matrixes contain all of the information from the original four discrete channels but with less separation.

At the receiving or listening end of the system, the two matrixed signals must be dematrixed, or decoded, by an opposite process, thereby deriving the original signal components for each of the four original channels of information with acceptable fidelity. The distribution may not be identical, but the overall quality should be equal to that of the original.

In similar systems used by other manufacturers, the best distribution of power to the four (simulated) channels at the output may be adjusted with individual gain-control circuits by means of techniques and equipment described elsewhere in this book.

In these systems, there is a reduction in the degree of separation available in the four-channel sound output, but the accumulated separation among all four channels should be equal to the separation between the two channels in the system in which the signals are processed. This means that if you can achieve 30-dB separation between two channels on a medium such as tape, you should be able to obtain at least 15-dB separation between any two of the four channels decoded from two properly matrixed channels on this medium.

COMPATIBILITY

It is likely that monophonic equipment will be used to some degree for some time to come. It is therefore important that stereo systems be compatible with monophonic systems. By "compatible" we mean that one should be able to play a monophonic record or broadcast tuner on a stereo system and be able to play a stereo record or tuner on a monophonic system. If any part of the system—the record, the tuner, or the amplifier—is limited to monophonic performance, the result is monophonic reproduction. However, we say a system is "compatible" if monophonic and stereophonic reproduction is practical and normal in the same system. Design for compatibility has been mainly one involving records, phono pickups, and broadcasting techniques; it is relatively easy to combine or separate the two channels of the remainder of a stereo system to produce monophonic or stereo output.

The Westrex stereo recording method, now standard, is designed for near-perfect compatibility in record reproduction, as will be shown in Chapter 3. Unfortunately, in practice, true compatibility does not yet exist because of mechanical limitations of the older monophonic pickups. As a result, stereo records should never be played with monophonic pickups. However, monophonic records can be played with stereo pickups, and often the reproduction is more pleasing than when the record is played on a monophonic system. Of course, there is no stereo effect. Details about compatibility and difficulties in attempting to play stereo records with monophonic pickups are discussed in Chapter 3.

Compatibility is the chief obstacle to the development of commercially successful four-channel sound. All problems in this regard could be solved eventually, given enough time and money. But, until the cost and time of such developments can be reduced and agreements are made between the various segments of the industry and the Federal Communications Commission, four-channel sound is likely to continue to develop in the present direction: specialized according to use rather than compatible in all regards.

The main reason for the present problems centers around the success of the disc in its present form and the difficulty and costs involved in developing a new process and system for making and using discs that could be played on monophonic and two-channel equipment as well as on four-channel equipment. It is interesting to note that arrangements have been proposed that would provide compatibility between two- and four-channel discs, but without the capability for monophonic playback. This would be a satisfactory solution for the future, but at present has the disadvantage that much home equipment is still monophonic.

3

Stereo Techniques

The desirability of stereo reproduction, and what had to be done to realize it, has been known for many years. However, it was only recently that definite techniques were developed and applied to hi-fi equipment. In this chapter, these techniques are discussed.

Stereophonic sound equipment utilizes all the circuitry contained in monophonic equipment. In addition, the circuits and devices necessary to add one or more channels and provisions for two, three, or four different outputs must be included in stereophonic equipment.

In the past decade, the reproduction obtained from monophonic and two-channel stereophonic hi-fi equipment has reached a high state of development. The quality of single-source sound and stereo sound, with the best systems, is such that the average human is unable to detect any distortion.

BASIC COMPONENTS OF STEREO SYSTEMS

The basic components of a monophonic system are shown in Fig. 1-2. This arrangement may be adapted to multiple-channel stereo by adding another channel or channels of amplification and additional speaker systems, which will permit stereo reproduction of sound. Then, by adding a stereo phono pickup, tape deck, or multiplex adapter, the input devices can be converted for stereo operation.

Improvements in the form of adapters are available to provide one control for balancing the outputs of the separate channels, and other conveniences are available for patching up the old single-channel system to produce stereo. However, this kind of arrangement is not competitive in convenience of operation, flexibility, simplicity, or other refinements when compared with equipment designed especially for stereo operation.

STEREO RECORDING

To us, as hi-fi listeners, the manner in which recorded or broadcast programs are picked up at the source is only of academic interest. With discs and tuners, we concentrate on reproducing the music available to us, and the studio techniques over which we have no control do not interest us. However, with the availability of multitrack tape recorders for the home, there is growing interest in amateur recording, and the addition of stereo makes it much more interesting. Also, a knowledge of microphone placement problems for pickup helps us in deciding speaker placement for our reproducing system.

The most important general type of stereo pickup technique is called *time-intensity pickup.* This means that the signals in the two stereo channels will vary in both intensity and time according to the difference in direction and in distance of the sound source from the two microphones.

As explained in Chapter 2, the stereo effect can be obtained by picking up the sound with two microphones spaced a certain distance apart and located in front of the sound source. But how far should the microphones be separated? If they are too close together, the differences in time and intensity of the two signals are negligible, and the stereo effect is lost. Experiments have been made with microphones as much as 30 feet apart. At such distances, the reproduced sound seems to the listener to divide into two sources, especially if he is close to the speakers.

Experiments with the relative spacings involved have led to the conclusion that there is a definite optimum relation between the separation of the microphones and their distance from the source. A similar relation exists between the separation of the listener's speakers and their distance from him. This relation can be illustrated by the triangle formed by the two speakers and the listener in Fig. 3-1A. In practice the angle at the listener should not be less than 30 degrees nor more than 45 degrees. This relation keeps the distance of the listener from the speakers approximately the same as the distance between the speakers, and ensures sufficient time-intensity variation to provide stereo effect.

It is generally agreed that in the ideal two-channel system, the listener's speakers should be the same distance apart as the studio microphones. However, if we consider the size of a full symphony orchestra, and how far the microphones need to be spaced to have a good part of the orchestra between them, we realize that the same spacing at both ends of the circuit is not always practical. In practice, compromises are made. A good rule to follow is to spread the microphones as far apart as you can and still pick up an appreciable amount of sound from the middle portion of the source. Usually the pattern will be such that the microphones can be placed along the sides of a triangle as in Fig. 3-1B.

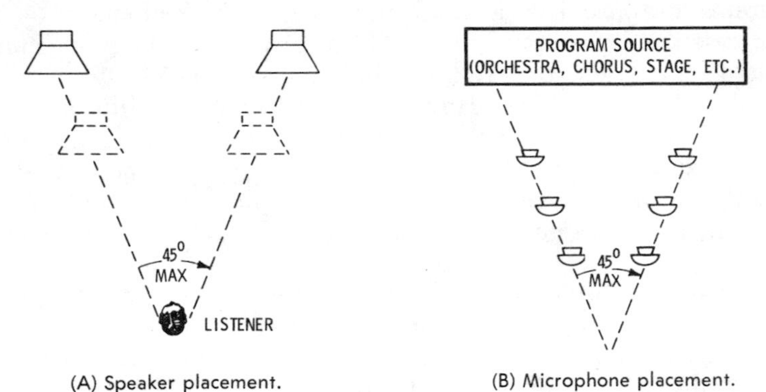

(A) Speaker placement. (B) Microphone placement.

Fig. 3-1. Suitable locations for stereo speakers and microphones.

Another method, referred to as the *intensity-difference system,* is sometimes used in stereo recording. In this system, the microphones are not spaced; however, they have a directional characteristic (Fig. 3-2). The stereo effect is obtained by proper orientation of the microphones. One method is to mount the microphones at 90 degrees to each other and 45 degrees to the center of the source, as shown in Fig. 3-3A. Another method is to use one microphone aimed directly at the source for the main signal, and a second microphone to pick up side signals. This is illustrated in Fig. 3-3B.

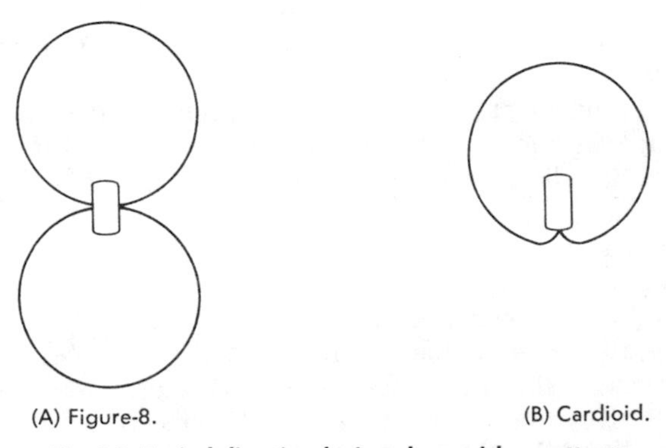

(A) Figure-8. (B) Cardioid.

Fig. 3-2. Typical directional-microphone pickup patterns.

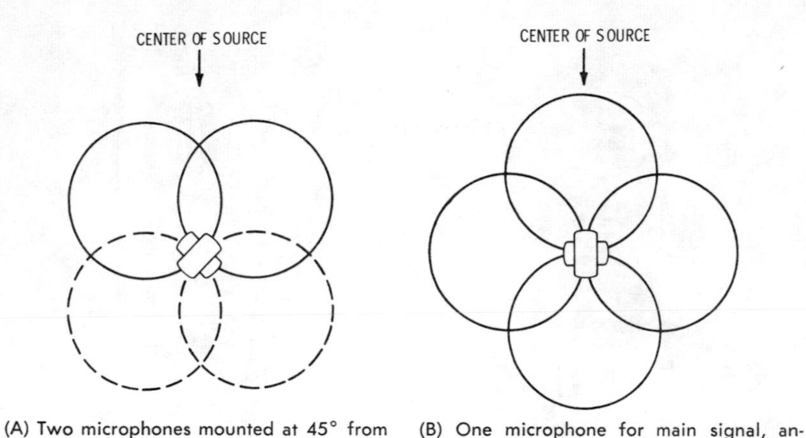

(A) Two microphones mounted at 45° from center of source.

(B) One microphone for main signal, another for side signals.

Fig. 3-3. Placement of two directional microphones for stereo pickup.

A directional microphone especially designed for stereo pickup is pictured in Fig. 3-4A. This microphone contains two directional elements, as shown in the cutaway side view of Fig. 3-4B. Each element has a cardioid pickup pattern, and the two elements are aimed 90 degrees apart. Thus, each element is aimed 45 degrees from the center of the sound source, producing the response given in Fig. 3-4C.

DISC RECORDING

It was not until practical methods of stereo disc recording were developed that standardized stereo could reach our homes. Although other good methods have been considered, the Westrex method of stereo recording is the one adopted by the recording industry. We shall, therefore, confine our discussion here to that method.

Requirements

Let us review briefly the general requirements of a stereo disc recording method. They are as follows:

1. Two completely independent left and right signals must be recorded separately in such a way that there is a minimum of interference or mixing between them, and so that, in playback, they can be separately recovered in the same form as that in which they were recorded.
2. The system must be compatible with monophonic recording and playback systems. A stereo playback system should play monophonic records without loss of fidelity; likewise, a mono-

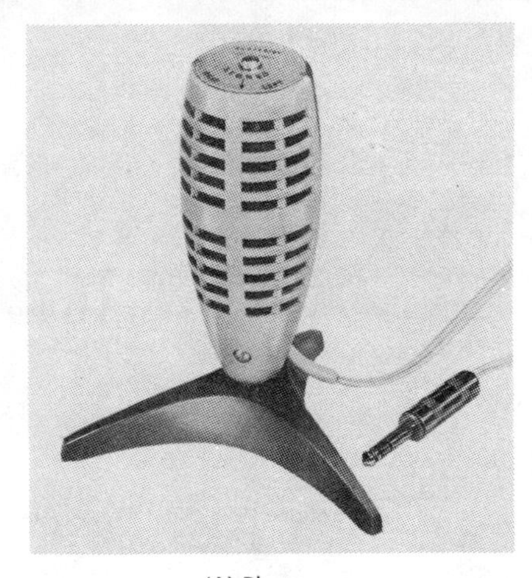

(A) Photo.

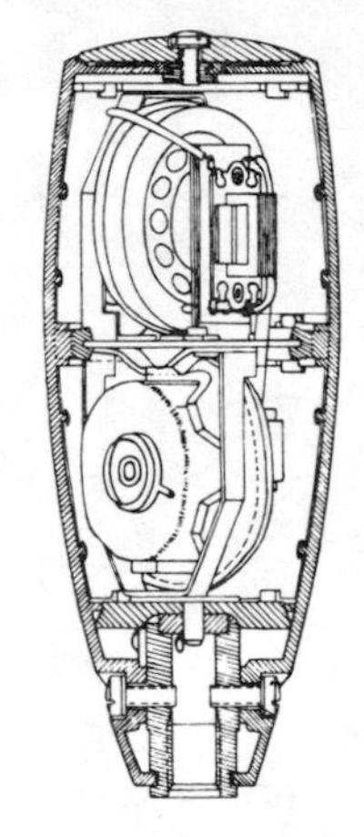

(B) Construction.

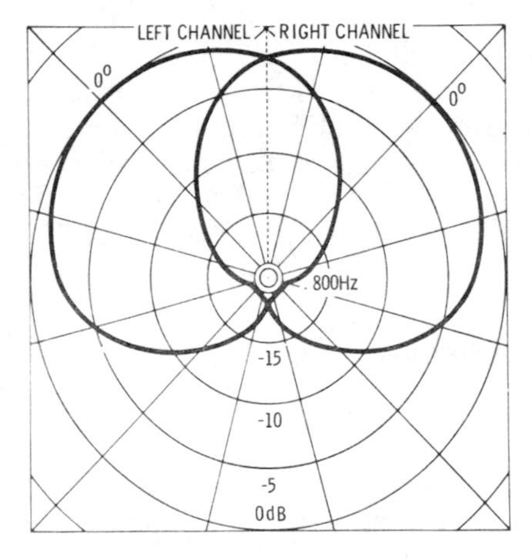

(C) Response pattern.

Courtesy North American Philips Corp.

Fig. 3-4. A directional microphone designed for stereo pickup.

phonic system should play stereo records and produce as good a monophonic output as when it is playing monophonic records.

When Edison first developed phonograph recording, he used vertical motions of the recording stylus to record the sound vibrations. This came to be known as the "hill-and-dale" method of recording. However, in such a system, the "hills" in the record groove had to lift the playback arm and cartridge. Since early arms and cartridges were relatively heavy, this motion caused excessive wear. Because of this, the record industry adopted the lateral recording method, in

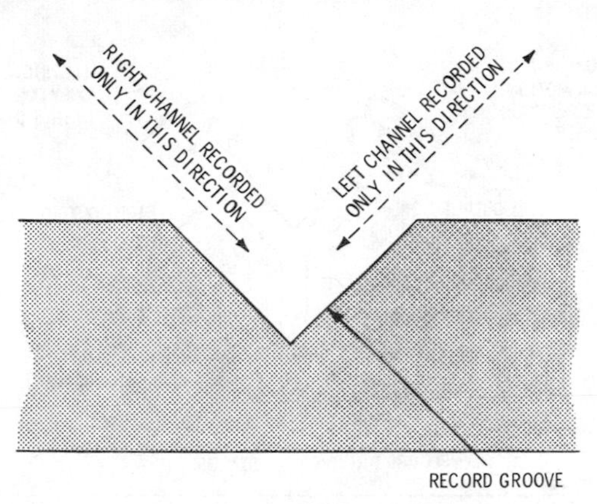

Fig. 3-5. Basic principle of Westrex system of stereo recording.

which the needle goes from side to side in accordance with the sound vibrations. All commercial monophonic records for home use are laterally recorded, and monophonic pickups are designed accordingly.

The Westrex System

In the Westrex stereo recording system, both lateral and vertical motions are employed. Modern pickup and arm design is such that vertical motions are usable under the proper conditions. The basic principle of the Westrex system is illustrated in Fig. 3-5. The sound signal for each stereo channel is recorded along a direction at 45 degrees to the horizontal. In other words, each channel is recorded on one side of the record groove in such a way that its undulations will move a playback stylus back and forth along one of the 45-degree directions. The two directions are at right angles (90 degrees) to each other, and motions along either one alone do not affect motion along the other. Thus, the motions of one channel do not interfere with the motions of the other, and two separate output signals, each corresponding to the motions for its respective channel, are obtained from the stereo cartridge.

A symbolic representation of how the axial motions in the two 45-degree directions can generate two separate signals is given in Fig. 3-6. Imagine the assembly of the stylus and the two bar magnets free to move in any direction but not to rotate. Then, if the stylus is pushed to the left, it goes into a position like that of Fig. 3-6B. The magnet at the right merely moves across the diameter of the coil perpendicular to its axis. There is no axial motion in or out of the right coil, so no current is generated in this coil.

Now suppose that the stylus-magnet assembly is pushed to the right, as shown in Fig. 3-6C. Conditions are now the reverse of those

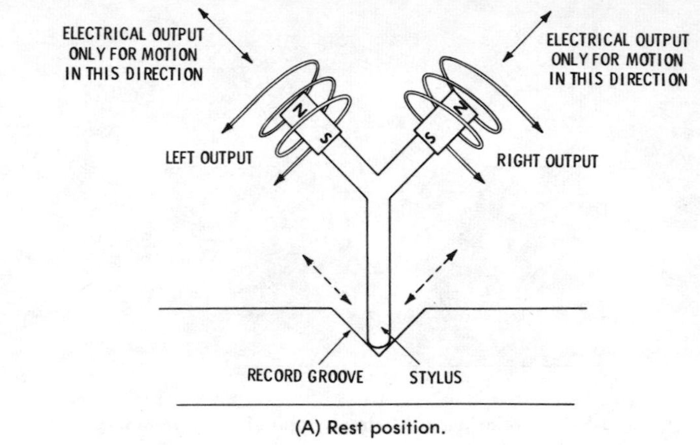

(A) Rest position.

(B) Movement to left.

(C) Movement to right.

Fig. 3-6. Principle of operation of stereo phonograph cartridge.

in Fig. 3-6B. A current is generated in the right coil, but none is generated in the left coil.

Thus, vibrations recorded in the groove so that they push the stylus at 45 degrees toward the left produce pickup output current only in the left coil. Likewise, vibrations recorded in the groove so that they push the stylus at 45 degrees toward the right produce pickup output current only in the right coil.

It is difficult to imagine any case in which both left (L) and right (R) signals would not be present. Because of this, you may wonder how the playback needle can move in both 45-degree directions at once. Naturally, it cannot do this. Instead it moves in a direction and with an amplitude dictated by the vector resultant of the two forces along the 45-degree paths. This means the needle can move as a whole in any resultant direction, including side to side and straight up and down. Five combinations of relative L and R signals are analyzed in Fig. 3-7. Only one signal is present in Figs. 3-7A and 3-7B. Accordingly, in each of these cases the needle moves along the direction of the single channel present. Thus if the motion is along one of the 45-degree sides of the groove, the pickup "senses" that it is receiving signal from only one channel. Fig. 3-7C illustrates the special case of equal signals from L and R. In this case, the resultant motion is always vertical, and the record "looks" to the needle like a "hill-and-dale" recording. The vertical motion indicates that both signals are equal in amplitude, and the amount of motion indicates their amplitude. Fig. 3-7D indicates the motions when the L signal is much stronger than the R signal. The larger the L signal with respect to the R signal, the farther the resultant is to the left of vertical. The angle of resultant motion "tells" the pickup the ratio between the two amplitudes, and the amplitude of the resultant indicates their amplitudes. The situation depicted in Fig. 3-7E is the same except that the R signal is stronger than the L signal. The examples in Fig. 3-7 show that for any combination of resultant needle angle and amplitude of motion, there is a distinct amplitude for both the L signal and the R signal.

Since the amplitudes of the L and R signals are usually approximately equal at any instant, the conclusion to be drawn from Fig. 3-7 and the preceding discussion is that the groove modulation in stereo records is predominantly vertical. However, vertical groove modulation tends to be more distorted and produce more record wear than lateral modulation. To change the Westrex recording system to one

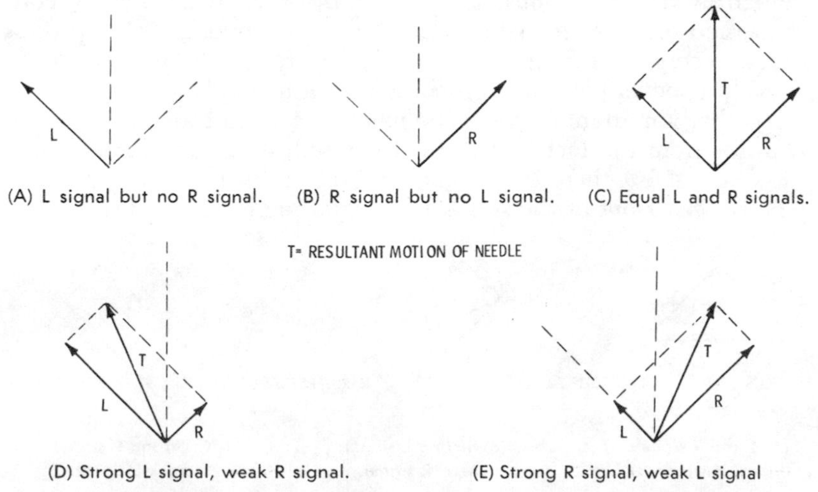

(A) L signal but no R signal. (B) R signal but no L signal. (C) Equal L and R signals.

(D) Strong L signal, weak R signal. (E) Strong R signal, weak L signal

Fig. 3-7. Resultant motion for various combinations of L and R components.

in which groove modulation is predominantly lateral, the phase of one of the signals driving the stereo record cutter is reversed. This changes the resultant needle force from a predominantly vertical motion to one which is predominantly lateral. This method is illustrated in Fig. 3-8. Note that Figs. 3-8A, 3-8B, and 3-8C are the same as Figs. 3-7C, 3-7D, and 3-7E, respectively, except that the R vector is reversed. The Record Industry Association of America (RIAA), which accepted the Westrex system, has stipulated that "equal in-phase signals in the two channels shall result in lateral modulation of the groove." This requirement is satisfied by the phase reversal of one of the signals to the cutter as shown in Fig. 3-8; but a compensating reversal must take place at the pickup or later in the system for playback. However, this is a simple matter, because it involves only reversing the connections of one of the pairs of leads from the pickup.

Special Factors in Stereo Disc Systems

Although the phase reversal of one of the signals to the recording head makes the modulation of a stereo record predominantly lateral, the fact still remains that vertical motions play a vital part in recording and playback. This fact brings with it some special problems in stereo pickup design.

The stereo pickup must have high vertical *compliance;* that is, it must allow for easy motion of the needle in the vertical direction. If it does not, the needle exerts undue pressure on the record, causing rapid wear. A good stereo pickup is designed to have the necessary vertical compliance, but monophonic pickups are not. Thus, although as far as motions required of the pickup needle are con-concerned there is compatibility, use of a monophonic pickup on a stereo record will ruin the record after relatively few playings. Therefore, never play stereo records with a monophonic pickup.

Another important factor is the *pinch effect* illustrated in Fig. 3-9. It arises from the fact that the cutting stylus has a flat front edge. At A in Fig. 3-9, there is no signal and the stylus is laterally motionless; the flat front of the stylus is at right angles to the direction of

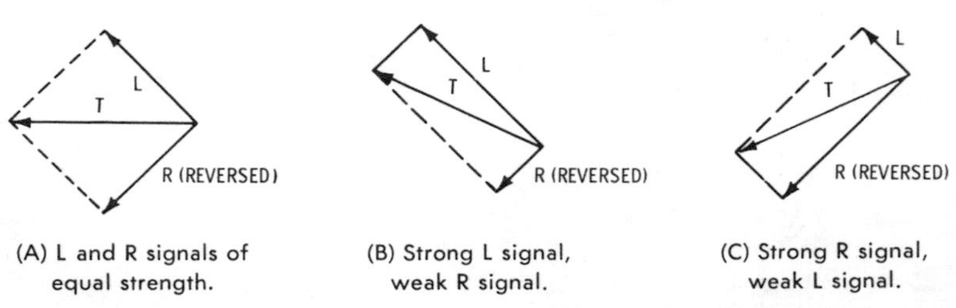

(A) L and R signals of equal strength.

(B) Strong L signal, weak R signal.

(C) Strong R signal, weak L signal.

Fig. 3-8. Vector addition of L and R components with one channel electrically inverted.

Fig. 3-9. Cause of the pinch effect.

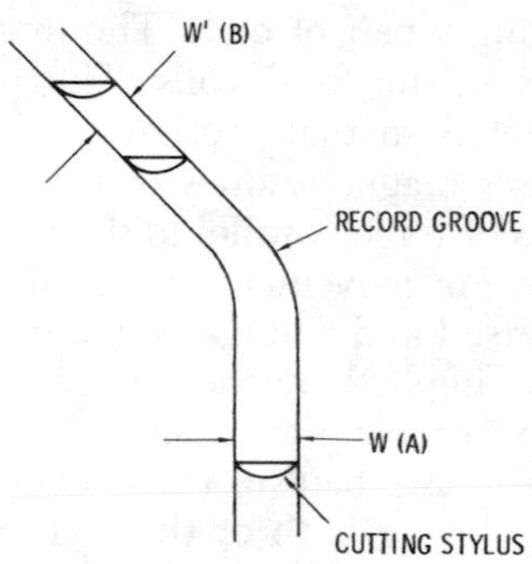

the groove, which at this point has its greatest width (W). However, at B modulation has been applied; the stylus still has the motion it had at A, but now it also has the lateral motion imparted by the signal applied to the cutting head. The flat front of the stylus, with its cutting edges, now plows sideways through the record material, so the width of the groove here is W′, which is considerably less than W, the width at A. Unless the tip of the playback needle is extremely fine, the narrowing at B tends to lift it out of the V-shaped groove. This lifting motion is the same as might be experienced with vertical groove modulation. It does not affect a monophonic pickup (except to make the record wear faster), but with the vertical-sensitive stereo pickup it introduces an unwanted signal and distortion.

To attempt to overcome the effects of pinching, the tip radius of stereo pickup needles has been reduced from the over 1-mil (0.001 inch) monophonic microgroove value to 0.5-0.7 mil. However, this reduction in tip diameter introduces another disadvantage—the tip pressure on the record is increased. Pressure is force per unit area, so even if the force of the needle on the record has not increased, a decrease in its area causes a corresponding increase of pressure. Pressure is a main factor in wear on records, so the smaller needle tip results in greater wear, if the other factors remain equal. To keep wear down, stereo pickups are operated at very low needle force. Four grams is usually considered a reasonable maximum.

The same types of pickup devices are used for stereo as for monophonic reproduction. They include crystal cartridges, ceramic cartridges, and moving-coil and moving-magnet magnetic cartridges. The suspensions used for the needle and its associated mechanical vibrating assembly are very critical, and many clever arrangements have been worked out by the engineers.

The principle of how moving magnets are used to generate the two stereo outputs was illustrated in Fig. 3-6. Although that diagram does not correspond to any manufactured cartridge, the manufactured ones do have a small magnet (or two) moving near a core having a pair of coils. The motion of the magnet(s) generates voltages in the fixed coils. The pole pieces and moving magnets are oriented so that motion of the needle in one 45-degree direction causes magnetic lines of force to be cut by one coil but moves the lines of force parallel to the other coil so that no voltage is generated in it. For movement in the other 45-degree direction, the situation is reversed and voltage is generated in the other coil.

As implied by the name, in the moving-coil magnetic-type (dynamic) pickup, the coils move and the magnet stands still. A special linkage mechanism with jeweled bearings imparts rotational motion to either or both of the coils as the needle is moved by the record undulations. Each coil responds to needle motion in one of the 45-degree directions. The coils operate in a magnetic field, so an audio voltage is generated in them when they turn.

Ceramic elements can also be used in a stereo pickup, as illustrated in Fig. 3-10. As a ceramic crystal is bent, a voltage is generated. As the needle moves to the left, one ceramic element is bent

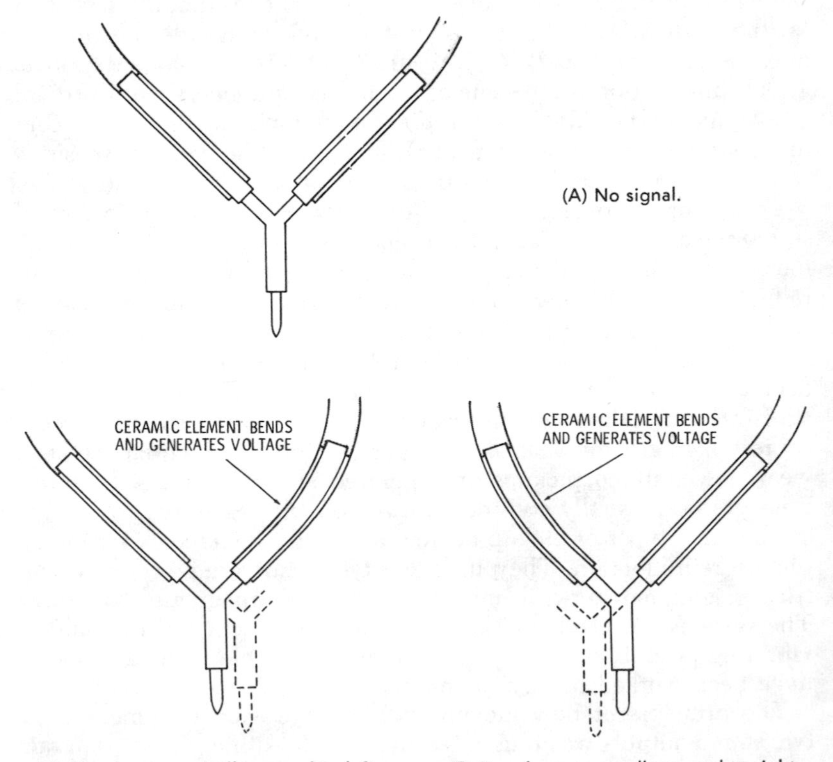

Fig. 3-10. Stereo playback with ceramic elements.

(Fig. 3-10B); similarly, movement to the right bends the other element (Fig. 3-10C). Thus, two isolated stereo outputs are provided.

Four-Channel Record Discs

The Westrex recorded disc contains two separate signals on the 45°-45° sound groove. A new system has been introduced to record and play back four discrete signals (four channels) on a similar groove. This system, however, requires considerably expanded capability of components such as the record cutter, disc, pickup cartridge, and needle, all of which must be able to operate together with a flat response curve up to 50 kHz.

This system utilizes direct recording of frequencies up to 20 kHz for the left and right channels (similar to the standard stereo process), and a modulated 20- to 45-kHz subcarrier to carry encoded difference information for the other two channels.

The bearings (positions) of the channel inputs (microphones) and outputs (speakers) are as follows:

Channel One:	Left Front
Channel Two:	Left Rear
Channel Three:	Right Front
Channel Four:	Right Rear

The sum of channels one and two is recorded on one side of the record groove, and the sum of channels three and four is recorded on the other side of the groove. Thus, when this four-channel stereo disc is played on standard two-channel stereo equipment, full compatibility of two-channel reproduction from a four-channel disc is obtained.

To reproduce four-channel sound from this same record, a special decoding device is required. The four independent channels are extracted through the use of dematrixing techniques (addition and subtraction of signals), and then the four decoded signals are fed to a four-channel stereo system for reproduction.

Tracking Problems

Trackability distortion is produced when a phono stylus-cartridge transducer does not track, or "trace," the grooves of a record in the same manner as originally cut by the cutting stylus. This action causes all kinds of distortions to be produced, and parts of the original sound are not reproduced. As techniques of record and tape making improve, wider frequency ranges and greater transients are accurately stored in the recording mediums. These improvements create a requirement all the way through the hi-fi system for improved pickup, amplification, and reproducing techniques, and for the equipment to keep pace. However, the essence of a quality system is the ability of its pickup device to provide precise dynamic coupling to the medium and accurately transduce the program material as it was recorded.

A number of factors must be just right to ensure proper *tracking*, that is, to ensure that the needle stays in the groove and does not cause excessive wear. Good tracking is achieved when the needle follows both sides of the groove with equal pressure on each side at all times. Tracking is much more important in stereo playback than in monophonic playback because the needle will wear one side of the groove more than the other if it is forced against the side of the groove. In stereo, wear on one side of the groove reduces the amplitude of one stereo channel with respect to the other stereo channel; therefore, if the wear is appreciable, a stereo record would become useless much sooner than a monophonic record under the same conditions.

This wear on the sides of the grooves is another reason why the needle force of a stereo pickup must be less than that of a monophonic pickup. But, reduction of needle force means more difficulty in maintaining tracking because the needle will sometimes have a tendency to "skate" across the record. Fortunately, the reduction of the needle-tip radius to 0.5 to 0.7 mil to overcome pinch effect (as explained before) also improves tracking.

It can be seen from the preceding discussion that the pickup arm for stereo cannot be too carefully adjusted. It must have an absolute minimum of resistance to lateral motion so that it tracks smoothly when balanced for the recommended needle force.

Adjustment of the pickup arm and needle is easier with a "single-play" turntable than with a record changer because of the absence of the complex change mechanism. With a changer, the needle must rest heavily enough in the grooves to allow the pickup arm to trigger the change mechanism. This takes a great deal more force than is necessary to move the arm across the record. In addition, because the records are stacked on the turntable during operation of a changer, the needle has a different angle for each record. Thus distortion is introduced in the reproduction, and record wear is much greater than with single-play mechanisms. This does not mean that changers cannot give top-quality reproduction, but, everything else being equal, the single-play setup is simpler to adjust and operate.

Turntables

At first thought, it would seem that the requirements for turntables and turntable drives would be the same for stereo as for monophonic systems. However, this is not so. Stereo turntable and drive requirements are more exacting because of the greater inherent sensitivity of stereo pickups to vertical vibration and the susceptibility (in the case of magnetic pickups) to hum pickup from the motor.

Rumble—The effect produced in the sound output by low-frequency signals generated by vibrations in the motor and drive systems is called *rumble*. These signal components usually have frequencies of from 30 to 60 Hz. Therefore, if the system as a whole does not have extended low-frequency response, rumble is not such a great problem. Thus, if your speaker system cuts off at about 100 Hz, you can stop worrying about the fine points of rumble production. However, any high-fidelity system worthy of the name reproduces signal components down to 50 Hz or below, and turntable rumble is an important factor.

A stereo pickup is more sensitive to rumble than a monophonic pickup because it is sensitive to vertical vibrations, and the vertical vibrations are usually two or three times as strong as the horizontal vibrations in a turntable drive assembly. Therefore, special measures must be taken for stereo-system turntables to minimize vibration and its effects. Otherwise, severe rumble is present in the output signal of the pickup. Although there is no standard, rumble is usually measured with respect to a fairly strong 1000-Hz signal obtained from a standard test record. Low-level passages of music may be as much as 40 dB below the test-record output, so rumble should be at least 45 dB down. At −60 dB, rumble usually is completely inaudible, so this is a desirable objective.

In record players designed for stereo, rumble should be minimized by damping in the drive system and by the use of motors which deliver power as smoothly as possible.

Hum Pickup—The reason for the greater susceptibility of some stereo pickups to hum is the fact that there are two channels instead of one. The hum currents in the two coils of a magnetic-type stereo pickup combine in the output during operation. When a monophonic record is being played, the hum currents can be made to cancel by connecting the coils in parallel.

The source of most hum pickup is the turntable drive motor. Induction and synchronous motors have coils carrying alternating current from the power line, and thus they radiate hum. In general, the higher the motor power, the more hum is radiated. Because of this, it might seem that the motor power should be made as low as possible. However, the lower the motor power, the more difficult it is to get good speed regulation and the more likely it is that wow will be introduced. It is general practice to make the turntable relatively heavy to provide the inertia for good speed regulation. But more motor power is required to drive the heavier turntable; hence more hum is produced. Record changers require more drive power than single-play systems, so changers tend to be more subject to hum.

TAPE RECORDING

Stereo recording and playback from tape may soon be as common as stereo records, because of technical improvements, price reduction, and greater convenience in use. Many audiophiles believe that tape offers the greatest opportunity for the ultimate in high-fidelity stereo reproduction.

Although stereo tapes were available long before stereo records came into general use, certain basic problems have slowed them from enjoying a wide distribution. The first of these problems is price. At present, stereo tapes cost more than comparable discs, and they must be played on a machine of relatively high quality, which also costs more than record players providing comparable reproduction. The second basic problem, now overcome, was the inconvenience of handling. A disc can easily be put on a turntable, the pickup placed on it, and music obtained with little delay. When a tape is to be played, ordinary rolls must be carefully keyed into position, and the tape threaded through the guides and past the heads of the machine to the pickup reel. Magazine-type tapes are now available which eliminate such time-consuming operations, but they are still expensive.

In spite of the price disadvantage, tape does have many advantages. It is practically immune to wear and deterioration of quality with playing. A tape can be played thousands of times without noticeable degradation, providing reasonable care is used in its storage. The transfer from storage on the tape to an electrical signal in the amplifier is accomplished without the necessity of mechanical parts that vibrate at sound frequencies as in phonograph pickups. Thus wear and resonance effects are minimized.

The difference between a stereo tape system and a monophonic tape system is that the stereo system uses two, four, or eight tracks recorded on the tape. The recording and playback heads each have two or four units, one for each channel. The tracks and the gaps in the head are separated by a guard band so that there is no interaction between the two signals.

The arrangement in Fig. 3-11, used with open-reel systems, allows two tracks to be recorded or played back simultaneously for two-channel stereo. Four-track tape permits additional playing time for the same length of tape. The tape can be played in one direction first and then turned over and played in the other direction. The four tracks are recorded on the tape as illustrated in Fig. 3-12A. The guard bands do not need to be as large as for the two-track tape in Fig. 3-11, because in Fig. 3-12A only alternate tracks are used during tape travel in a given direction. Fig. 3-12B shows the open-reel tape format when the tape is intended for application to

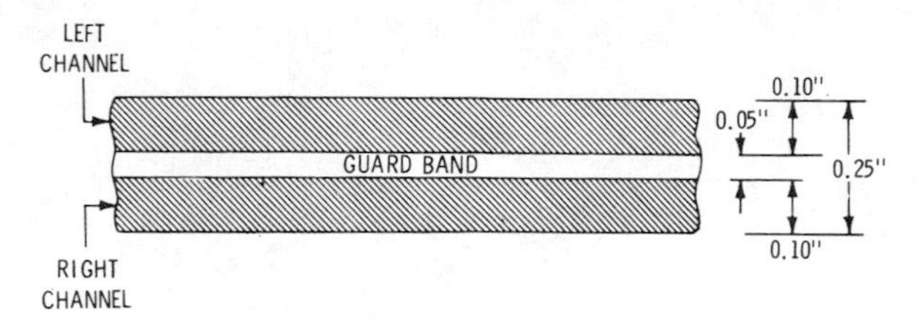

Fig. 3-11. Tracks on two-track stereo tape.

four-channel stereo. The tape is played in one direction only, and all four tracks are used simultaneously.

The open-reel tape system has inconveniences in handling, loading, threading, playing, and storage, and the tape is susceptible to breaking and tangling. To overcome these problems, the cartridge and the cassette have been developed. These are permanent containers in which internal spools hold the tape in a protected position, ready to play at any time.

The mechanical arrangement of the cartridge is shown in Fig. 3-13. The cartridge is primarily intended for playback of mass-produced prerecorded programs, although recorders for cartridge tapes are available. Fig. 3-14 shows the tape formats for two and four-channel cartridges. The cartridge tape operates at 3¾ inches per second, in one direction only, and has eight tracks. For two-channel stereo operation, the tape carries two channels of information in each of four pairs of tracks (Fig. 3-14A). The first pass plays tracks 1 and 5. Then the tape head indexes to pick up tracks 2 and 6, and this process is repeated through tracks 3 and 7 and, finally, 4 and 8. The heads then return to replay tracks 1 and 5; the machine plays continuously until shut off.

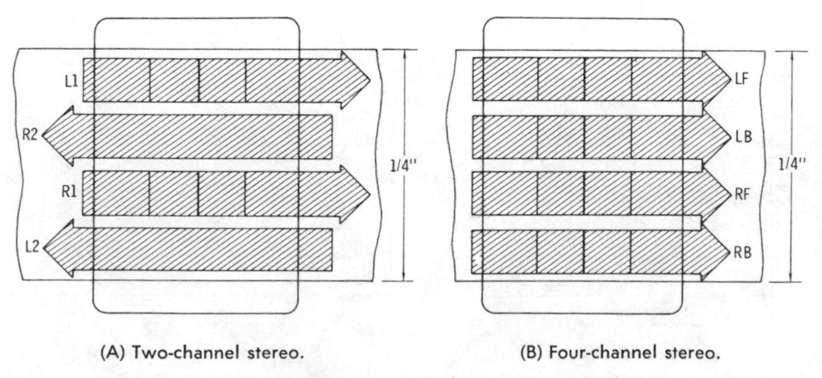

(A) Two-channel stereo.

(B) Four-channel stereo.

Fig. 3-12. Track arrangements on four-track open-reel tape.

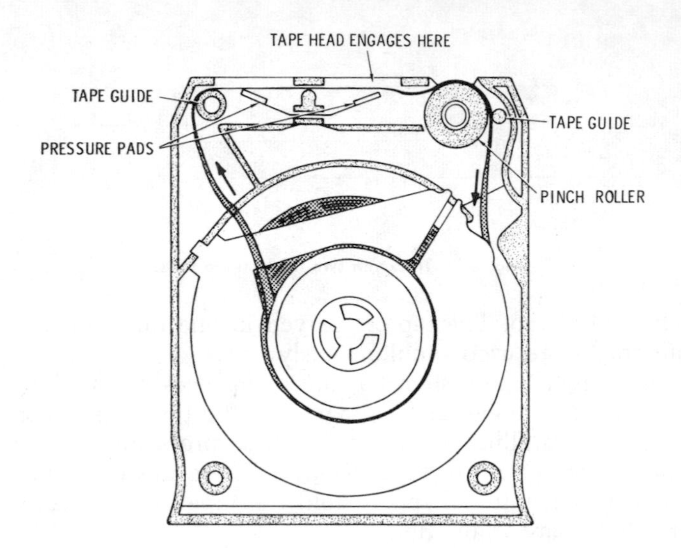

Fig. 3-13. Construction of a tape cartridge.

The four-channel cartridge format is shown in Fig. 3-14B. In this arrangement, alternate tracks—for example, tracks 1, 3, 5, and 7—are used in one pass. Then the tape head indexes, and tracks 2, 4, 6, and 8 are used in the second pass. Only two passes are required to play the entire tape, and the available playing time for a given length of tape is only half as great as for the two-channel system. The spacing between tracks, however, is sufficient to provide adequate separation between channels.

The cassette is a miniature reel-to-reel tape device, as shown in Fig. 3-15. Tapes are threaded permanently, and each end is coiled

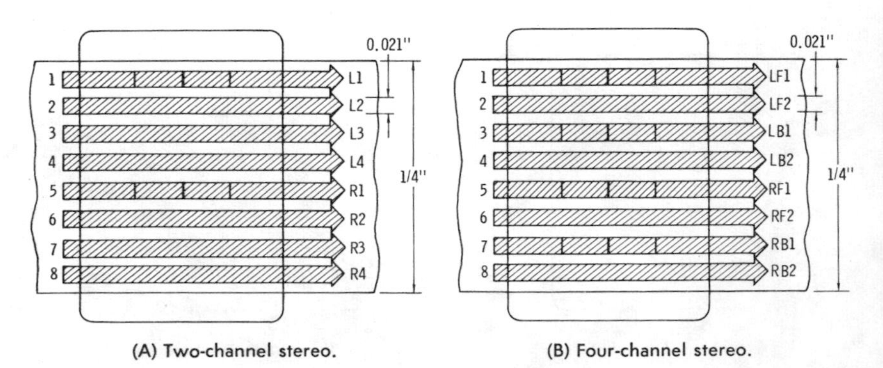

(A) Two-channel stereo.

(B) Four-channel stereo.

Fig. 3-14. Track arrangements for cartridge tape.

and attached to a hub. The capstan, pinch roller, and heads of the cassette deck can pass through holes in the plastic tape housing to operate the tape to record and play back.

The cassette is suitable for use as a home or business recording and playback medium, or it may be used for prerecorded programs. Cassette track formats are shown in Fig. 3-16. Fig. 3-16A shows the standard licensed format for two-channel stereo; it is compatible with mono use. Figs. 3-16B and 3-16C show proposed formats for use in four-channel applications.

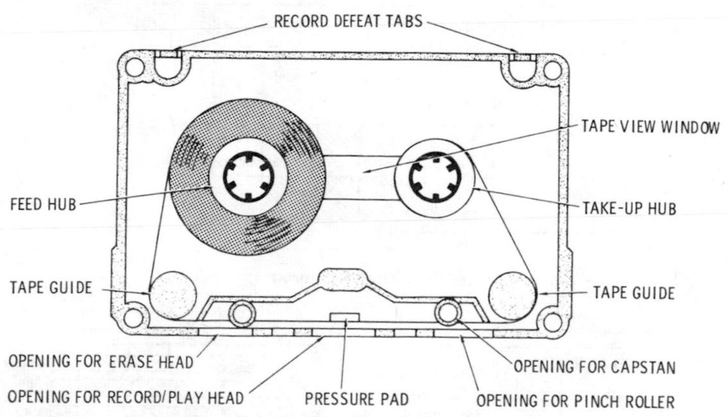

Fig. 3-15. Construction of a tape cassette.

The cassette is recorded from one end of the tape to the other in one direction. Then the cassette is turned over for recording in the opposite direction. (Some recorders reverse the head arrangment so that it is not necessary to remove the cassette, and continuous play is provided.) In order to get two sets of four-channel stereo tracks (four discrete channels) on the narrower tape, smaller track widths are used (Fig. 3-16C).

The location of the tracks on the cassette tape is different from the track positions for open-reel or cartridge tape. Head gaps are spaced to give the least possible cross talk between the two adjacent tracks while maintaining adequate stereo separation. Since the tracks of a stereo pair are side by side, a single, double-width monophonic head can pick up and blend the two stereo channels. Cassette decks can incorporate rewind and fast-forward modes. Because a very low tape speed ($1\frac{7}{8}$ ips) and thin tapes are used, the cassette can hold up to two hours of information (one hour in each direction). Four basic cassette recording times are available for program material: 30, 60, 90, and 120 minutes.

To prevent accidental erasure of a cassette, you may knock out the record-defeat tab (Fig. 3-15) for either or both sides. This prevents most recorders from switching into the record mode. Prerecorded cassettes are delivered with the tabs removed.

Recent improvements in techniques for reducing tape noise and in chemical formulas for the manufacture of tape have opened the way for cartridges and cassettes to provide sound quality equal to that of discs. There are several new mediums such as Crolyn, titanium, Coboloy, and other formulations to replace ferric oxide.

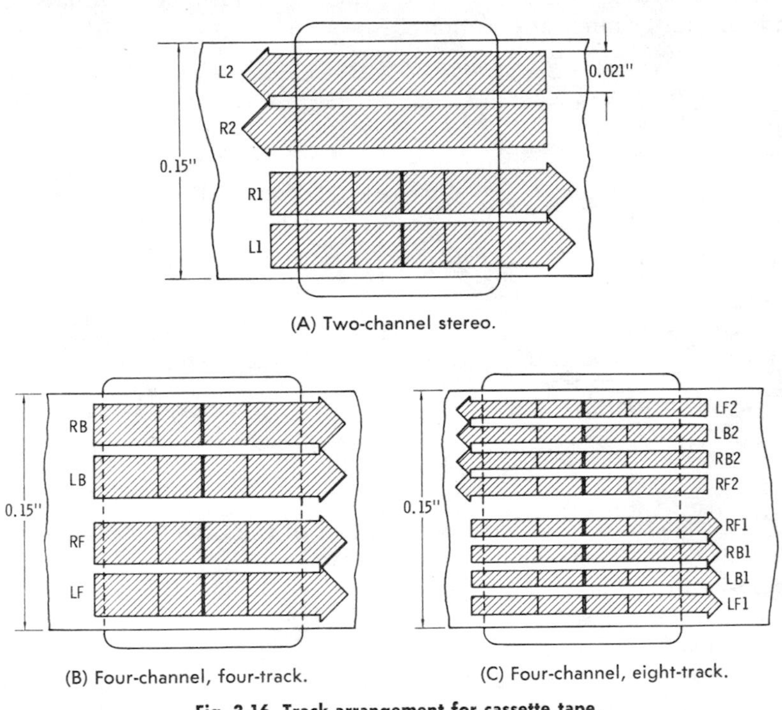

(A) Two-channel stereo.

(B) Four-channel, four-track.

(C) Four-channel, eight-track.

Fig. 3-16. Track arrangement for cassette tape.

Also, several manufacturers are improving the results obtainable with present formulations by increasing their density in the application process. These materials and techniques have demonstrated an ability to hold more "bits" of information on a given area of tape than do the former ferric-oxide applications. This ability is related to the signal frequency response, the level of output, and the signal-to-noise ratio. However, use of some of these new formula tapes requires a change in the level of bias necessary to be applied in the recorder. Therefore, a recorder must have a bias control or switch to be compatible with all kinds of tapes currently available. These improvements apply to all tapes, including those used in open-reel, cartridge, and cassette systems.

All stereo tape machines now use "in-line" heads, in which the gaps for adjacent channels are exactly centered along the same vertical line. However, there were some machines made with staggered heads; in these the tape passed over first one head, then the other. Tape recorded for this arrangement is obviously not playable on the in-line–head machine. It was at first thought that staggering was necessary to prevent cross talk between the channels, but the desired isolation is now obtained by proper spacing between the in-line gaps.

One of the advantages of tape as a stereo medium is the fact that isolation between channels is inherently much better than for phonograph pickups. Even with the small spacing between the tracks of a two-track system, 40 dB of separation is normally obtained. The desired channel separation is obtained much more easily in the four-track system because in this system the head gaps have more physical separation.

STEREO BROADCASTING AND RECEPTION

We have already considered two sources of stereo high-fidelity music for the home listener: disc recordings and tape recordings. A third source is the signal of a radio broadcast station. Instead of a record player or tape machine, a tuner is used. Tuners for stereo are shown in Chapter 4. They are similar to single-channel tuners except for the technique of multiplexing, which will be described in this chapter.

Two-Station Stereo Broadcasts

The first method used for stereo broadcasting and reception is illustrated in Fig. 3-17. Two complete transmitters were used, one sending out the left (L) stereo signal, the other the right (R) stereo signal of the same program. All combinations of a-m broadcast, fm broadcast, and television sound-channel transmitters have been tried. Most popular were the "fm/am" (one fm station and one a-m station) and the "fm/fm" (two fm stations) methods.

The two-transmitter method of transmitting stereo was convenient because it utilized existing transmitting and receiving equipment with no circuit changes. However, it did have the following disadvantages:

1. It wasted frequency spectrum space. Two station channels had to be used for each program.
2. Differences in propagation characteristics of the waves radiated by the two transmitters led to variation in signal amplitude and quality within the separate channels. This was particularly

true of fm/am combinations, where the carrier frequencies are so widely separated.

3. In some cases, especially with fm/fm, the duplication of complex receiving equipment represented an excessive expense. Ninety percent of the programs were nonstereo, requiring only one receiver; but to receive stereo, two receivers were needed. Also, a number of am/fm tuners could not receive both bands simultaneously, and so they had to be supplemented with additional receiving equipment.

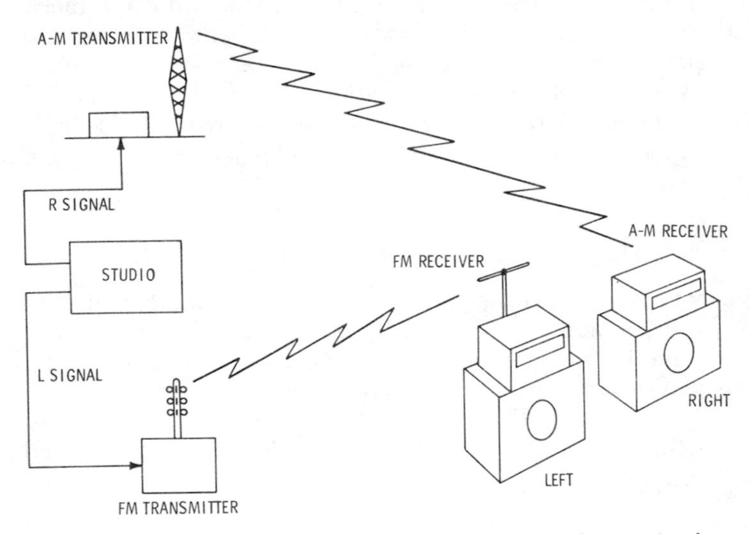

Fig. 3-17. Two-transmitter method for radio transmission of stereo signals.

Stereo Multiplex

The disadvantages of the two-station method made it imperative that a method of transmitting both stereo signals on one carrier be devised. The result is the method called *stereo multiplex*. "Multiplex" means "a method or arrangement for sending two or more messages simultaneously on one carrier or circuit." This is exactly what stereo multiplex does—it transmits both channels on the same fm carrier.

The standard fm broadcast system is based on an audio frequency response of approximately 50 Hz to 15 kHz. There is nothing about the basic modulating system to prevent extension of this audio-frequency range to 75 kHz (further extension would require more bandwidth than is currently allotted). Therefore, if at some point above the limit of human hearing (but still within the 75-kHz bandwidth) another signal is added, it will be amplified and detected by the regular receiver circuits. However, being above the range of hearing (supersonic), it will not interfere with the regular audio signal. Thus, the carrier is modulated by the regular audio signal plus the supersonic signal.

This supersonic modulation component is known as the *subcarrier.* It, in turn, is modulated by another audio signal. As long as the bandwidth of the subcarrier is not allowed to extend downward into the range of the regular carrier, no interference will occur. This complete modulated subcarrier signal is part of the composite signal which modulates the main carrier. One of the stereo signals is transmitted as modulation of the main carrier, and the other signal as modulation of the subcarrier.

In receiving, two demodulating circuits are needed. First the main carrier is demodulated, to get the first stereo signal and the modulated subcarrier. Then the modulated subcarrier is separated and demodulated, to get the second stereo signal.

There will always be some listeners who don't care about being equipped for stereo reception. Also, for portable use, and where cost is to be minimized, a conventional single-channel receiver is called for. For these reasons, it is important that the stereo system be compatible. By "compatible" we mean that while stereo multiplex transmission is in progress a listener with a conventional receiver (one not equipped for stereo) should be able to receive the transmission as a full monophonic signal.

If one channel of the multiplex system is used for the L signal and the other for the R signal, compatibility is not provided. The owner of the conventional receiver hears only the modulation of the main carrier, which in this case would be just the L or R signal; the subcarrier frequency is above audibility, so neither the subcarrier nor its modulation is used. For a truly compatible system, the nonstereo listener should be able to hear the combination of the signals of both the L and R channels.

To meet this compatibility requirement, the main carrier can be modulated with a full monophonic signal (L plus R), and the subcarrier channel can be modulated by the difference between the two signals ($L - R$). The $L + R$ modulation signal provides the nonstereo listener with his full monophonic reception; his receiver does not respond to and thus ignores the subcarrier.

For stereo reception, the L and R signals are recovered by demodulating the subcarrier signal and adding or subtracting the difference signal to or from the main carrier ($L + R$) signals. Electrical addition of the signals can be accomplished simply by applying both signals across a common impedance. Subtraction can be accomplished by first inverting one of the signals (changing its phase by 180 degrees), then adding it to the other signal.

STEREO VERSUS HIGH FIDELITY?

One point should be emphasized: Stereo reproduction is part of high fidelity, not an "additional feature." The impression is sometimes wrongly given that "hi-fi" and stereo are two separate stages of development of audio reproduction. This is a fallacy, unfortunately sometimes encouraged by a statement that some equipment features "high fidelity and stereo."

Stereo is a "fine point" in high fidelity; that is, if you do not have all the good qualities of high-fidelity monophonic systems to start with, other forms of distortion will prevent appreciation of the benefits of the stereo effect. If the reproduction of your system is clean and clear of harmonic, intermodulation, and transient distortion, it is likely that improvements in spatial sound effects through stereo will be appreciated.

4

Program Source Equipment

Any hi-fi installation requires one or more pieces of equipment to provide program material. Those commonly used may be described as follows:

1. *Record Players:* There are two popular types. The most popular is the automatic record changer; for optimum results, the preferred type is the professional but plain turntable, manually operated. Both types are available in models that will play records of all sizes and speeds.
2. *Tuners:* These are a-m and fm radio receivers especially designed to receive, amplify, and rectify signals of wide bandwidth to provide a hi-fi audio signal.
3. *Tape:* Playback equipment can provide several hours of recorded program material as does the record player. The primary advantage of tape is that one may easily record, store, and play back one's own recorded material of music, entertainment, or information for any purpose.
4. *Television:* The audio components of a television signal may be picked up by a special television audio tuner or from the television receiver and fed through a system to provide hi-fi audio to accompany the television picture program.
5. *Microphones:* These are electromechanical transducers designed to convert sound waves into electrical impulses. Microphones are used occasionally in hi-fi to record, entertain, or announce, or for other special purposes.

All record players have a motor, turntable, pickup arm (tone arm), pickup cartridge, and needle. The professional high-quality reproduction systems generally have a manually operated turntable. Record changers with automatic devices to change records at the end of play of a record, providing continuous operation for long periods, are more popular. We will cover the component parts of both the manually operated and automatic types, starting with the needle.

Reproducing Needles

To minimize wear and secure the maximum useful life from a recording, a properly shaped needle must be used. A properly formed reproducing needle will also minimize background noise.

Fig. 4-1 illustrates a group of playback needles seated in the grooves of a recording. The needle at (1) is of a theoretically ideal shape. At (2) and (3) the needles are too sharp and will gouge the bottom of the groove. The one at (4) is too blunt and will cause excessive wear on the walls of the groove, resulting in their eventual breakdown. At (5) is a needle of satisfactory shape. However, the cutting stylus that makes the groove that the playback equipment must track is shaped like a diamond. The difference in shape between the cutting stylus and playback needle causes trackability distortion and pinch distortion.

When a cutting stylus cuts a sine-wave groove, it moves back and forth in one plane, and the points of contact in the groove are

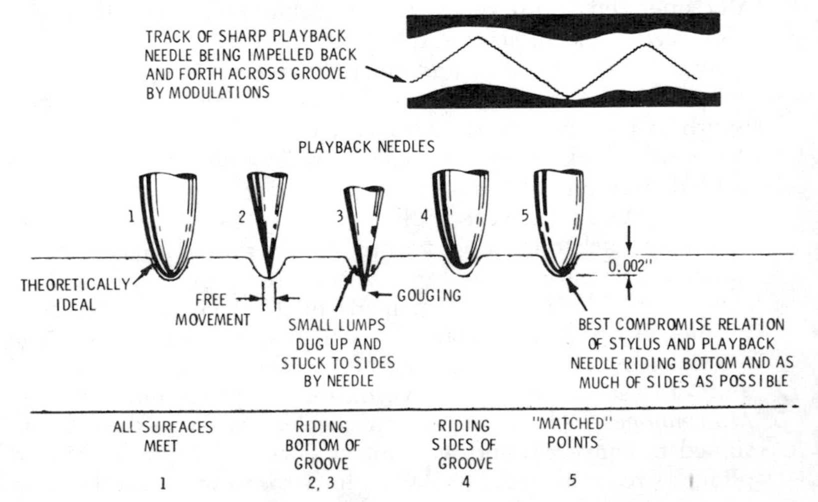

Fig. 4-1. Properly and improperly shaped playback needles.

in a line parallel to the radial from the center of the record being cut. However, when a round needle follows this groove, the only time the points of contact are in a line parallel to the radial from the center of the record is at the peaks of the sine-wave excursion (Fig. 4-2A). At the mid-point, or steepest-slope point, of the sine-wave groove, the points of contact of the cutting stylus are still parallel to the radial, but the points of contact of a round stylus are

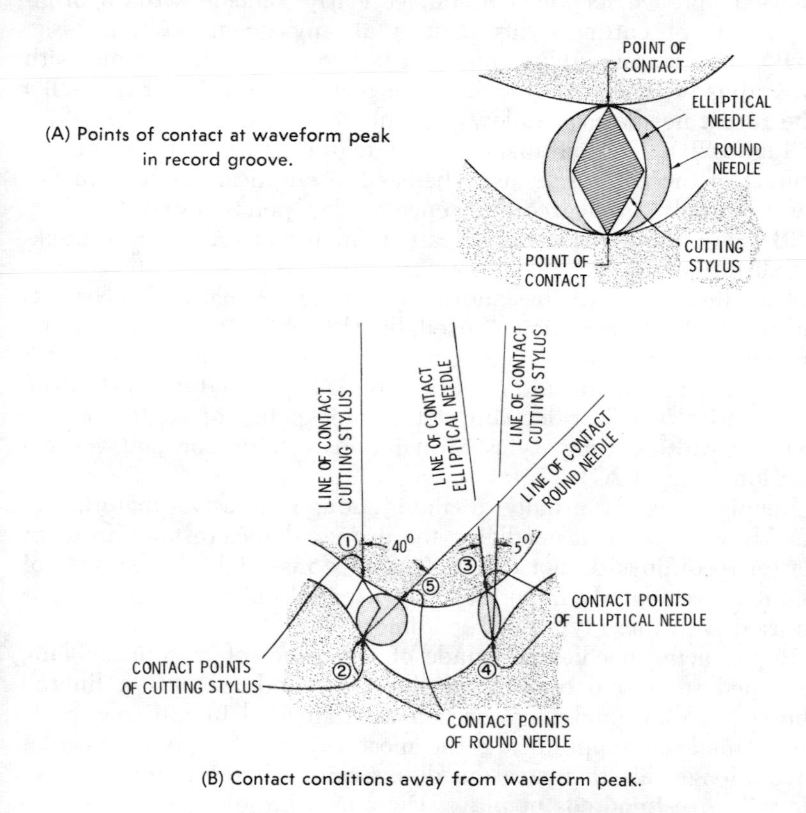

(A) Points of contact at waveform peak in record groove.

(B) Contact conditions away from waveform peak.

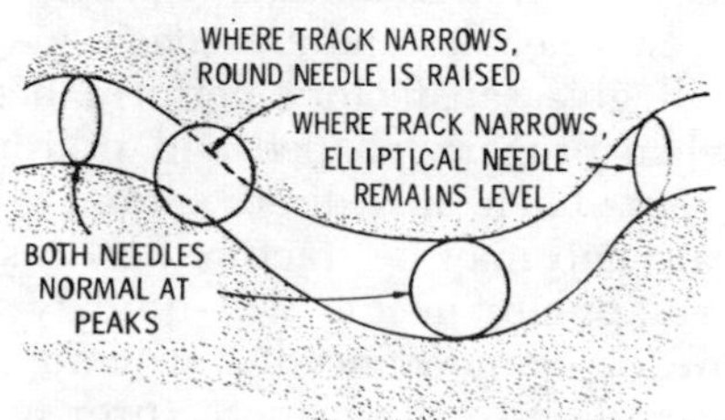

(C) Pinch effect for round and elliptical needles.

Fig. 4-2. Needle shape and performance.

inclined nearly 45 degrees to the radial passing through the center of the record. The resulting tracking distortion is comparatively large.

This condition is mostly overcome by the use of an elliptically shaped needle. The needle is mounted in a position so that the long axis of the ellipse is nearly parallel to a radial passing through the center of the record. Thus, when the elliptically shaped needle tracks the groove, its points of contact nearly coincide with the original points of cutting-stylus contact at any given position (Fig. 4-2B). The cutting stylus cuts at a 0-degree operating angle with the radius of the groove circle. (See points 1 and 2 of Fig. 4-2B.) The round needle may follow, making contact on points (2 and 5 of Fig. 4-2B) of a line making a 40-degree angle with the line of contact of the cutting stylus, whereas the elliptical needle will follow with only a 5-degree difference angle (points 3 and 4 in Fig. 4-2B) with a proportionate increase in fidelity or reduction of tracking distortion.

The elliptical needle has another advantage in that it also corrects for pinch-effect distortion caused by the changing width of the groove made by the moving cutting stylus. As the groove width narrows, the round needle rises because it is pinched upward. Since the elliptical needle rides almost the same points of contact as the constant-width cutting stylus, it maintains a fairly constant vertical position (Fig. 4-2C).

Regular records usually contain enough abrasive material to quickly wear a metal needle to the proper shape. Instantaneous or master recordings do not contain abrasive material, and because of this it is particularly important that a needle of the correct shape be used with these types of recordings.

Reproducing needles are made of a number of materials. Plain, hardened steel, and osmium needles are good for only a limited number of plays and therefore are seldom used in hi-fi playback. Those made of sapphire are the most common. Sapphire needles have a longer life than steel needles. A properly used sapphire needle will give hundreds of plays. Diamonds are also used in reproducing needles and are considered to be the best. Diamond needles will give satisfactory performance for thousands of playings, and when properly shaped and polished they are superior to all other types. Their only disadvantage is that they are expensive and delicate and may be fractured by a slight impact. A chipped diamond or sapphire needle will quickly ruin any type of recording. The life of diamond needles in comparison with other materials varies according to the manufacturer and the user. A life expectancy for a diamond needle many times longer than that of competitive materials is not unusual. Wearing qualities of different needles are

Fig. 4-3. Needle wearing qualities.

shown in Fig. 4-3. Prices of diamond needles are dropping to reasonable levels, making them best buys for top hi-fi playback.

The needle-tip size for monophonic (single-channel) records may be over 0.001 inch (1 mil), whereas needle tips for stereo records are either 0.5 or 0.7 mil.

Pickup Cartridges

The construction and the electrical characteristics of the various available pickups vary greatly. They can be classified into five distinct groups. These are: crystal, ceramic, magnetic, dynamic, and capacitance.

Units of varying quality can be obtained in each group. The tastes and desires of the user and the use to which the pickup is to be put may govern its choice. With systems designed to reproduce the voice for reference purposes only, an inexpensive pickup with a comparatively narrow frequency range is suitable. Where the highest quality of reproduction is required, a special highly damped pickup having a frequency characteristic flat to beyond 15,000 Hz is necessary.

Crystal and Ceramic Pickups—Crystal and ceramic pickups are the lowest-cost units mentioned in the foregoing. They are simple in design and construction, have fair frequency response characteristics, and hi-fi types have low distortion content.

Some crystalline substances possess the ability to produce an electrical charge under certain conditions. When they are stressed mechanically, a charge is produced on their surfaces. If a voltage is applied to the surfaces of a crystal with piezoelectric properties, a mechanical deformation of the crystal will take place.

The *piezoelectric crystal* acts as a generator and converts mechanical motion into an electrical charge. Crystal microphones and phonograph pickups can be thought of as piezoelectric generators. A crystal is also similar to a motor. When a potential is applied to a crystal, it moves. It converts electrical energy into mechanical motion. Crystal headphones and record cutters are piezoelectric motors.

A piezoelectric crystal as used in microphones and pickups is a formation of crystalline Rochelle salt. The Rochelle-salt crystal possesses the piezoelectric property to a high degree. It is approximately 100 times as active as a regular quartz crystal.

Rochelle salt crystals are formed in large bars. These bars are cut into slabs or plates for use in the manufacture of crystal elements. The two commonly used crystal plates are usually referred to as *expander* and *shear plates,* as shown in Fig. 4-4. The crystal is either a shear or expander plate, depending on the way it is cut from the bar.

A crystal plate is said to have three axes. The latter are the electrical (*AA*), the mechanical (*BB*), and the optical (*CC*) axes. An expander plate is cut at a 45-degree angle to the optical and mechanical axes of a crystal bar. A shear plate is cut with its edges parallel to the mechanical and optical axes of the crystal bar.

When a potential is applied to the two large faces of each plate, mechanical motion is developed at an angle of 45 degrees from that of the mechanical and optical axes. (When a force is applied, an electrical potential will be developed at the faces.) Therefore, the expander plate will increase its length and, at the same time, decrease its width. If the polarity of the faces of the crystal is changed, the crystal will decrease its length and increase its width.

The same action takes place when a potential is applied to a shear plate, except that expansions and contractions occur along the diagonals of the plate instead of parallel to the edges, as in the case of the expander plate. When mechanical pressure is applied, a potential voltage is produced.

In order to form a crystal element for use in a crystal cartridge or other device, a number of crystal plates are cemented together. This makes possible more effective utilization of the properties of the crystal.

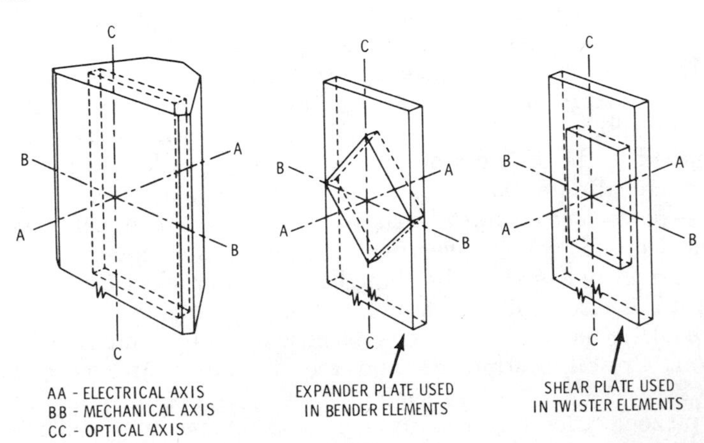

Fig. 4-4. Types of crystal plates.

An element which consists of a number of expander plates cemented together is referred to as a *bender* element, while an element formed from a number of shear plates is known as a *twister* element. The names "bender" and "twister" refer to the action which takes place when an electrical potential is applied to the element.

The multiple crystal has a number of important advantages over a crystal employing a single plate. It greatly decreases the undesirable effects of saturation and hysteresis and reduces the effects of temperature on the impedance and sensitivity of the unit. Fig. 4-5 shows the construction of a bender and a twister crystal element. The faces of each crystal plate are milled smooth, and foil or graphite electrodes are applied. Leads are connected to the electrodes, and after they have been properly oriented, the plates are bonded together with cement.

(A) Bender element. (B) Twister element.

Fig. 4-5. Construction of crystal elements.

The completed crystal element is coated with a special moistureproof material to protect it against deterioration under very dry or damp conditions. The crystal element is mounted in a nonconducting case and held at one end by a metal clamp. The other end of the crystal is free, permitting it to move torsionally. A bearing and chuck are mounted on the free end of the crystal. The bearing usually consists of rubber or a similar synthetic material. The chuck usually consists of a light metal such as aluminum. To restrain the crystal from vibrating at more than one mode, it is customary to restrain it slightly along its axes of motion. This is accomplished by cementing a strip of damping material along the length of the element. This strip also gives some damping effect in the other modes and helps to reduce the amplitude of the resonant peak of the crystal. Because the crystal is very stiff, its resonant frequency is normally not in the frequency range to be recorded.

As a circuit component, a crystal cartridge acts in the same way as does a capacitor and can be considered as such. Extremes of temperature drastically affect the operation of a crystal. The maximum sensitivity of a crystal is usually at about 75 degrees. As the temperature rises above or falls below 75 degrees, the sensitivity of the crystal falls off slowly. Cartridges using barium titanate have much improved temperature characteristics.

At temperatures in the neighborhood of 130 degrees, Rochelle salt crystals may permanently lose their piezoelectric properties. As a rule, temperatures slightly below this, that is, from 110 degrees to 120 degrees will not injure a crystal.

Ceramic pickups have overcome this difficulty and are now commonly used in place of crystals. A ceramic pickup has the advantage that it is quite stiff, and, because of this, variations in mechanical load do not greatly affect the performance. Ceramic pickups have operational disadvantages in that they have some distortions and resonances which are almost always present. Also, most have rolloff of the high frequencies. Types such as the cartridge shown in Fig. 4-6 have overcome these faults to a major degree.

Magnetic pickups—The magnetic pickup is a current-operated device. The construction of magnetic pickups varies greatly. Essentially they consist of a coil and magnet and another magnet to which the needle is attached. The movable magnet to which the needle is affixed is damped. This is accomplished in a number of ways, depending on the construction of the particular cartridge.

The coil is connected directly to the input of the preamplifier. Current through it varies with the change in the density of the flux about the fixed magnet. This variation in flux is produced by variation in the force exerted on the movable magnet, for each change in position. Since the reproducing needle is connected to this magnet, it produces output in proportion to the movement.

The frequency range of a magnetic pickup is greater than that of the crystal type. The finest magnetic pickups have a frequency

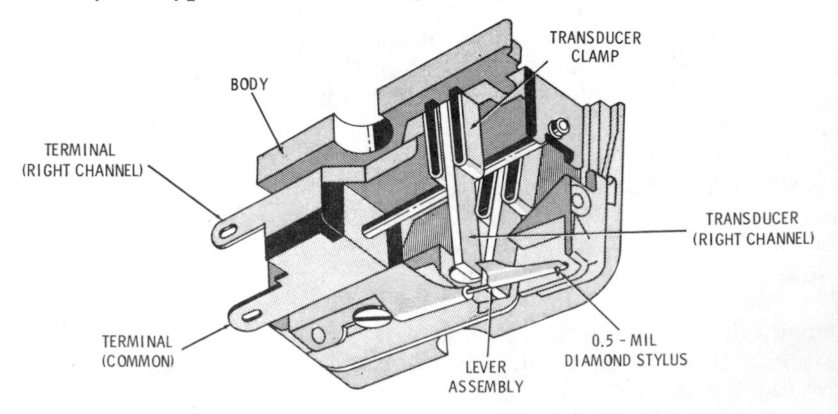

Courtesy CBS Electronics, Div. of CBS, Inc.

Fig. 4-6. Construction of the Columbia Professional 55 ceramic stereo cartridge.

range of from 50 to over 15,000 Hz. The distortion content of a fine magnetic pickup may be as low as 0.1 percent at 400 Hz, and 1 percent to 4 percent at high frequencies. A typical hi-fi reluctance-type magnetic cartridge is shown in Fig. 4-7.

Dynamic Cartridges—The dynamic cartridge, as the name implies, is of the moving-coil type. The unit consists of a movable coil, to

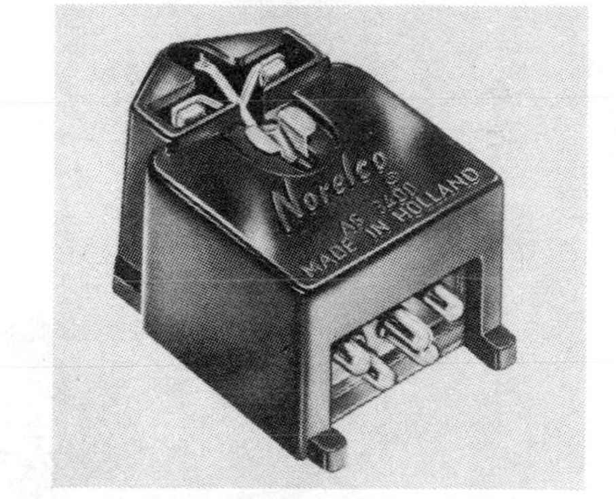

(A) Pickup.

Courtesy North American Philips Corp.

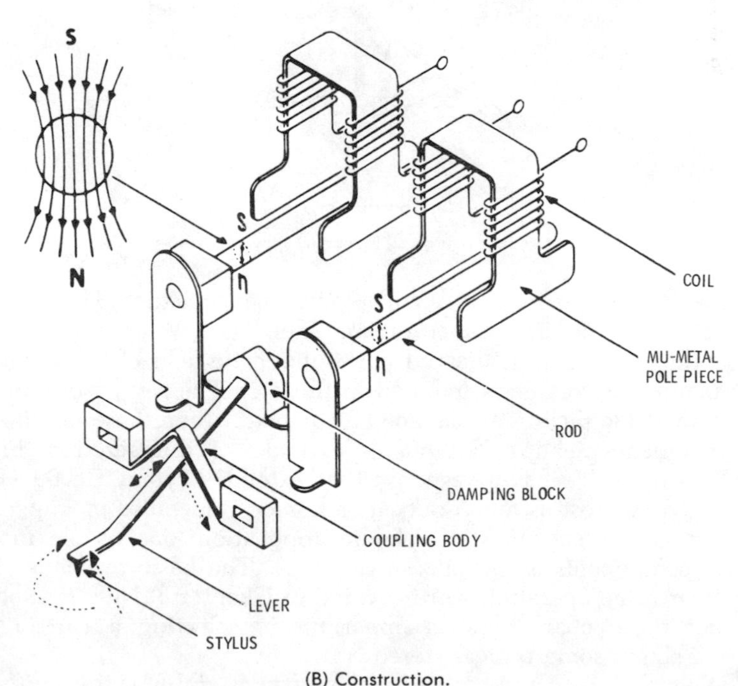

(B) Construction.

Fig. 4-7. A moving-magnet type of stereo pickup.

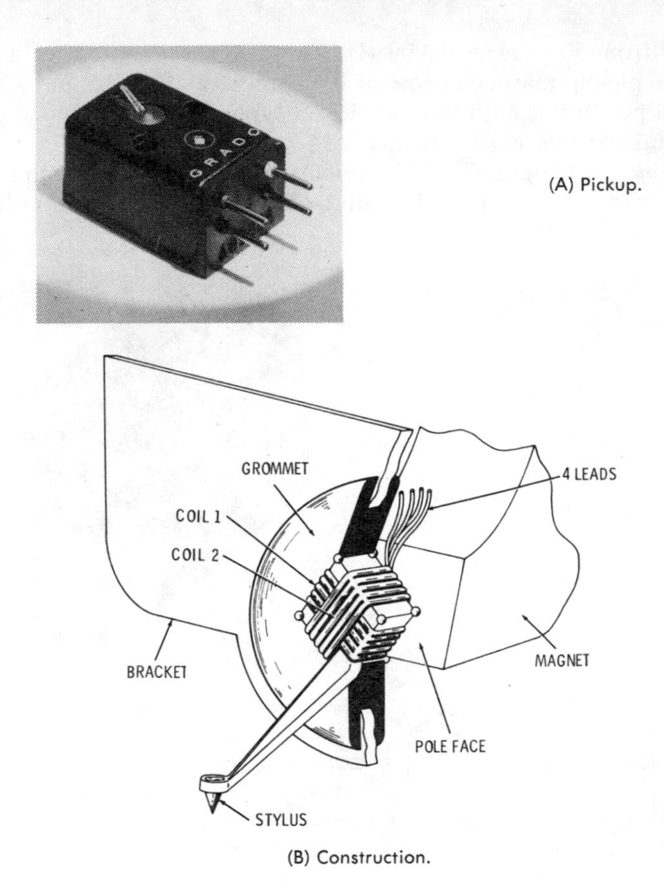

(B) Construction.

Fig. 4-8. The Grado moving-coil (dynamic) type of stereo pickup.

which the needle is mounted, and a permanent magnet. The coil is connected to a low-impedance preamplifier input. When the needle moves the coil, which is located in the magnetic field of the permanent magnet, a voltage is induced in the coil. The results obtained with a dynamic pickup are as good as or better than those obtained with magnetic pickups. A dynamic cartridge is illustrated in Fig. 4-8. This cartridge has a response that extends beyond 20,000 Hz, and it gives almost as much output as a good reluctance pickup.

Stereo Cartridges—In stereophonic application, one of the most critical components is the pickup cartridge. The basic principles of stereo cartridge operation were covered in Chapter 3. Now we shall consider those characteristics important in selecting a cartridge. Fig. 4-9 shows some typical stereo cartridges.

The weight of a cartridge and pickup arm assembly is transmitted to the record as needle force. The maximum allowable needle force without excessive record wear is related to the size of the needle tip. This is because wear is dependent on pressure, which is force per unit area; thus the smaller the needle tip, the less the needle force must be to prevent wear. If needle-tip size were made too small, the necessary reduction in needle force would result in loss of tracking; that is, the needle would skate over the record instead of staying in the groove. Thus the minimum needle-tip size is kept to 0.5 mil, and the range of needle force is roughly from ½ to 6 grams.

As is the case with all high-fidelity system elements, we are interested in the frequency response of a cartridge. Response is usually stated in terms of frequency range and the deviation, in dB, of the response over that range. We are naturally interested in the widest possible frequency range, but it is doubtful if response below 20 Hz makes much difference. However, the high limit of the range should extend to near 20,000 Hz or beyond to take full advantage of the better recordings.

The output voltage of a cartridge is important, not only because it is used to calculate how much amplification is needed, but also because the higher the output, the more chance there is of having a good signal-to-noise ratio. However, sometimes it is desirable to sacrifice some signal strength to ensure minimum distortion, best frequency response, and minimum record wear. Output voltage does not establish criteria unless the level at which the needle is driven is also specified. Standard records are used to provide the drive for output voltage tests. Some cartridge manufacturers state output for a needle velocity of 5 centimeters per second (cm/s), and others for a 10-cm/s needle velocity. Naturally, the output should be higher for the greater velocity. In Fig. 4-10, the needle velocity is shown in relation to needle force and frequency. For total outputs below 10 millivolts, a separate preamplifier may be necessary, depending on the gain of the system used.

Channel separation is the indication of how well the left signal is kept out of the right channel, and the right signal out of the left channel. Good separation is necessary for good stereo effect, since the difference between the two signals is what produces the spatial effect. If a cartridge were not carefully designed with separation in mind, one channel would affect the other, and the outputs of the two channels would tend to become the same. Tests have shown that a minimum of 15-dB separation should be maintained.

To operate properly, a cartridge must be connected to the proper amplifier input impedance. Generally speaking, ceramic and crystal pickups must work into a relatively high resistance (15 kilohms to several megohms) compared with magnetic cartridges (5 to 100 kilohms). Each cartridge manufacturer specifies the proper load characteristics for his models.

Compliance is a measure of how easy it is to move the needle in the directions it must be driven during playing. It is ordinarily measured in millionths of a centimeter per dyne (10^{-6} cm/dyne), i.e., the distance, in millionths of a centimeter, that the needle can be pushed by a force of 1 dyne in that direction. For stereo cartridges, it is important that they not only have a high compliance,

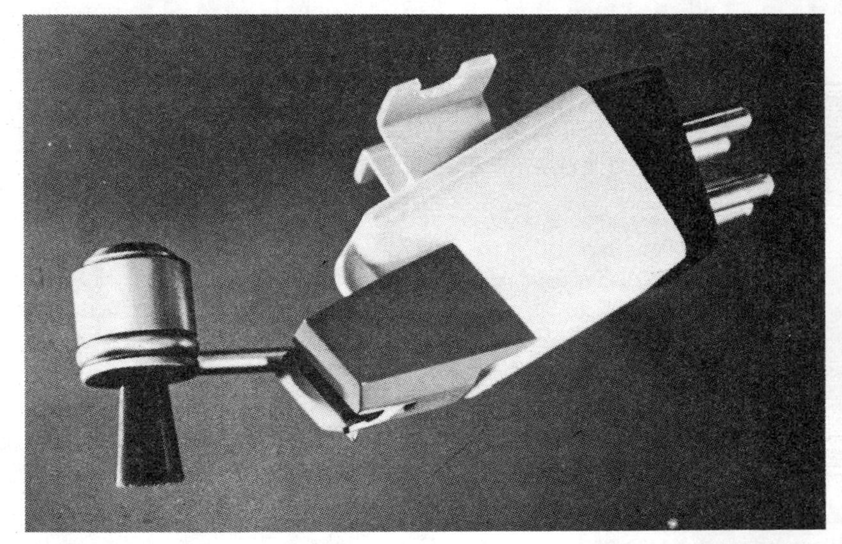

Courtesy Pickering & Co., Inc.

(A) Pickering Dustamatic.

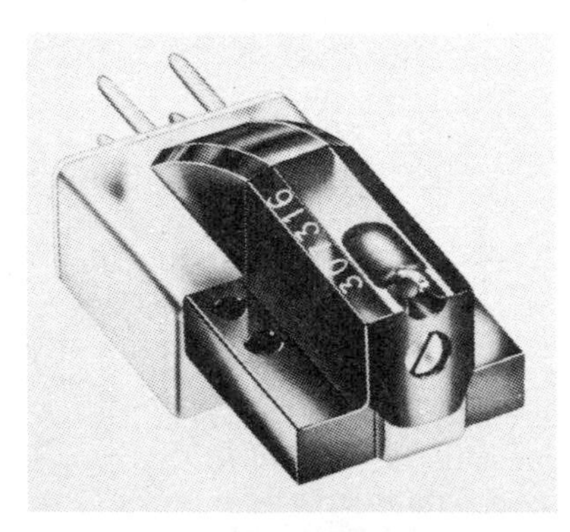

Courtesy Ortofon Div., Elpa Marketing Industries, Inc.

(B) Ortofon SPE/T.

Courtesy Shure Brothers Inc.

(C) Shure V-15 Type II.

Fig. 4-9. Typical

but that this high compliance apply in all directions of motion. This is why many stereo cartridge manufacturers specify both lateral and vertical compliance. It will be shown that it is the low vertical compliance of most monophonic pickup cartridges that makes it necessary that stereo records never be played with monophonic pickups.

Another characteristic that is sometimes specified by cartridge manufacturers is channel balance. It is given as the deviation from

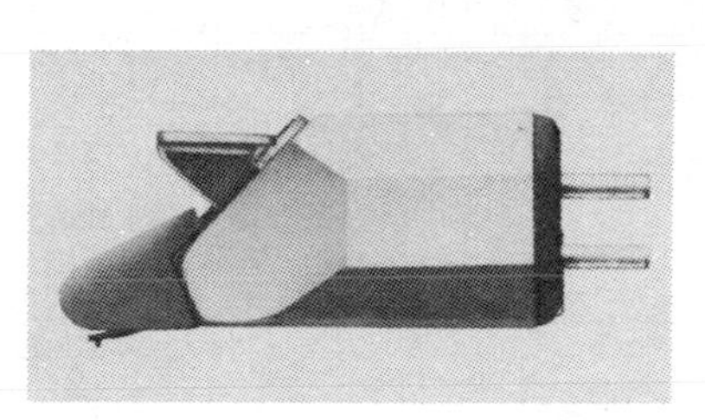

Courtesy Empire Scientific Corp.

(D) Empire 888P.

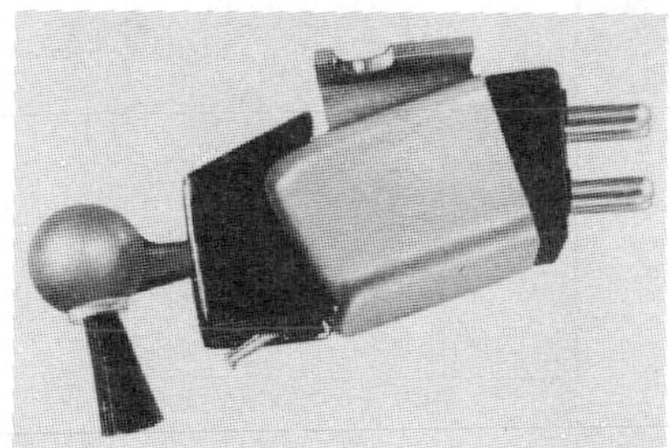

Courtesy Stanton Magnetics, Inc.

(E) Stanton Longhair.

(F) Dusting action of brush.

stereo cartridges.

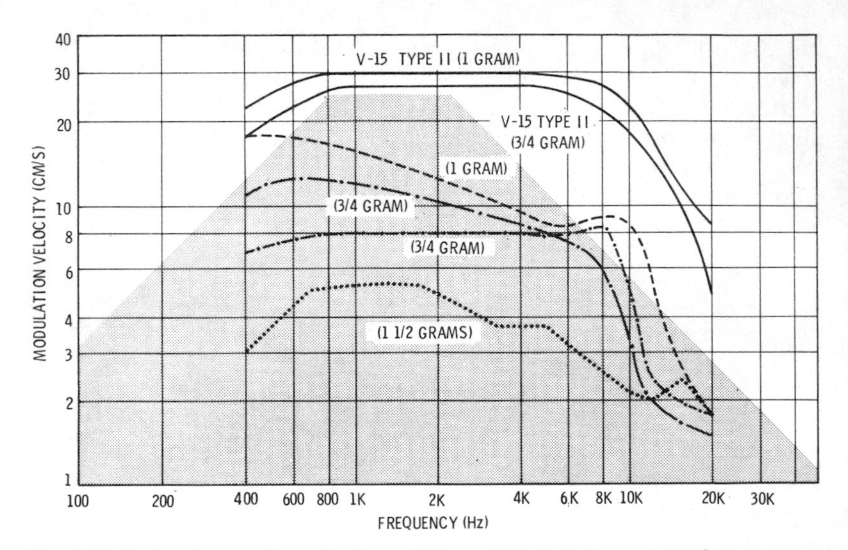

Fig. 4-10. Performance and needle force of Shure high-performance cartridges compared with characteristics of earlier models.

balance, that is, the difference in dB between the outputs of the two channels for the same drive. Actually, this characteristic is not very critical because just about all stereo systems include a channel-balance adjustment to compensate for reasonable differences in output. However, a relatively large unbalance, such as more than 3 dB, is an indication of a defective cartridge. The optimum arrangement is a system that is naturally balanced throughout, and balance deviation of 1 dB or less should be considered good.

Application—The pickup cartridge should be mounted in a suitably designed arm. The needle pressure should be left at the lowest value consistent with the design of the cartridge used. For optimum results, the arm should have an offset head to minimize tracking error, and side-to-side motion should be as free as possible.

Vertical sensitivity is an important characteristic of any cartridge. This sensitivity is mostly related to the vertical movement sensitivity of the cartridge and stylus. Low vertical sensitivity in a properly designed monophonic unit will reduce hum and other turntable noises without affecting the quality of pickup of the lateral groove modulation from the record. However, as explained previously, vertical sensitivity is important for proper reproduction of stereo records. Hum may be picked up by induction in magnetic- and dynamic-type pickups. Hum sensitivity is another important consideration in pickup selection.

In general, reluctance-type magnetic pickups are most widely used. Dynamic pickups reproduce extremely well, but they are less sensitive. Ceramic units are best for medium-quality work requiring high output from the pickup unit; these units will drive a basic amplifier directly without need for a preamplifier. In comparison with ceramic units, the response of the better reluctance types is more nearly flat, and they have the least distortions and resonances. While hi-fi crystal pickups are entirely suitable for most reproduction, the best crystal units usually have a number of small resonances and distortions. Capacitance units are suited to special applications requiring extremely wide frequency response. However, much of the operating quality of any cartridge depends on the pickup arm and the needle pressure.

The Pickup Arm (Tone Arm)

The pickup arm or tone arm is the carriage for the cartridge and needle. Its design and principle of operation are very important to high-fidelity reproduction. To permit the needle to respond without distortion to record groove deflections (tracking ability), the tone arm should have free movement in all directions. The force of weight of the arm applied vertically to the needle should be adjustable to provide for the needs of various kinds of cartridges. Optimum pressure will generally be found between ½ gram and 6 grams, with an average setting of 1 recommended for most cases. Several companies make a device available for measuring this weight near the point of application. The weight must be within the range required for a certain needle-cartridge combination. It must be heavy enough to keep the needle in the groove to follow the modulation, and light enough to allow free movement and low wear on the needle and the record. The pressure on the needle therefore should be adjustable and the adjustment made preferably with a calibrated scale.

The pickup arm should track as nearly as possible the original line of cut; otherwise, a tracking error will develop and cause distortion. See Fig. 4-11. The record is cut along line *AB*, and the usual pickup arm operates in an arc, as shown by line *CD*. Placement of the mounting for the pickup arm can vary the tracking to a large degree. When the playback is made with the arc of contact of needle to record too far off, the needle follows the modulation with a modified movement because the round needle tip is moving with a direction and amount per modulating element different from the direction and amount per modulating element of the original cutting stylus. This causes tracking distortion in the signal generated by the cartridge.

The arm is best designed to ride on precision aligned bearings for the greatest reduction of resistance to movement and for the lowest inertia. Some units use ball bearings.

Pickup arms that provide for different kinds of cartridges enable one to try and use different cartridges.

For stereo application, the pickup arm (tone arm) and turntable and drive arrangement must meet special requirements which are more rigid than those sufficient for monophonic reproduction. Some pickup arms designed particularly for stereo are illustrated in Fig. 4-12. Let us consider some of the features to be considered when one is selecting a pickup arm for stereo.

One of the characteristics most important in stereo pickup arms is the tracking error just described. When there is an angular difference between the direction of motion of the groove the needle is in and a line between the pickup-arm lateral pivot and the needle, tracking distortion and unbalance in the two outputs are produced. In other words, instead of the needle pulling "in line" with the pickup, at some parts of the motion across the record it pulls in a different direction, producing proportional unbalance. The lateral component of force caused by tracking error in a stereo record groove produces more wear on one side of the record groove than on the other. Hence, one stereo component signal will be reduced in relation to the other. Also, stereo cartridges operate at much lower needle force than mono cartridges and cannot tolerate as much lateral tracking-error force without jumping out of the groove.

Tracking error is minimized in pickup arms by orienting the pivot point and needle point locations with respect to the record and making the angular offset of the pickup with respect to the longitudinal axis of the arm optimum. By proper adjustment of both of these variables, arm manufacturers have been able to reduce tracking error to a fraction of a degree in some models.

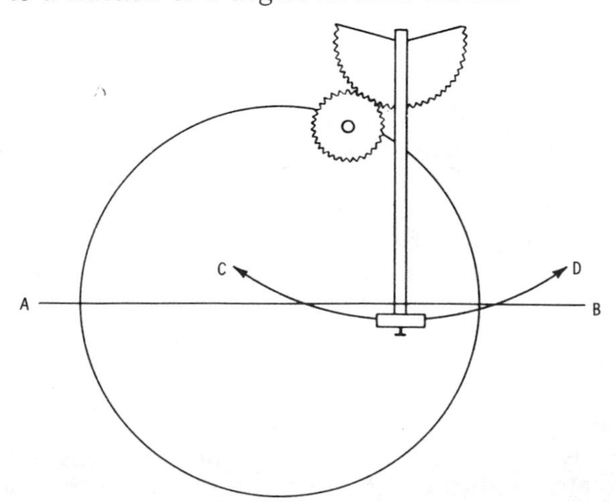

Fig. 4-11. Tracking error of tone arm.

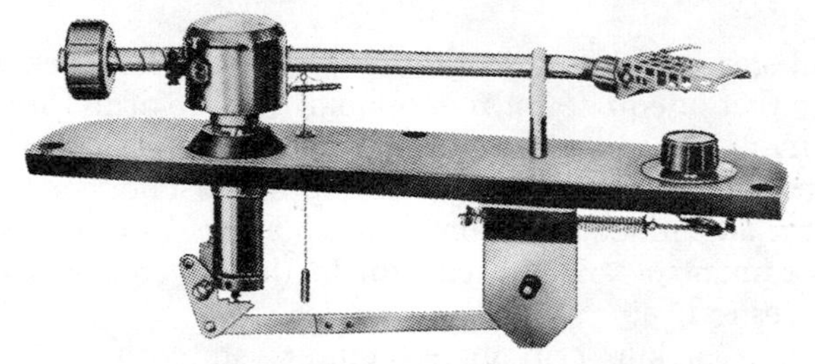

Courtesy Thorens Div., Elpa Marketing Industries, Inc.

(A) Thorens TP14.

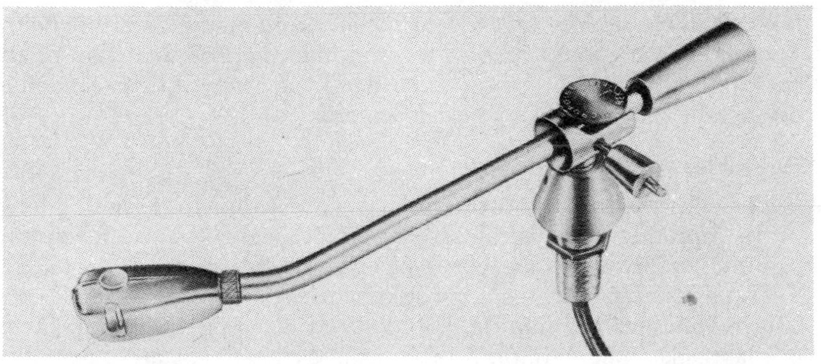

Courtesy Rek-O-Kut Company, Inc.

(B) Rek-O-Kut Micropoise.

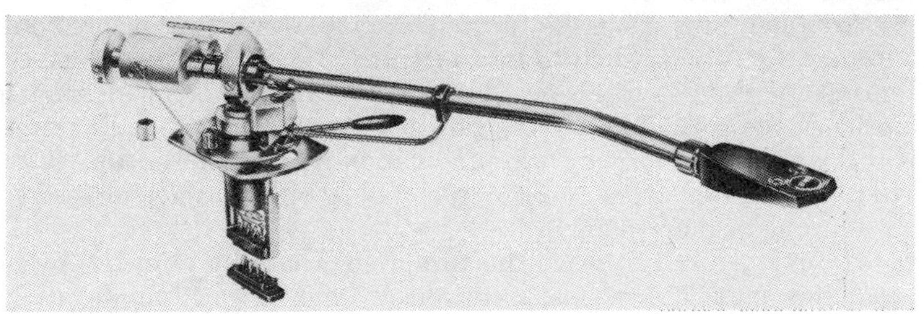

Courtesy Shure Brothers Inc.

(C) Shure SME Series 2.

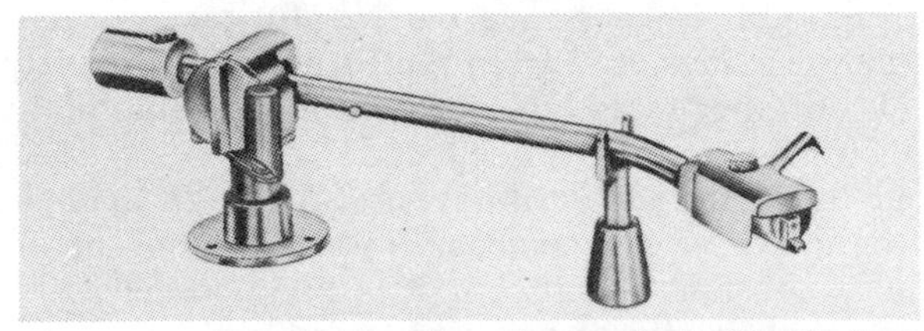

Courtesy Empire Scientific Corp.

(D) Empire 980.

Fig. 4-12. Typical pickup arms designed for stereo.

A good stereo pickup arm should have a needle-force adjustment finer than that adequate for monophonic pickup arms. Stereo models require more precision, because of the lower forces and smaller tolerances to which they must be adjusted. The normal arrangement is for adjustment first for perfect balance (zero needle force), then adjustment of another control to the exact number of grams of force desired.

Sometimes pickup arm manufacturers state the *tracking force*. This is the force that must be exerted laterally against the needle to move it in across the surface of the disc. In arms for stereo, the tracking force seldom exceeds 2 grams, and in many cases it is below 1 gram. The tracking force of an arm must be less than that of the cartridge used with it, or the cartridge rating would be exceeded as the needle moved inward on the record.

Turntables

The construction of a turntable has a great deal to do with fidelity of the reproduction. Wabble, wow, and other turntable defects show up unfavorably when a record is played back.

Three standard turntable speeds are used. They are 78.26, 45, and 33⅓ revolutions per minute. Fixed speed is a very important factor in playback, since variations of as little as 1 percent are detectable. Speed variations are caused by changes in the power source of the turntable and motor.

Wabble results when the turntable, spindle, and bearing assembly are not carefully aligned to be concentric. When wabble occurs, the surface of the record moves closer to and farther from the pickup head as the turntable moves. This results in changes in the strain on the needle, modifying the modulation as it is picked up. In extreme cases, the stylus may completely leave the surface of the record at one or more points.

The motor used to drive the turntable is usually powered by alternating current, and, as is universally the case, vibration at the power frequency is set up within the motor. If this vibration is transmitted to the turntable and pickup, it will result in hum modulation of the pickup. This is known as *rumble*. In good turntable assemblies, the motor is insulated from the turntable sufficiently to eliminate rumble. Rubber mats lying on top of the table and under the record will further reduce this effect.

To avoid the defects described, careful design and construction of the motor and drive system are necessary. Motors can be designed to give fairly constant speed, but it is not possible to design and build a motor inherently of constant speed without some degree of undesirable fluctuation. The fluctuation will cause distortion. Fluctuation should be reduced by the turntable drive system. This may be achieved by use of resilient frictional drivers or application of governors or other frictional drags to provide regulation of the speed. The inertia of a heavy turntable will also contribute to regulating the speed and to reduction of the effect of motor fluctuation.

Frictional drives and drags must be carefully designed and applied so that the frictional force applied is constant, or the table speed will vary through each 360-degree revolution. This causes the effect known as "wow."

Fig. 4-13 shows a number of driving methods commonly employed in record players and changers. The gear-train or direct motor drive (Fig. 4-13A) requires a very powerful motor free from vibration. If properly designed, this type of drive is very satisfactory. Because good units are very expensive, this drive system is rarely encountered.

The arrangement shown in Fig. 4-13B is the direct rim-drive method. This method provides a single speed. The turntable is driven by a rubber wheel attached to the motor shaft. The rubber wheel also serves to isolate the motor from the turntable, reducing the transmission of motor variations. Modern arrangements use one or more rubber idlers, further reducing transmission of motor fluctuations and vibrations.

One of the defects of this system arises from the deformation of the rubber wheels which results if the wheel is left in contact with the turntable rim when the equipment is not in use. The wheel is flattened at the point in contact with the turntable rim. In the beter units, provisions are available for removing the pulley from contact with the turntable when the equipment is not in use.

In Fig. 4-13C, a multiple-speed rubber-wheel drive system is shown. Two rubber wheels are provided, one giving a 78-rpm turntable speed, and the other a 33⅓-rpm turntable speed. The desired speed can usually be chosen by changing the position of a lever connected to the drive mechanism.

The single- and dual-speed rubber-wheel drives are often constructed so that the rubber wheel mounted on the motor shaft drives an idler which, in turn, drives the turntable. This greatly simplifies the design of facilities for removing the wheel from contact with

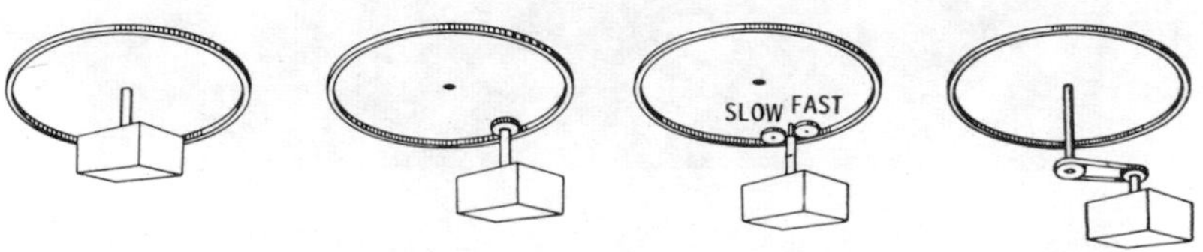

(A) Direct motor drive. (B) Rim drive. (C) Two-speed idler drive. (D) Belt drive.

Fig. 4-13. Turntable drive methods.

the turntable, and it permits better control of the pressures between the rubber wheel and the turntable rim. The possibility of slippage, which is a common fault of a direct rubber-wheel drive, is also greatly reduced. This system is also applied to three-speed players.

Courtesy Acoustic Research, Inc.

(A) AR turntable.

Courtesy Marantz Co., Inc.

(B) Marantz SLT-12 turntable.

Fig. 4-14. High-

Fig. 4-13D shows a belt-drive arrangement often encountered in playback equipment. A rubberized composition belt is connected between pulleys on the motor shaft and turntable spindle. This system is very good with respect to low transmission of vibration.

Courtesy Matsushita Electric Corporation of America

(C) Panasonic SP-10 turntable.

Courtesy RABCO

(D) RABCO turntable.

fidelity turntables.

Another drive method uses a conical driving member that is continuously variable. This provides for adjustment to any of the three popular speeds of 78, 45, and 33⅓ rpm and in addition allows for setting an exact speed around any one of the three standards to get a precise pitch from a particular record according to individual desire.

Record Changers Versus Turntables

Turntables have design and application advantages in that they are simple and have a limited function compared with an automatic record changer. It is generally conceded that the highest-quality reproduction can be obtained from a good turntable. Record changers, on the other hand, can provide convenience with excellent quality. However, the additional mechanisms, devices, and controls necessary to change the record may cause problems in speed variations, rumble, and other defects. Each has its place—the turntable for flawless reproduction, and the record changer for hours of excellent-quality continuous program without attention. Fig. 4-14 shows hi-fi turntables, and Fig. 4-15 shows typical record changers.

The cartridge, the pickup arm, and turntable requirements for stereo operation are so interrelated that many audiophiles prefer to buy the whole combination as a unit. Typical combinations are illustrated in Fig. 4-16. As explained in Chapter 3, variation of needle height, greater tracking force, and stronger rumble effects make record changers somewhat inferior to manual players for stereo. Changers that are used for stereo are very similar to those employed for monophonic reproduction. However, they contain refinements to reduce rumble, tracking force, etc.

The turntable combination shown in Fig. 4-16B is a new design using solid-state electronics for speed control. Voltage from the ac line is reduced through a transformer, changed to dc by a rectifier, and fed to a push-pull amplifier which powers the motor. Inasmuch as this amplifier can produce 20 watts and the motor needs only 5 watts, there is considerable margin for reliability. The amplifier is fed by a Wien bridge oscillator which determines the power frequency according to the speed of rotation selected by the user. The pitch, or fine speed, control in this system is based on the introduction of changes in the oscillator circuit by means of a switch, controlled by an adjustment at the top center of the unit.

The motor itself is a 16-pole synchronous type which locks to the frequency fed from the amplifier-oscillator circuit and thus maintains constant speed at each speed setting. With this motor-drive system, the turntable can be operated from power sources of 100 to 250 volts and at power-line frequencies of 50 to 60 Hz. With some

Courtesy Garrard, Div. British Industries Co.

(A) Garrard SL55B.

Courtesy JVC America, Inc.

(B) Nivico Model 5204.

Fig. 4-15. Typical record changers.

modification at the factory, the unit can even be made to run from a dc source such as batteries. The motor drives the platter through a rubber belt. The bearings are self-lubricating. The platter, a 12-inch nonferrous casting, weighs 7 pounds, 4 ounces; it is covered

Courtesy Garrard, Div. British Industries Co.

(A) Garrard Zero 100.

Courtesy Thorens Div., Elpa Marketing Industries, Inc.

(B) Thorens TD-125.

Fig. 4-16. Combination turntable, pickup arm, and cartridge assemblies for stereo.

with a thick rubber mat; and its center piece may be inserted upside down to accommodate 45-rpm records. There are three controls: the speed selector (left), the off/on switch (right), and (center) the fine-speed adjustment and an illuminated strobe marker to permit accurate speed adjustment.

APPLICATION OF ELECTRON TUBES, SEMICONDUCTORS, AND INTEGRATED CIRCUITS

Over the first half of this century, electron tubes reached a high state of development in rf and audio applications. Semiconductors and integrated circuits are relatively new, but they have been developed to the point where they now can provide performance equal to that of tubes with the advantages of reduced weight, smaller size, improved reliability, and lowered cost of operation.

Semiconductor devices are small but versatile units that can perform an amazing variety of control functions in electronic equipment. Like electron tubes, they have the ability to control almost instantly the movement of charges of electricity. They are used as rectifiers, detectors, amplifiers, oscillators, electronic switches, mixers, and modulators.

Semiconductor devices have many important advantages over other types of electron devices. They are small and light in weight (some are less than an 1/8 inch in length and weigh only a fraction of an ounce). They have no filaments or heaters, and therefore require no heating power or warm-up time. They are solid in construction, extremely rugged, free from microphonics, and can be made impervious to many severe environmental conditions. The circuits required for their operation are usually relatively simple.

The simplest type of semiconductor is the diode. Crystal, silicon, and tunnel diodes are used in hi-fi equipment. Diodes have two elements. When another layer (element) is added to the semiconductor diode to form three layers (with two junctions), the capability of amplification is added. The resulting device is called a *bipolar transistor;* it has an emitter, base, and collector (similar in effect to the cathode, grid, and plate of a vacuum triode).

Later transistor developments, the field-effect transistor (FET) and the integrated circuit, have contributed to improved performance in tuners, receivers, and amplifiers, and at the same time have made possible the mass manufacture of smaller equipment with efficiency equal to the most sensitive and powerful tube-type receivers. Some of the most recent improvements in the application of transistors have been provided for rf amplification, conversion, i-f amplification, detection, and control circuits in stereo equipment.

In the bipolar transistor, performance depends on the interaction of two types of charge carriers, holes and electrons. Field-effect transistors are unipolar devices (i.e., operation is basically a function of only one type of charge carrier, holes in p-channel devices and electrons in n-channel devices).

Early models of field-effect transistors used a reverse-biased semiconductor junction for the control electrode. In the metal-oxide-semiconductor field-effect transistor (MOSFET), a metal control *gate* is separated from the semiconductor *channel* by an insulating oxide layer. One of the major features of the metal-oxide-semiconductor structure is that the very high input resistance of MOS transistors (unlike that of junction-gate field-effect transistors) is not affected by the polarity of the bias on the control (gate) electrode. In addition, the leakage currents associated with the insulated control electrode are relatively unaffected by changes in ambient temperature. Because of their unique properties, MOS field-effect transistors are particularly well suited for use in such applications as voltage amplifiers, rf preamplifiers, i-f amplifiers, and other circuits used in hi fi.

The distinguishing feature of an integrated circuit is that all components required to perform a particular electronic function are combined and interconnected on a common substrate. The constituent elements of integrated circuits lose their identities as discrete components, and the devices assume the appearance of "microminiaturized" function blocks. In comparison to their discrete-component counterparts, integrated circuits offer enhanced performance and new plateaus of reliability, at reduced costs. In addition, the availability of complete solid-state circuits in packages no larger than those of conventional discrete transistors makes possible further reductions in the size and weight of electronic equipment.

TUNERS

Tuners are available to receive fm or a-m signals separately or in one combined unit. Some tuners provide for the preamplifier-equalizer functions because in some installations no program-material source other than the tuner is required. In other cases, a separate preamplifier control is not as desirable as having all controls on the same panel with the tuning controls. The performance of such a tuner preamplification-compensation system should not be expected to be as good as a top-quality separate control preamplifier, but the design of this arrangement usually provides as much quality as is usable for fm or a-m reception or other average pickup. For amplitude-modulation reception, the upper limit of frequency response is usually 10,000 Hz or less, because a-m broadcast stations are separated only by this amount, and reception of two nearly equal signals only 10 kHz apart will produce a strong 10-kHz beat note. This note is unpleasant to sensitive ears if not filtered out. Frequency-modulation tuners can provide a greater range of reproduction and high fidelity over the entire audible range, if all elements of the systems involved are designed, arranged, and operated to achieve maximum performance.

A-M Tuners

Amplitude-modulation tuners, covering the broadcast band from 540 to 1600 kHz, are usually of the superheterodyne type, having an rf preamplifier to reduce image interference, a first detector and oscillator to convert the signal to a lower frequency, where there are better rf amplification conditions, a second detector to remove the sound signal from the carrier signal, and an audio amplifier and output arrangement to bring the signal to a level sufficient to drive a basic audio amplifier or control unit of hi-fi type. In addition, there may be the usual control features, such as automatic volume control, noise limiting, and others.

Amplitude Modulation—Amplitude modulation is defined as the process of changing the amplitude of an rf carrier in accordance with the intelligence to be transmitted. When there is no modulation, the radio-frequency carrier portion of an amplitude-modulated wave is of constant frequency and constant amplitude, as shown in Fig. 4-17A. An audio modulating frequency is superimposed on this carrier in a manner that causes the amplitude of the carrier signal to vary as illustrated in Fig. 4-17B, leaving the carrier frequency unchanged. The pattern shown in Fig. 4-17B is commonly referred to as a *modulation envelope*.

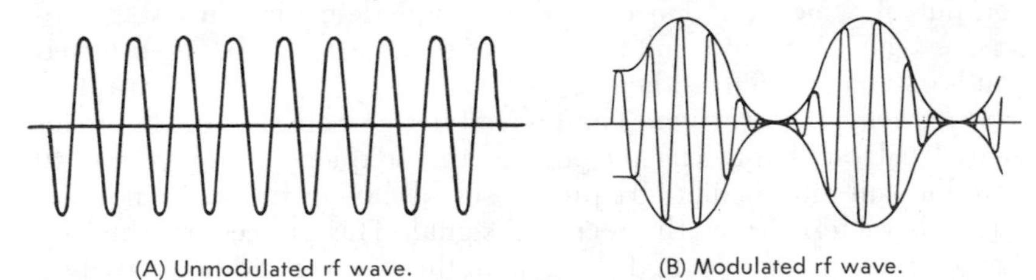

(A) Unmodulated rf wave. (B) Modulated rf wave.

Fig. 4-17. Amplitude modulation.

Sidebands—An amplitude-modulated wave is composed of a number of frequencies: the radio frequency of the carrier wave, the modulating audio frequency or frequencies, and combinations of these frequencies. These combination frequencies are called the *sideband frequencies* and are the result of mixing the radio frequency and the modulating frequencies. Whenever any two frequencies are mixed together, two new frequencies are produced. One of these is the sum of the two frequencies, and the other is the difference between the two original frequencies. Thus, for a modulating frequency of 5000 Hz and a carrier frequency of 1000 kHz, sideband frequencies of 995 kHz and 1005 kHz are produced. If the modulating frequency is increased to 10,000 Hz, sidebands will be produced at 990 kHz and 1010 kHz.

It is these sideband frequencies that carry the intelligence in an amplitude-modulated wave. When an rf carrier is modulated by many audio frequencies, such as occur in speech or music, the side frequencies consist of a band of frequencies above and below the carrier frequency. The width of each of these bands is determined by the highest modulating frequency. For this reason, hi-fi a-m signals must have an available bandwidth equal to twice the highest frequency to be reproduced.

A-M Superheterodyne Receivers

A *superheterodyne receiver* is one in which the desired signal is mixed with a locally generated signal to produce an intermediate-frequency signal. This intermediate-frequency signal is then amplified and detected to produce the audio frequency. Fig. 4-18 is a simplified block diagram of a typical superheterodyne receiver.

The rf amplifier stage receives the weak signal intercepted by the antenna, amplifies it, and passes it on to the mixer. In the mixer stage, the received signal is heterodyned with the output of the local oscillator. The output of the mixer stage is an intermediate-frequency (i-f) signal which has the same modulation characteristics as the received signal. The i-f signal then passes through a number of amplifiers, referred to as intermediate-frequency amplifiers, the output of which is applied to the second detector. This stage removes the i-f component from the signal, leaving the undistorted audio signal, which is then amplified and applied to the speaker.

Frequency Conversion—The converter stage consists of the mixer and local oscillator. The purpose of the frequency converter is to produce an intermediate-frequency signal having the same modulation characteristics as the received signal. This is accomplished by generating an unmodulated rf signal in the receiver and heterodyning it with the received signal. By this method, a third signal is generated, the frequency of which is equal to the difference between the locally generated and incoming signal frequencies.

Two circuits are required to generate the i-f signal, an oscillator and a mixer. Tubes of special design have been developed so that

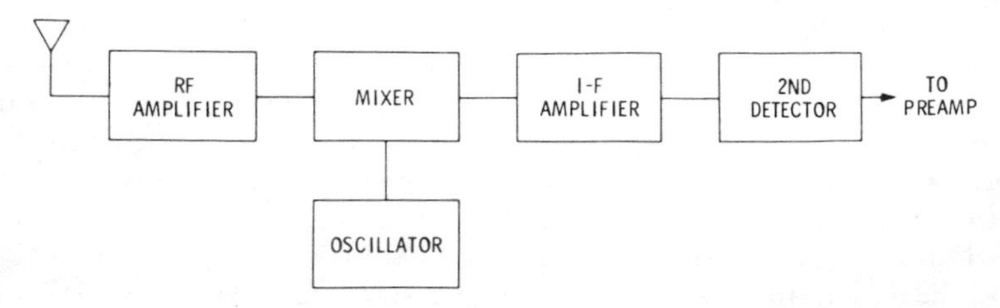

Fig. 4-18. Block diagram of superheterodyne a-m receiver.

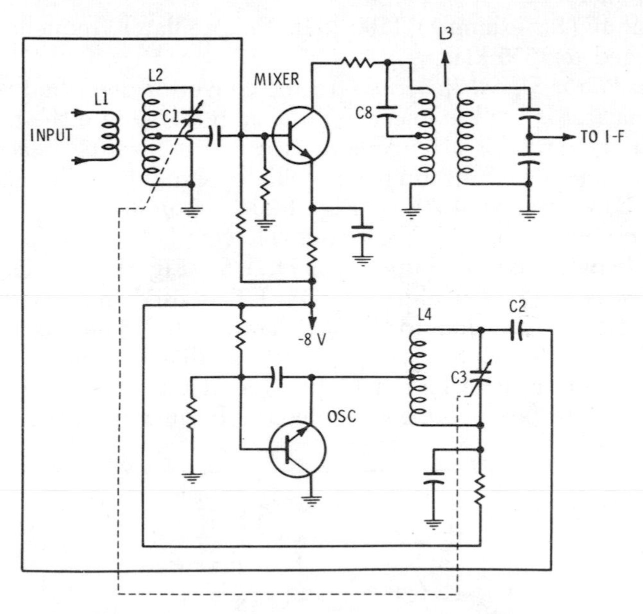

Fig. 4-19. Converter with separate mixer and oscillator.

both functions can be accomplished by one tube. Many receivers, however, employ separate mixer and oscillator stages. A typical transistor-type converter using a separate mixer and oscillator is shown in Fig. 4-19. The rf input is coupled to the mixer tuned circuit, L2 and C1, by means of coupling coil L1. This circuit (L2-C1) is tuned to the frequency of the incoming signal and applies this signal to the mixer base. The oscillator operates at a frequency equal to the incoming signal frequency plus the intermediate frequency. Output from the oscillator is coupled to the mixer base through capacitor C2. The signal at the collector of the mixer is thus the result of both the incoming signal and the oscillator signal. Signals at the oscillator frequency, the receiver signal frequency, the difference frequency, and several others appear in the mixer output. The circuit of L3 and C8 is tuned to the difference frequency, and this signal builds up to a high amplitude while other signals are largely eliminated.

Capacitors C1 and C3 are ganged so that when the mixer circuit is tuned to the frequency of an incoming signal, the oscillator is tuned so that its frequency remains equal to the incoming signal plus the intermediate frequency. The most common intermediate frequency is 455 kHz. If the receiver signal is at a frequency of 1000 kHz, the oscillator frequency must be 1455 kHz to produce an intermediate frequency of 455 kHz. If the mixer is tuned to a new signal at (for example) 1100 kHz, the oscillator frequency must be changed to 1555 kHz.

Oscillator Signal Injection—In the converter described previously, a capacitor is used to inject the oscillator signal into the base circuit of the mixer. This arrangement is called *capacitive injection.* Two other methods of injecting the oscillator signal into the mixer circuit are shown in Fig. 4-20. In Fig. 4-20A, *inductive injection* is used. The oscillator coil (L4) is inductively coupled to the mixer cathode circuit by means of coupling coil L3. In Fig. 4-20B, *electronic injection* is used. A dual-gate MOSFET is used in this circuit, providing a separate gate for the oscillator signal. The incoming signal applied to the control gate and the oscillator signal applied to a second control gate both act upon the electron stream through the transistor to produce the intermediate frequency in the output cir-

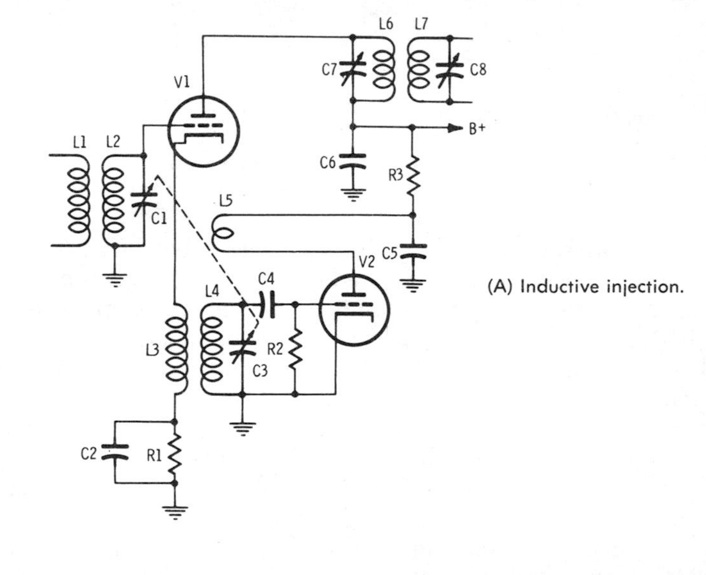

(A) Inductive injection.

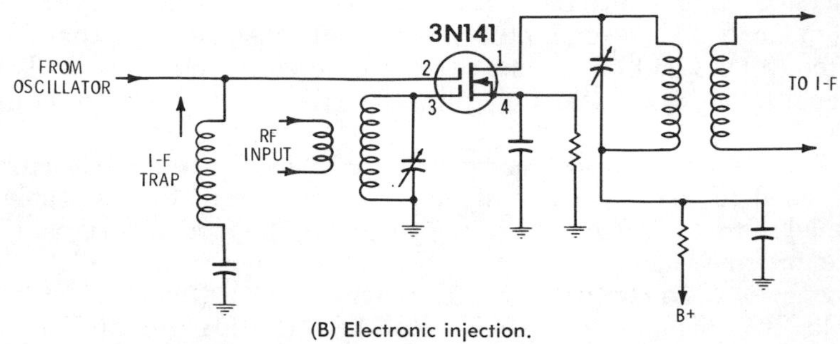

(B) Electronic injection.

Fig. 4-20. Methods of injecting oscillator signal into mixer.

cuit. The injection method illustrated in Fig. 4-20B is superior to those of Figs. 4-19 and 4-20A in that it reduces interaction between the mixer and oscillator circuits.

I-F Amplifiers—The i-f amplifiers provide most of the voltage amplification of the signal of a superheterodyne receiver. One or two and sometimes three i-f amplifier stages are used. A typical tube-type i-f amplifier circuit is shown in Fig. 4-21. The input and output circuits are inductively coupled by means of i-f transformers T1 and T2. The primaries and secondaries of the transformers are tuned. Since the incoming signal is always heterodyned to the same intermediate frequency, the four tuned circuits are operated at the same frequency at all times. This makes it possible to design and adjust the circuits to obtain maximum gain. The i-f transformers are mounted in small metal cans and are adjusted to the proper frequency by means of variable capacitors, as shown in the figure, or by means of movable powdered-iron cores. The capacitor-tuned types are often provided with fixed powdered-iron cores to increase gain and selectivity.

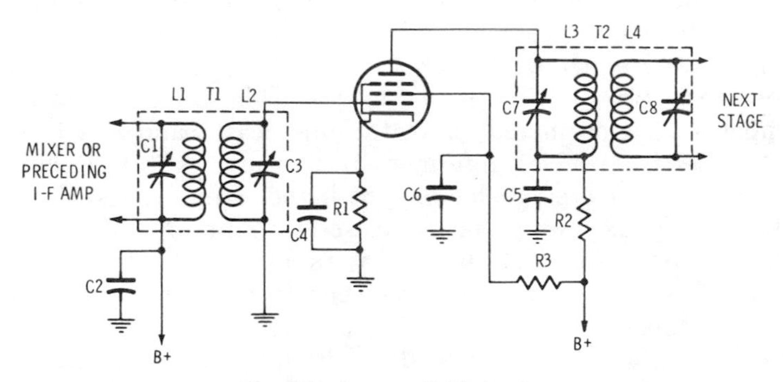

Fig. 4-21. A pentode i-f circuit.

Because of the high gain of i-f amplifiers, coupling between input and output circuits must be kept to a minimum. This is accomplished by careful shielding and placement of parts and by providing suitable decoupling networks in plate, screen, and grid circuits. Decoupling networks usually consist of a resistor and capacitor connected as shown in Fig. 4-21. The plate decoupling network consists of R2 and C5, while R3 and C6 provide screen decoupling.

Hi-Fi Bandpass and Image Rejection—The two most important factors influencing the choice of an intermediate frequency are bandpass and image rejection. For several reasons it is possible to obtain greater bandpass as the intermediate frequency is raised. Therefore, when maximum bandpass is desired, the intermediate frequency is made as high as possible, consistent with other factors.

If the oscillator of a superheterodyne is tuned to 1455 kHz and the intermediate frequency is 455 kHz, signals at 1000 kHz (oscillator minus intermediate frequency) and 1910 kHz (oscillator plus intermediate frequency) may be received by tuning the mixer to the desired signal. This is possible because both frequencies when heterodyned with the oscillator signal will produce the same difference frequency. In practice, the mixer is tracked so that it is always tuned to either the oscillator frequency plus the intermediate frequency or the oscillator frequency minus the intermediate frequency. If the mixer frequency is equal to the oscillator frequency plus the intermediate frequency, then the oscillator frequency minus the intermediate frequency is referred to as the *image frequency*. If the mixer is tuned below the oscillator frequency, then the higher frequency is called the image frequency. Regardless of which frequency the mixer is tuned to, some signal energy will appear in the mixer output if a strong image-frequency signal is present. This difficulty occurs because the mixer circuit is not selective enough to reject the image signal. Suitable image rejection is obtained by choosing an intermediate frequency high enough to provide sufficient separation between the received-signal and image-signal frequencies. As the intermediate frequency is increased, the image frequency moves farther away from the frequency to which the mixer is tuned, and the image rejection increases. In the broadcast band and at somewhat higher frequencies, intermediate frequencies in the neighborhood of 455 kHz are satisfactory; at higher frequencies, the intermediate frequency must be increased to obtain suitable image rejection. Generally, it is necessary to make a compromise and choose a frequency somewhere between that which gives optimum image rejection and that which gives the greatest selectivity.

Transistor Tuned Amplifiers—In radio-frequency (rf) and intermediate-frequency (i-f) amplifiers, the width of the band of frequencies to be amplified is usually only a small percentage of the center frequency. Transistors may be used in these applications effectively to select the desired band of frequencies and to suppress unwanted frequencies. The selectivity of the amplifier is obtained by means of tuned interstage coupling networks. A typical transistor i-f amplifier section is shown in Fig. 4-22.

Application of Integrated Circuits to 12-MHz I-F Amplifier for A-M Receiver—Fig. 4-23 illustrates the use of integrated circuits in an i-f amplifier for an a-m receiver. The amplifier is encased in a metal box, and adequate shielding and supply decoupling are provided. The i-f amplifier has three stages, each of which is designed to provide a gain of 25 dB. The source resistance to the input circuit was selected to provide a satisfactory compromise for gain, noise figure, and modulation-distortion performance. The input and output transformers, T1 and T4, have high unloaded Q's to preserve good noise performance and to maximize the output power. The interstage transformers, T2 and T3, have low unloaded Q's to achieve the required gain. The second detector has a bandwidth of 10 kHz. Typical overall performance characteristics are as follows:

Power drain = 83 milliwatts
Power gain (from input to second detector output) = 76 dB
Agc range (first stage) = 60 dB
Noise figure = 4.5 dB

RF Amplifiers—An rf amplifier is not absolutely necessary in a superheterodyne receiver; in fact, many receivers do not include such a stage. However, the incorporation of an rf amplifier greatly improves the performance of a receiver. The purpose of an rf amplifier is to improve the image rejection and the sensitivity of the receiver. As explained in the discussion of if amplifiers, the mixer stage

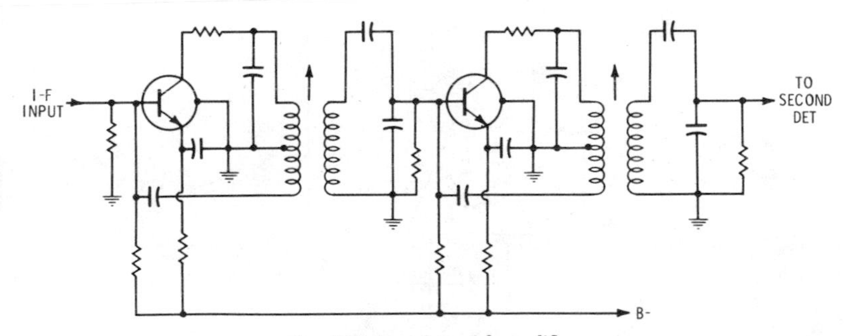

Fig. 4-22. Transistor i-f amplifier.

does not have sufficient selectivity to reject strong signals at the image frequency completely. The rf stage increases the image rejection by amplifying the desired signal. The image signal is not amplified, and thus image interference is reduced. Some receivers use as many as three rf stages to secure optimum image rejection in combination with an intermediate frequency low enough to permit high selectivity.

Considerable noise is generated in converter circuits. This noise is superimposed on the signal and appears in the output of the receiver. To be received, a signal must have an amplitude greater than the noise generated in the converter stage. An rf amplifier increases the amplitude of the incoming signal before it reaches the converter stage. Since the converter noise remains constant, the additional signal amplification makes it possible to receive signals which would otherwise be lower than the converter noise level. Some noise is also generated by rf amplifiers, and, when such stages

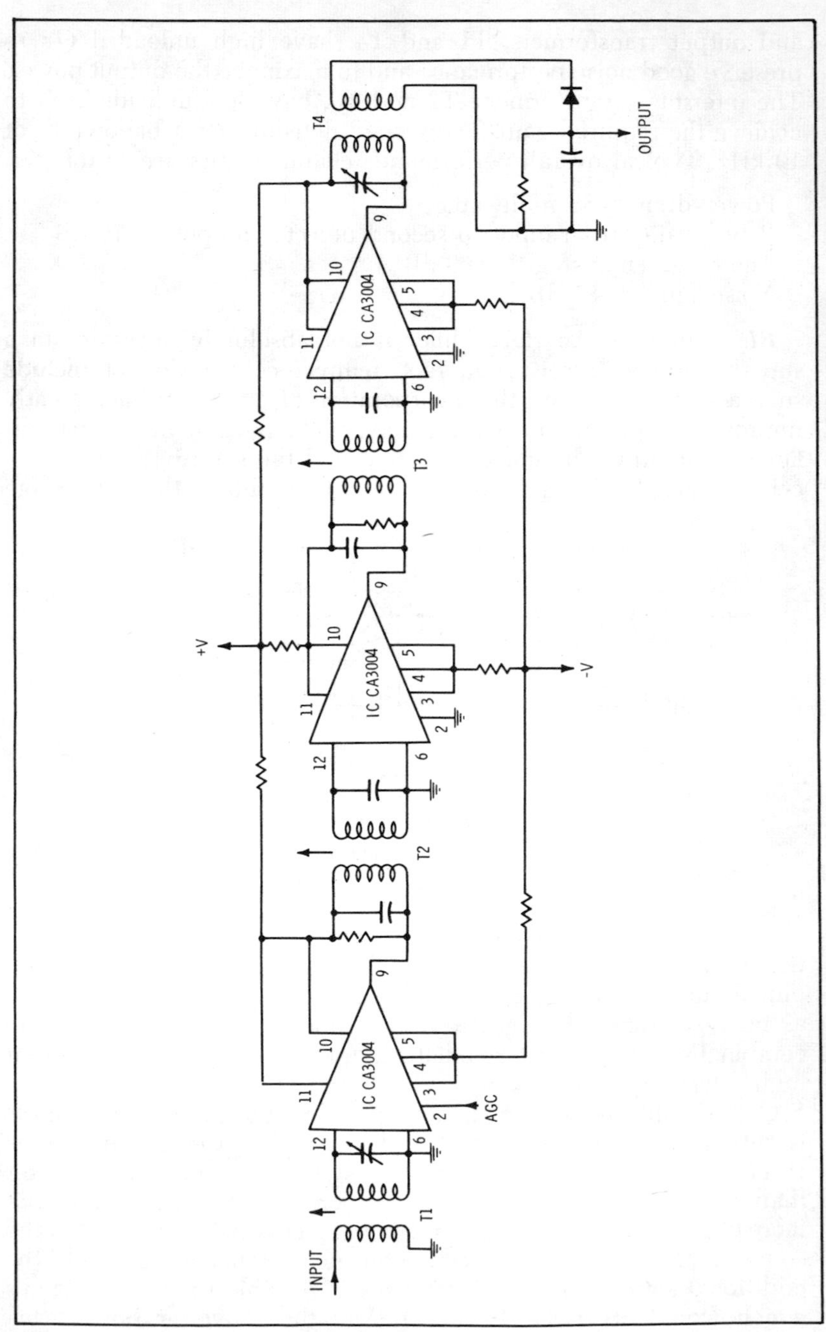

Fig. 4-23. Integrated-circuit i-f amplifier for a-m receiver.

are employed, the absolute sensitivity of the receiver is determined by the noise in the first rf stage. Radio-frequency amplifiers, however, generate much less noise than converters. The ability of an rf stage to improve the sensitivity of a receiver is particularly important at frequencies above 10 MHz. Below 10 MHz, man-made noise is too great to make very high sensitivity useful.

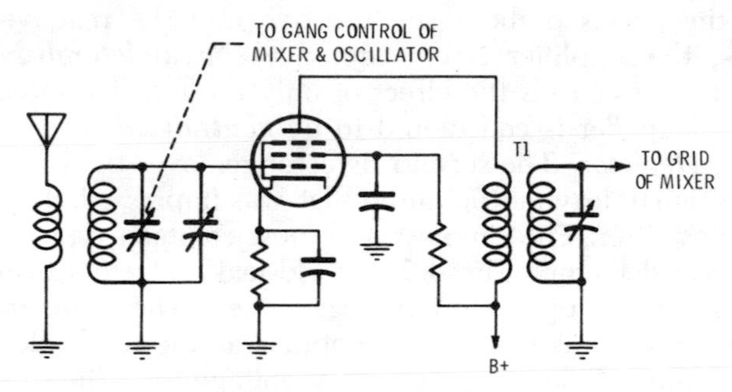

(A) Using pentode tube.

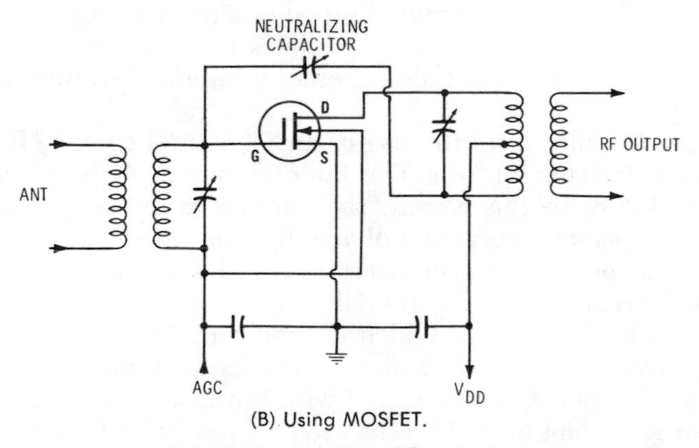

(B) Using MOSFET.

Fig. 4-24. Typical rf amplifier circuits.

A typical rf amplifier circuit is shown in Fig. 4-24A. It consists of a pentode tube with a tuned grid circuit and an impedance load. Pentodes are generally used because of their high gain and low interelectrode capacitance. Because of their high gain, rf amplifiers must be carefully shielded and decoupled to prevent oscillation.

A typical neutralized rf amplifier circuit using an n-channel MOS transistor is shown in Fig. 4-24B. The transistor shown is intended for operation at frequencies up to 60 MHz, and therefore it is highly suitable for use in a-m broadcast receiver circuits. Typically, its forward transconductance does not drop 3 dB until approximately 150 MHz. The stage shown in Fig. 4-24B has a typical power gain of 10 to 18 dB at 60 MHz. Cross modulation typically is less than one percent for interfering signal voltages up to 200 millivolts.

Since transistors do not have internal shielding, external feedback circuits are often used in tuned coupling networks to counteract the effects of the internal transistor feedback and thus provide more gain or more stable performance. If the external feedback circuit cancels the effects of both the resistive and the reactive internal feedback, the amplifier is considered to be *unilateralized.* If the external circuit cancels the effect of only the reactive internal feedback, the amplifier is considered to be *neutralized.*

Second Detector—The second detector removes the i-f component from the signal, leaving the audio that was impressed on the carrier at the transmitter. The simplest and most common type of detector is the diode detector (Fig. 4-25). Grid-leak detectors overload too easily for use in superheterodyne receivers. The plate detector is sometimes used, but it is not as popular as the diode detector because it is more difficult to obtain avc voltage from the former.

The diode detector is used in many forms. It has the advantage of a cathode-follower output. Cathode-follower circuits have lower impedance, allowing use of longer lines to feed the amplifier unit. Cathode-follower circuits also operate with low distortion and high stability.

Fig. 4-26A illustrates the use of an integrated circuit (IC) as an envelope detector for a-m. The internal circuit of the IC is shown in Fig. 4-26B. In this circuit, the emitter of the output transistor (Q6) is operated at zero voltage by connection of an external resistor in the bias loop of constant-current transistor Q3.

The current in the differential-pair transistors (Q2 and Q4) is increased to the point at which common-collector output transistor Q6 is biased almost to cutoff. For this current increase, constant-current transistor Q3 is operated with terminal 4 open, and emitter resistor R is shunt loaded by the external resistor at terminal 3.

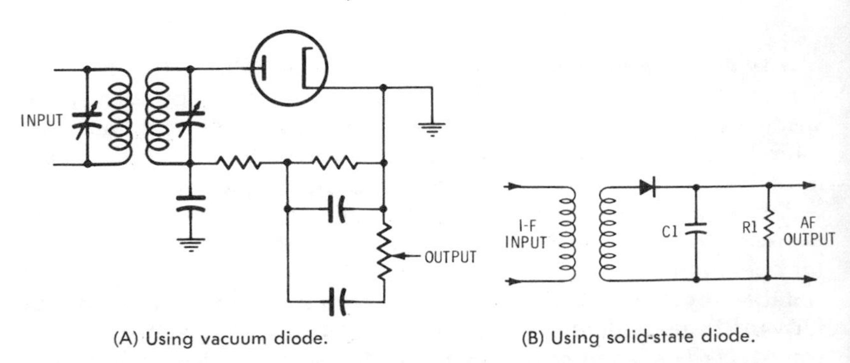

(A) Using vacuum diode.

(B) Using solid-state diode.

Fig. 4-25. Diode detectors.

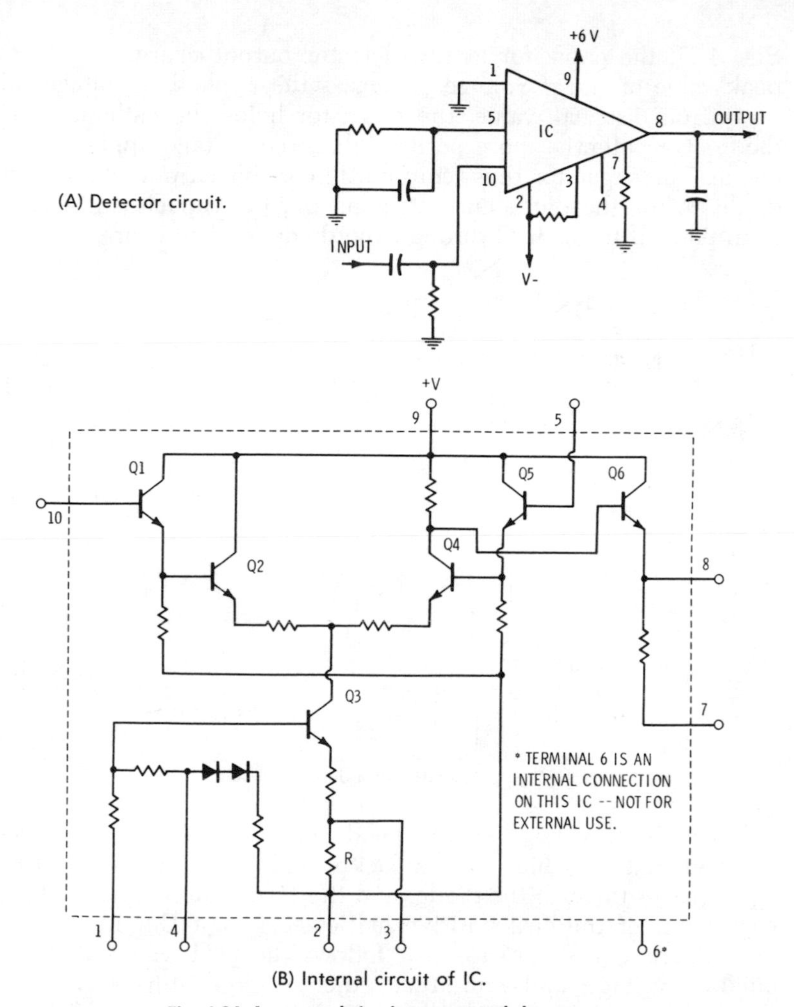

Fig. 4-26. Integrated-circuit a-m second detector.

Although the output transistor is nearly cut off, all the other active devices are operating in their linear regions. For small ac signals, therefore, the circuit provides linear operation except for Q6, which is turned on only by a positive signal. The maximum acceptable input signal depends on the linear range of the differential amplifier. An external filter capacitor is connected between terminal 8 and ground to remove the rf signal from the detected audio output.

The purpose of a detector is to eliminate alternate half-cycles of the waveform and detect the peaks of the remaining half-cycles to produce the output voltage (Fig. 4-27). Between points A and B in

Fig. 4-27, the capacitor at the detector output charges up to the peak value of the rf voltage. Then, as the applied rf voltage falls away from its peak value, the capacitor holds the cathode of the diode at a potential more positive than the voltage applied to the anode. The capacitor thus temporarily cuts off current through the diode. While the diode current is cut off, the capacitor discharges from point B to point C through the diode load resistor.

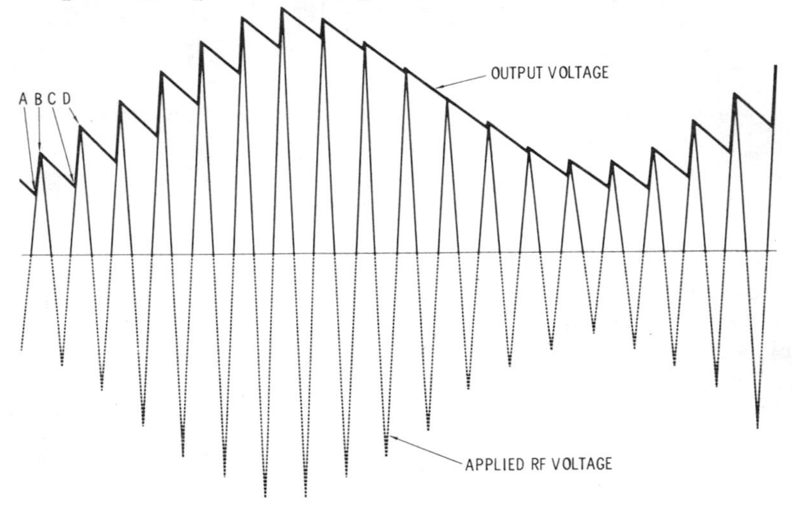

Fig. 4-27. Action of a detector circuit.

When the rf voltage on the anode rises high enough to exceed the potential at which the capacitor holds the cathode, current again passes through the diode, and the capacitor charges up to the peak value of the next positive half-cycle (point D). In this way, the voltage across the capacitor follows the peak value of the applied rf voltage and reproduces the af modulating signal. The jaggedness of the curve in Fig. 4-27, which represents an rf component in the voltage across the capacitor, is exaggerated in the drawing. In an actual circuit, the rf component of the voltage across the capacitor is small. When the voltage across the capacitor is amplified, the output of the amplifier reproduces the speech or music that originated at the transmitting station.

Another way to describe the action of a diode detector is to consider the circuit as a half-wave rectifier. When the signal on the anode swings positive, the diode conducts, and rectified current is delivered to the capacitor and load resistor. The voltage across the capacitor varies in accordance with the rectified amplitude of the carrier, and thus reproduces the af signal. The capacitor should be large enough to smooth out rf or i-f variations, but should not be so large as to affect the audio variations. (Although two diodes can be connected in a circuit smiliar to a full-wave rectifier to produce full-wave detection, in practice the advantages of this connection generally do not justify the extra circuit cost and complication.)

In the circuits shown in Fig. 4-25, it is often desirable to forward-bias the diode almost to the point of conduction to improve performance for weak signal levels. It is also desirable that the resistance of the ac load which follows the detector be considerably larger than the diode load resistor to avoid severe distortion of the audio waveform at high modulation levels.

Automatic Volume Control—The function of automatic volume control (avc), also called automatic gain control (agc), is to maintain constant output from a receiver when the amplitude of the incoming signal changes. This is accomplished by rectifying part of the received signal, at the output of the i-f amplifier, and developing a voltage across a suitable resistor. The magnitude of the voltage is proportional to the amplitude of the incoming signal. This voltage may be applied as bias to the grids of remote-cutoff tubes in the rf and i-f stages of the receiver to vary the gain of the receiver inversely as to signal strength.

A typical avc circuit is shown in Fig. 4-28. Tube V1 operates as a conventional diode detector. Tube V2 is the avc rectifier. Signal voltage is fed from the detector plate to the plate of V2 through

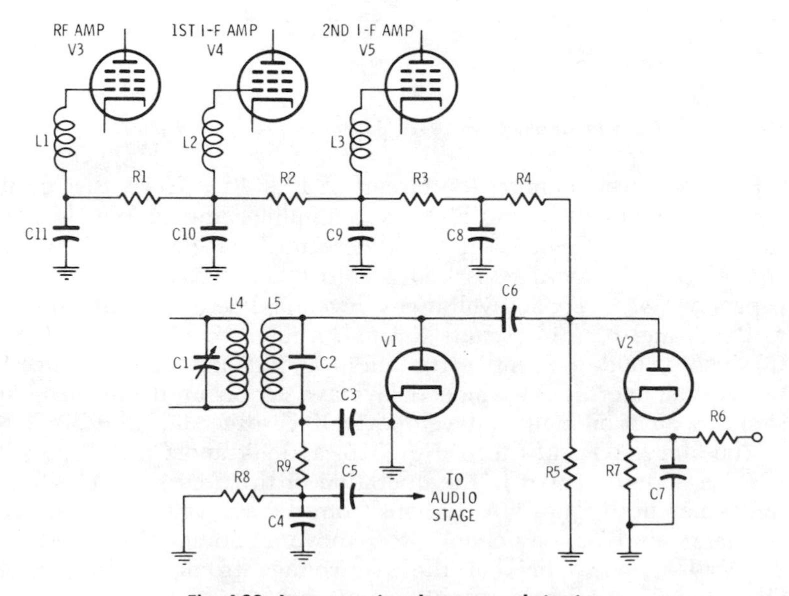

Fig. 4-28. An automatic-volume-control circuit.

coupling capacitor C6. The rectified signal current produces a voltage across diode load resistor R5 such that the upper end of R5 is negative with respect to ground. This negative voltage is applied to the grid circuits of the rf and i-f amplifiers through a filter and individual decoupling networks. An increase in the amplitude of the incoming signal increases the avc bias and reduces the gain of the receiver to maintain constant output. If the signal amplitude decreases, the avc bias decreases, and the receiver gain is raised. Resistors R6 and R7 form a voltage divider operating from the receiver B+ supply. The voltage divider places a positive potential on the cathode of the avc diode. This potential delays the development of avc voltage until the signal reaches a predetermined minimum value. On weak signals, there is no avc bias, and the receiver operates at full gain.

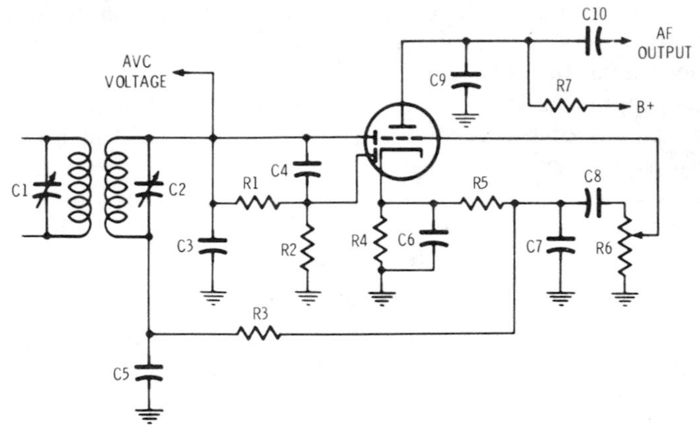

Fig. 4-29. Combined avc, second detector, and audio amplifier.

Some receivers employ the circuit of Fig. 4-29. Here, the second detector, avc rectifier, and first audio amplifier are combined in one tube. The upper diode is the signal detector. The lower diode, which acts as the avc rectifier, is coupled to the detector plate through capacitor C4. A rectified voltage is developed across R2 and applied to the rf and i-f grid circuits through a filter consisting of R1 and C3. The cathode current of the audio amplifier section produces a bias voltage across R4 which delays avc action until the incoming signal is great enough to develop a voltage exceeding the bias.

The filters, R1 and C3 in Fig. 4-29, and R4 and C8 in Fig. 4-28, play an important part in the operation of these circuits. The filters remove audio-frequency variations from the avc voltage. Their time constants must be long enough to remove all audio fluctuations but not so long as to prevent the avc voltage from following rapid changes in input signal amplitude.

A simple method of producing reverse avc for transistor circuits is shown in Fig. 4-30. On each positive half-cycle of the signal voltage, when the diode anode is positive with respect to the cathode, the diode passes current. Because of the diode current through R1, there is a voltage drop across R1, which makes the upper end of the resistor positive with respect to ground. This voltage drop across R1 is applied, through the filter consisting of R2 and C, as reverse bias on the preceding stages. When the signal strength at the antenna increases, therefore, the signal applied to the avc diode increases, the voltage drop across R1 increases, the reverse bias applied to the rf and i-f stages increases, and the gain of the rf and i-f stages is decreased. As a result, the increase in signal strength at the antenna does not produce as much increase in the output of the last i-f amplifier stage as it would without avc.

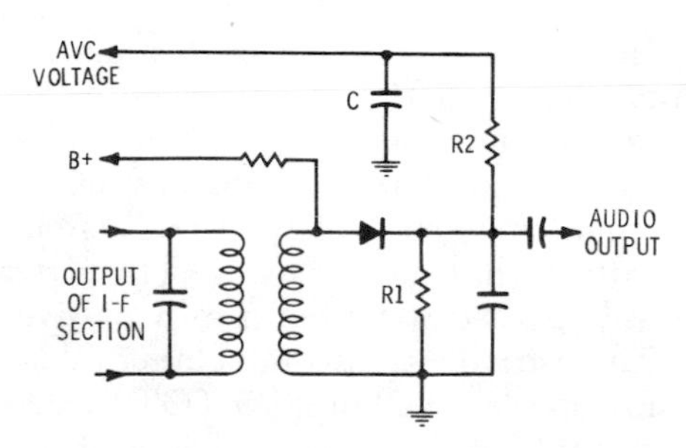

Fig. 4-30. Simple circuit for reverse avc.

When the signal strength at the antenna decreases from a previous steady value, the avc circuit acts in the opposite direction, applying less reverse bias and thus permitting the rf and i-f gain to increase.

The filter composed of C and R2 prevents the avc voltage from varying at an audio frequency. This filter is necessary because the voltage drop across R1 varies with the modulation of the carrier being received. If agc voltage were taken directly from R1 without filtering, the audio variations in avc voltage would vary the receiver gain so as to reduce the modulation of the carrier. To avoid this effect, the avc voltage is taken from capacitor C. Because of the resistance (R2) in series with C, the capacitor can charge and discharge at only a comparatively slow rate. The avc voltage therefore cannot vary at frequencies as high as the audio range, but can vary rapidly enough to compensate for most changes in signal strength.

There are two ways in which automatic gain control (another name for avc, as previously stated) can be applied to a transistor. In the reverse agc method, agc action is obtained by decreasing the collector or emitter current of the transistor, and thus its transconductance and gain. The use of forward agc provides improved cross-modulation characteristics and better signal-handling capability than reverse agc. For forward agc operation, however, the transistor used must be specially designed so that its transconductance decreases with increasing emitter current. In such transistors, the current-cutoff characteristics are designed to be more remote than the typical sharp-cutoff characteristics of conventional transistors. (All transistors can be used with reverse agc, but only specially designed types can be used with forward agc.)

Reverse agc is simpler to use, and provides less bandpass shift and tilt with signal-strength variations. The input and output resistances of a transistor increase when reverse agc is applied, but the input and output capacitances are not appreciably changed. The change in the loading of tuned circuits is minimal, however, because considerable mismatch already exists, and the additional mismatch caused by agc has little effect.

In forward agc, however, the input and output resistances of the transistor are reduced when the collector or emitter current is increased, and thus the tuned circuits are damped. In addition, the input and output capacitances change drastically and alter the resonant frequency of the tuned circuits. In a practical circuit, the bandpass shift and tilt caused by forward agc can be compensated to a large extent by the use of passive coupling circuits.

The variable-transconductance characteristic of the operational transconductance amplifier (OTA) integrated circuit is useful in an agc amplifier. This circuit has all the generic characteristics of the operational voltage amplifier (OVA). The forward transfer characteristic is best described by transconductance rather than voltage gain. The output of the OTA is a current, the magnitude of which is equal to the product of transconductance and the input voltage. The output circuit of this amplifier, therefore, may be characterized by an infinite-impedance current generator, rather than the zero-impedance voltage generator used to represent the output circuit of an operational voltage amplifier. The low output conductance of the OTA permits the circuit to approach the ideal current generator.

When the OTA is terminated in a suitable resistive load impedance and provisions are included for feedback, its performance is essentially identical in all respects to that of an equivalent operational voltage amplifier. The electrical characteristics of the OTA circuits, however, are functions of the amplifier bias current. In the integrated-circuit OTA, therefore, access is provided to bias the amplifier by means of an externally applied current. As a result, the transconductance, amplifier dissipation, and circuit loading may be externally established and varied.

Internal details of a basic integrated circuit used in avc circuits are shown in Fig. 4-31. An understanding of this circuit is best obtained by analysis of voltages and currents with almost complete disregard for voltage gain and impedance levels.

Transistors Q1 through Q4 perform conventional functions, serving as a current mirror, a constant-current source, and a differential pair. An amplifier bias current is externally developed and applied to the current mirror, Q1 and Q2, to bias the differential pair, Q3 and Q4. The differential output signal currents of Q3 and Q4 are amplified by the beta of the differential pnp transistor pair, Q7 and

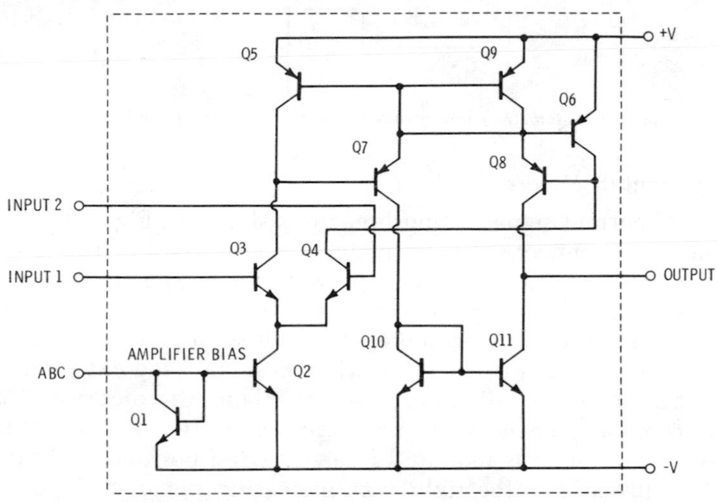

Fig. 4-31. Internal connections of integrated circuit.

Q8. Current mirror Q10 and Q11 then transforms the double-ended output of the pnp transistor network, Q5 through Q9, into a single-ended output. The entire circuit functions in a class-A mode. The amplifier bias current (abc) level establishes bias for all transistors in the amplifier.

Ideally, there is no need for a signal ground because the input signal is differential and the output signal is a current. The input and output terminals may operate at most ac and dc potentials within the range of the supply voltages. When the OTA operates in the open-loop condition, the transconductance, and thus the amplifier gain, can be varied directly by adjustment of the abc level. Therefore, an excellent agc amplifier is obtained by rectifying and storing the amplifier output and applying this signal to the bias terminal. Fig. 4-32 shows a functional diagram of such a system. Low-frequency feedback is provided around the gain-controlled stage to balance the amplifier. As the input signal increases, the amplifier bias current decreases and reduces the transconductance and therefore the system gain.

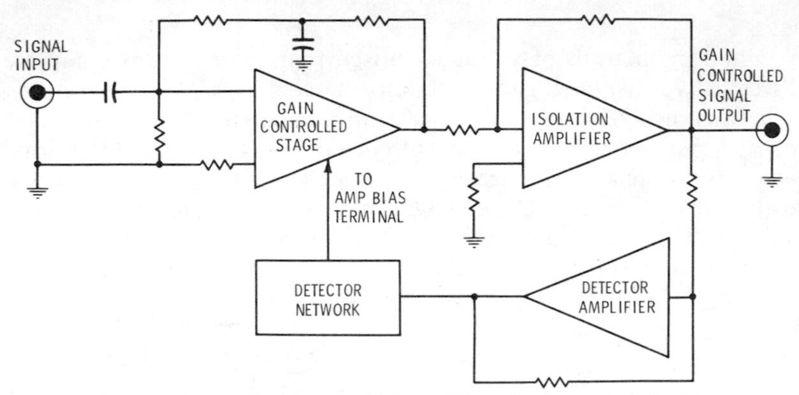

Fig. 4-32. Automatic-gain-controlled amplifier using integrated circuits.

Signal-Strength Meters

A plate-current signal-strength meter is shown in Fig. 4-33. A milliammeter is connected in the plate lead of several of the rf or i-f tubes which have avc voltage applied to their grids. As the signal increases, the avc voltage becomes more negative, and the plate current through the meter decreases. Resistor R is adjusted so that the milliammeter reads full-scale with no signal (highest plate current). This point is called "zero signal." Thus the meter indicator moves counterclockwise with increasing signal. In many commercial receivers, the meter is mounted in an inverted position, so that the pointer will move to the right with increasing signal strength.

A bridge-type signal-strength meter is shown in Fig. 4-34. Tube V1 is used to amplify the avc voltage. The current through R1, M, and R3 tends to cause the meter needle to move to the right, while the current through R2, M, and the tube tends to make the needle move to the left. At zero signal, these currents are made equal by

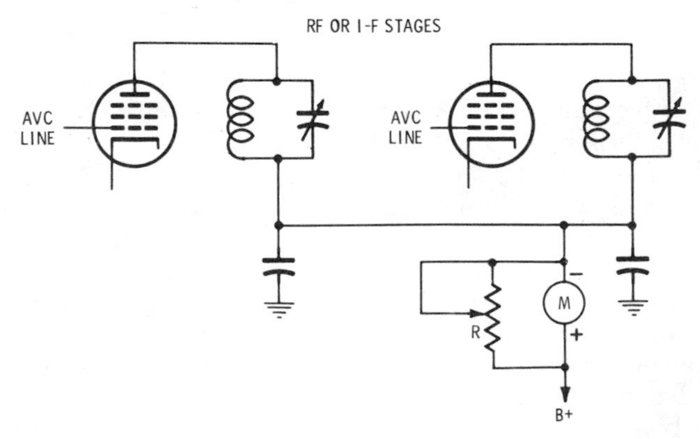

Fig. 4-33. A plate-current signal-strength meter.

Fig. 4-34. A bridge-type signal-strength meter.

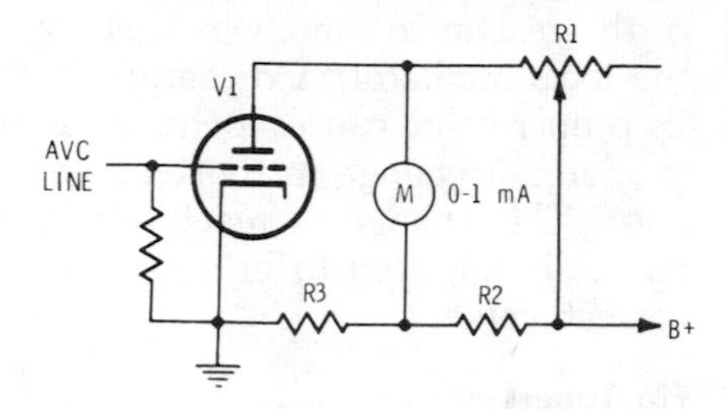

adjusting the resistance of R1. The operation of this circuit is based on the fact that a change in grid bias will cause a variation in the dc plate current of the tube. As the received signal amplitude increases, the avc voltage becomes more negative. The voltage is applied to the grid of V1, and the dc plate current decreases. Thus, the meter needle moves to the right with increasing signal strength.

Automatic Frequency Control

Automatic frequency control (afc) circuits are used in many superheterodyne receivers to compensate for frequency drift. This drift may be due to such factors as small changes in the oscillator or carrier frequencies. It is compensated for by automatically adjusting the oscillator frequency.

An afc system consists of two basic parts: a frequency detector and a variable reactance circuit. Fig. 4-35 shows a typical circuit. The discriminator (frequency detector) is of the Foster-Seeley type and is excited by the i-f signal from the final i-f amplifier stage. The discriminator output is a dc voltage the polarity of which depends on whether the intermediate frequency has deviated above or below its correct value, and the magnitude of which is proportional to the amount of deviation. This dc voltage is applied to the control grid

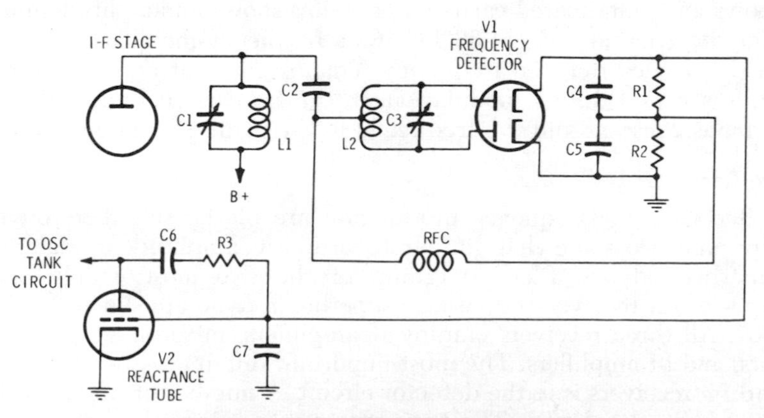

Fig. 4-35. An automatic-frequency-control circuit.

of the reactance tube, which is connected across the tank circuit of the local oscillator. A deviation in the intermediate frequency from its proper value causes a change in the dc grid voltage of V2, which produces a change in the reactance V2 presents to the local oscillator. This change in reactance is such that the oscillator is automatically adjusted to bring the intermediate frequency back to its correct value.

FM Tuners

Frequency-modulation tuners operate in the very high frequency (vhf) region and consequently have operating requirements different from those of a-m tuners. The rf circuits and components must be of optimum design to provide sufficient sensitivity to pick up fm stations under all conditions, and the audio aspects of all circuitry must be of sufficient quality of design and construction to receive and pass audio frequencies to 15 or 20 kHz.

Frequency Modulation

High-fidelity a-m reception is usually limited by the presence of high-level man-made noise and atmospheric disturbances which a-m receivers cannot reject without loss of fidelity. Also, because of the propagation characteristics of frequencies used for standard a-m broadcasting, out-of-area broadcast station signals can interfere with local reception (especially at night). At the frequencies used for fm reception, distant stations usually do not interfere, and the inherent noise-rejection characteristics of the fm receiver minimize the noise problem.

In frequency modulation, the frequency of the rf carrier is varied in accordance with the af or other signal to be transmitted. Amplitude and frequency modulation are compared in Fig. 4-36. Fig. 4-36A shows an unmodulated carrier, Fig. 4-36B shows an amplitude-modulated carrier, and Fig. 4-36C shows a frequency-modulated carrier. In the a-m carrier, the frequency remains constant and the amplitude is varied during modulation; in the fm carrier, the amplitude remains constant and the frequency is varied during modulation.

FM Receivers

Receivers for frequency modulation are of the superheterodyne type and are somewhat similar to ordinary amplitude-modulation superheterodynes. Block diagrams of the two most widely used types of fm receiver and an a-m superheterodyne are shown in Fig. 4-37. All three receivers employ rf amplifiers, mixer stages, oscillators, and i-f amplifiers. The most important differences between a-m and fm receivers is in the detector circuit. A number of fm detectors have been developed. The ratio detector used in the receiver in

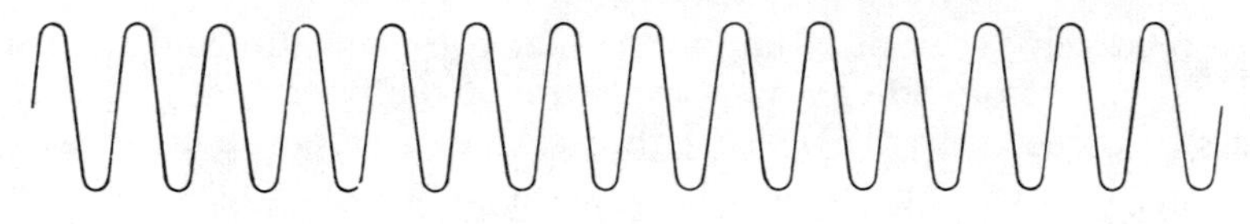

(A) Unmodulated carrier.

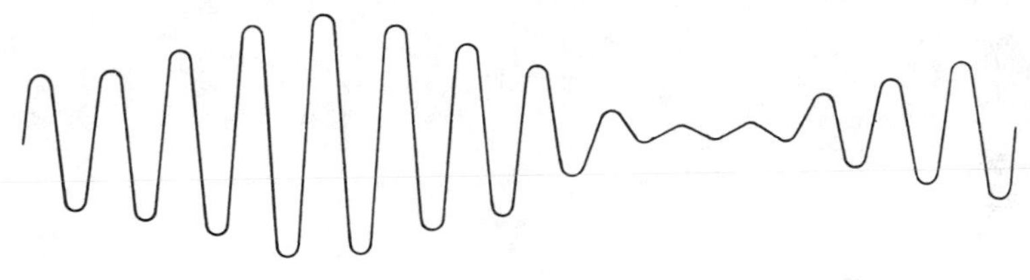

(B) Amplitude-modulated carrier.

(C) Frequency-modulated carrier.

Fig. 4-36. Comparison of amplitude and frequency modulation.

Fig. 4-37B removes the audio signal from the carrier and at the same time rejects amplitude impulses which may accompany it. The receiver in Fig. 4-37C employs a discriminator detector to remove the audio signal from the carrier. This detector is sensitive to amplitude impulses, and in order to eliminate them before detection, limiter stages must be provided. The limiter removes all amplitude fluctuations from the carrier before it is applied to the detector.

RF Amplifier—In fm receivers as in a-m receivers, radio-frequency amplifiers are used to secure improved signal-to-noise ratio, higher gain and selectivity, and improved image rejection. Improvement in signal-to-noise ratio is more important to secure in an fm rf amplifier than in an a-m rf amplifier because considerably more noise is generated in converter stages, and, as described, the addition of amplification before the converter increases the signal amplitude without increasing the noise.

When tubes are used in vhf rf amplifiers, they must have high mutual conductance, low interelectrode capacitance, and high input resistance. In the design of low-level solid-state tuned rf amplifiers, careful consideration must be given to the transistor and circuit parameters which control circuit stability, as well as those which maintain adequate power gain. The power gain of an rf transistor must be sufficient to provide a signal that will overcome the noise level of succeeding stages. In addition, if the signals to be amplified are relatively weak, it is important that the transistor and its associated circuit provide a low noise figure at the operating frequency. In stereo receivers, the noise figure of the rf stage determines the sensitivity of the receiver and is, therefore, one of the most important characteristics of the device used in the rf stage.

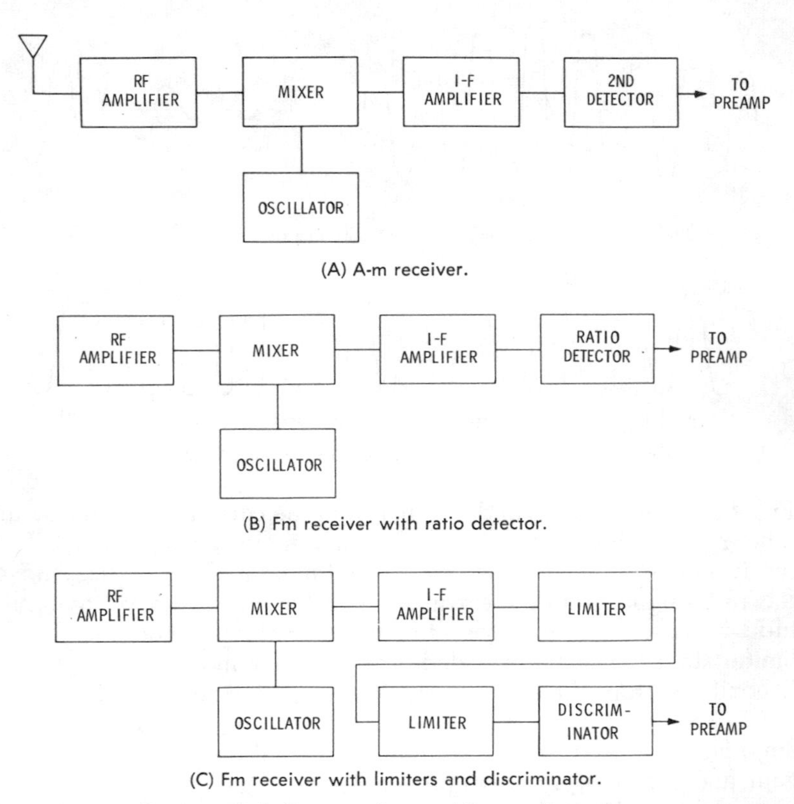

(A) A-m receiver.

(B) Fm receiver with ratio detector.

(C) Fm receiver with limiters and discriminator.

Fig. 4-37. Block diagrams of a-m and fm superheterodynes.

Field-effect transistors combine the inherent advantages of solid-state devices (small size, low power consumption, and mechanical ruggedness) with a very high input impedance and a square-law transfer characteristic that is especially desirable for low cross modulation in rf amplifiers. The output of the MOSFET in rf amplifier circuits is low in harmonics, and the MOSFET has a practical dynamic range capability about five times as great as for bipolar transistors. Also, MOSFET's provide less loading of the input signal, less change of input capacitance under overdrive conditions, and other characteristics that make the MOSFET particularly well suited to this application.

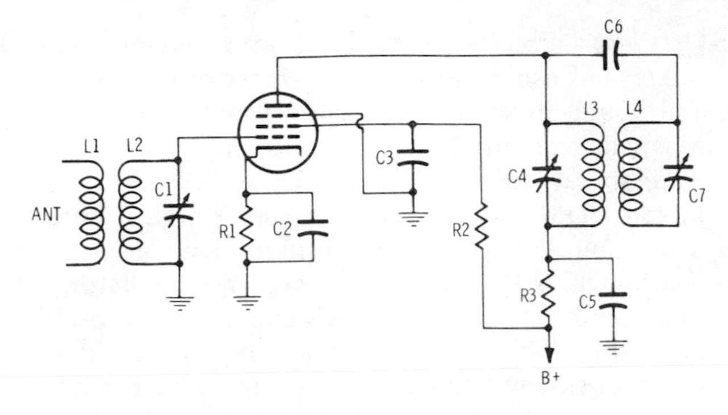

(A) Using vacuum tube.

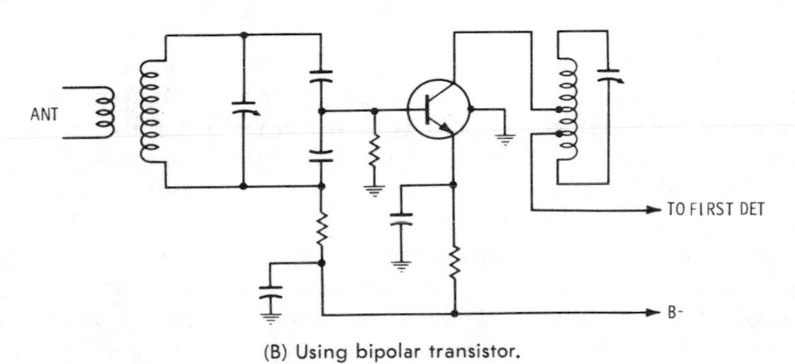

(B) Using bipolar transistor.

Fig. 4-38. Typical vhf rf amplifiers.

Components and circuits used in fm rf amplifiers must generate as little noise as possible. Because of the high frequencies at which these circuits operate, short leads, careful shielding, and high-quality insulation must be used.

The rf amplifier, mixer, and local oscillator taken together are called the *front end* of the receiver. Typical circuits used in fm front ends are shown in Figs. 4-38 through 4-41. In each circuit, the antenna transmission line is coupled to the input coil by means of a separate winding. This is required to match the high input impedance of the stage to the comparatively low impedance of the antenna transmission line. Most receivers are designed to match a 300-ohm line. To pass the complete fm signal, the rf amplifier must respond to a wide band of frequencies. This is accomplished by using low-*Q* coils to broaden the response curve.

Figs. 4-38A and 43B show the application of a highly sensitive electron tube and a vhf bipolar transistor, respectively, to vhf rf stages. The bipolar transistor can provide a sensitivity and signal-to-noise ratio sufficient to permit top-quality hi-fi stereo reception, but it presents difficulty in matching and cross-modulation effects.

Mixer-Oscillator—Frequency-modulation receivers generally use separate mixer and oscillator tubes or transistors to achieve greater efficiency, although in some cases these functions are combined in one device specifically designed for this application. The circuits employed are similar to those found in a-m receivers, with modifications to make them more suitable for use at high frequencies. The difficulties encountered in using a combined mixer-oscillator stage stem from interaction between the mixer and oscillator, which becomes troublesome at high frequencies and results in oscillator pulling and instability. These difficulties are largely avoided by using separate components and loose oscillator-mixer coupling.

It is much more difficult to minimize oscillator drift at the frequencies used for fm broadcasting than it is at a-m broadcasting frequencies. Heating, humidity, and B+ supply-voltage variations (regulation) all contribute to oscillator drift. The effects of changing humidity are minimized by coating circuit components with moistureproofing materials and by permitting a certain amount of temperature rise in the area surrounding critical components. Heating causes drift because it expands parts of critical components, which results in increased capacitance. It is minimized by using insulation materials with low temperature coefficients and by shunting tuned circuits with negative-temperature-coefficient capacitors to counteract the increase in capacitance taking place in other components. The effects of poor regulation in B+ voltage supplies are minimized by careful decoupling of the various circuits in the receiver.

The fm tuner circuit in Fig. 4-39 uses an MOS field-effect transistor in the rf amplifier stage and bipolar transistors in the mixer

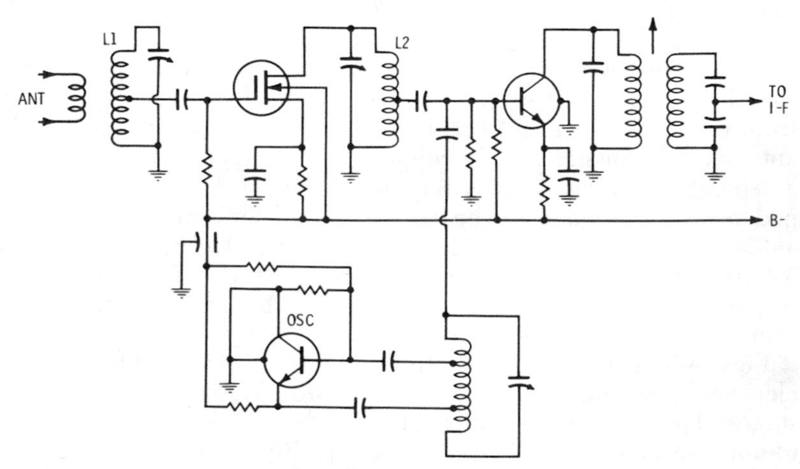

Fig. 4-39. Fm front end using MOSFET.

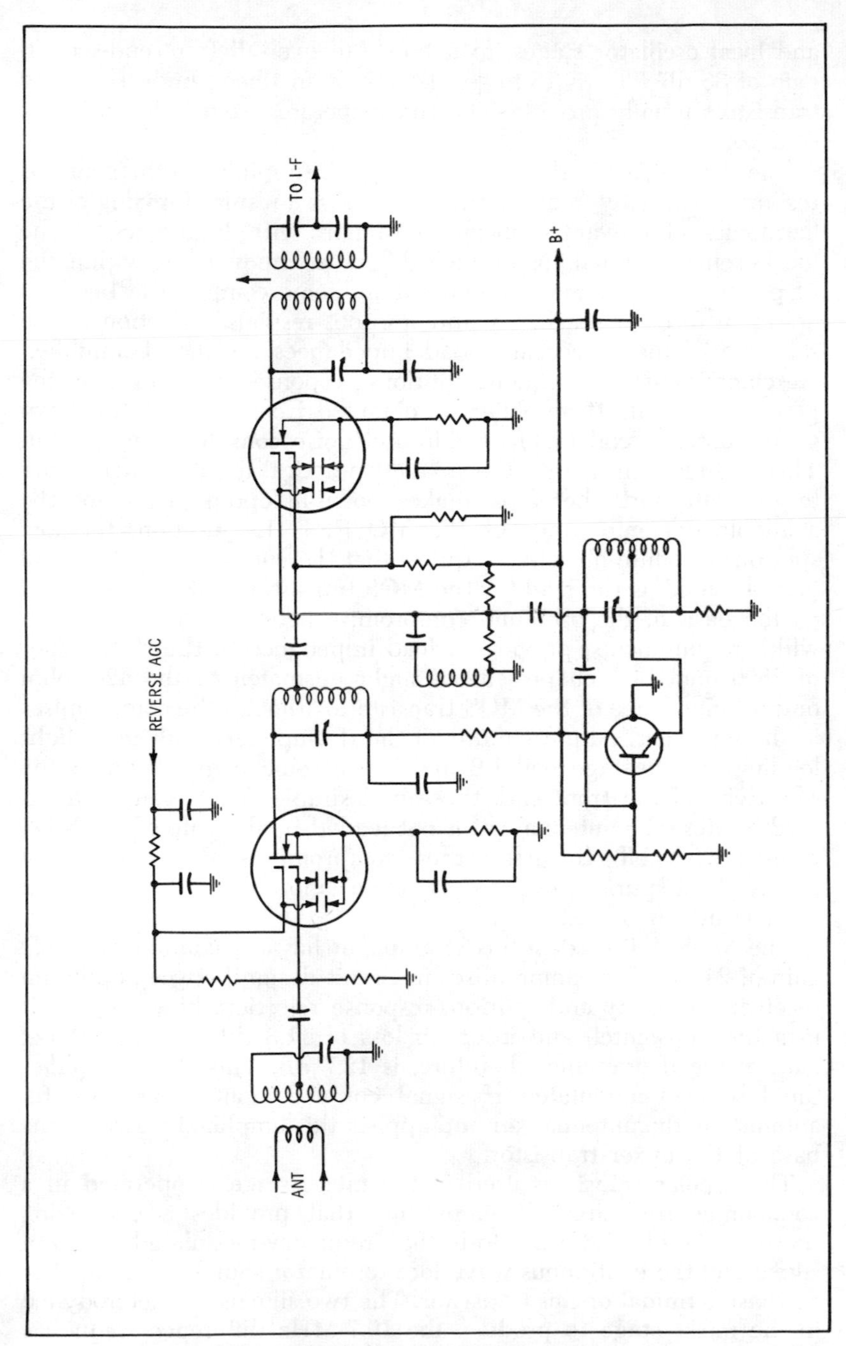

Fig. 4-40. Fm front end with dual-gate MOSFET's in rf and mixer stages.

and local oscillator stages, to achieve an over-all front-end-section gain of 35 dB. This is 15 to 20 dB more than silicon high-frequency transistors usually provide. The tuner operates from a dc supply of −15 volts.

The rf amplifier in the tuner is designed to minimize the spurious responses normally found in fm receivers as a result of mixing of the harmonics of unwanted incoming signals with harmonics of the local-oscillator signal to produce difference frequencies within the i-f passband. This objective necessitates some compromise between optimum receiver sensitivity and spurious-response rejection in the selection of the source and load impedances for the rf amplifier.

Achievement of minimum spurious responses requires that the gate input to the rf amplifier be obtained from a tap as far down on the antenna coil (L1) as gain and noise considerations permit. This arrangement assures the smallest practical input voltage swing to the gate and, therefore, makes possible optimum use of the available dynamic range of the MOSFET. In addition, the low spurious-response objective requires that the entire rf interstage coil (L2) be used as the load for the MOS transistor. This coil, selected on the basis of the optimum compromise between gain and bandwidth requirements, provides a load impedance to the rf amplifier of 3800 ohms, which presents a slight mismatch to the 4200-ohm output impedance of the MOS transistor. Although the compromises in the input and output circuits of the rf amplifier result in a slight loading of interstage coil L2 and cause some degradation in the selectivity of the front end, these undesirable effects can be tolerated because the antenna coil is not loaded by the gate of the MOS transistor. The effectiveness of these compromises is demonstrated by the excellent spurious-response rejection (more than 100 dB) which the circuit can provide.

The MOSFET used in this rf amplifier has a maximum available gain of 24 dB. The compromises in circuit design between optimum receiver sensitivity and spurious-response rejection, however, result in a total mismatch and insertion loss of 11.3 dB. The actual net gain of the rf amplifier, therefore, is 12.7 dB. This stage amplifies the frequency-modulated rf signal coupled from a 300-ohm fm antenna by the antenna coil and applies this amplified signal to the base of the mixer transistor.

The bipolar transistor used in the mixer stage is operated in a common-emitter circuit configuration that provides a conversion power gain of 21.8 dB. Both the frequency-modulated rf input signal and the continuous-wave local-oscillator signal are applied to the base terminal of this transistor. The two signals are heterodyned in the mixer stage to produce the 10.7-MHz difference frequency used as the intermediate frequency in fm receivers.

The bipolar oscillator transistor is operated in a common-collector circuit that generates an extremely clean oscillator waveform.

Fig. 4-40 shows applications of dual-gate protected MOSFET's in an arrangement that provides optimum use of the available dynamic range of the MOSFET in both the rf amplifier and mixer stages. The dual-gate MOSFET is very good for use as a mixer because the signals to be mixed are applied to separate gate terminals. This reduces oscillator radiation in the antenna circuits.

Fig. 4-41 shows application of integrated circuits to an fm stereo front end. For the optimum performance that can be achieved with such a circuit, the differential mode is used. The gain is about 40 dB overall, and the noise figure is 7.5 dB, which is higher than that achieved by other circuits shown in Figs. 4-38 through 4-40. However, where space is at a premium, this circuit will provide excellent performance, at lower cost.

To improve tuning operation of the modern hi-fi receiver, tuning diodes (varactors, or voltage-variable capacitors) have been introduced to replace variable capacitors and all their related mechanical apparatus. Tuning diodes are pn junction diodes in which the junction capacitance is varied by changing the applied reverse bias voltage. They are used in fm receivers as the variable element that changes the resonant frequency of series and parallel resonant circuits.

Tuning diodes offer the following advantages over mechanical capacitors:

1. Mechanical linkage and switching contacts are eliminated.
2. Channel or station changes can be made in less complicated arrangements by push-button, continuous-tuning, or signal-search systems or by sweep methods.
3. Precision automatic fine tuning is simplified.
4. Faster response time is provided.
5. Remote tuning is simplified.
6. Tuning components are many times smaller than the mechanical components they replace.
7. Circuitry can easily be adapted to modular or microcircuit types of packaging.
8. Miniaturization is simplified.

Fig. 4-42 shows a tuned circuit of this type as used in the rf circuit of an Altec Lansing stereo receiver. In the complete front end of this receiver, four tuning diodes (balanced varicap tuning) are all controlled by one external variable control voltage. This control voltage is varied by a potentiometer, but it could be controlled by voltages provided from automatic, remote, or other tuning subsystems.

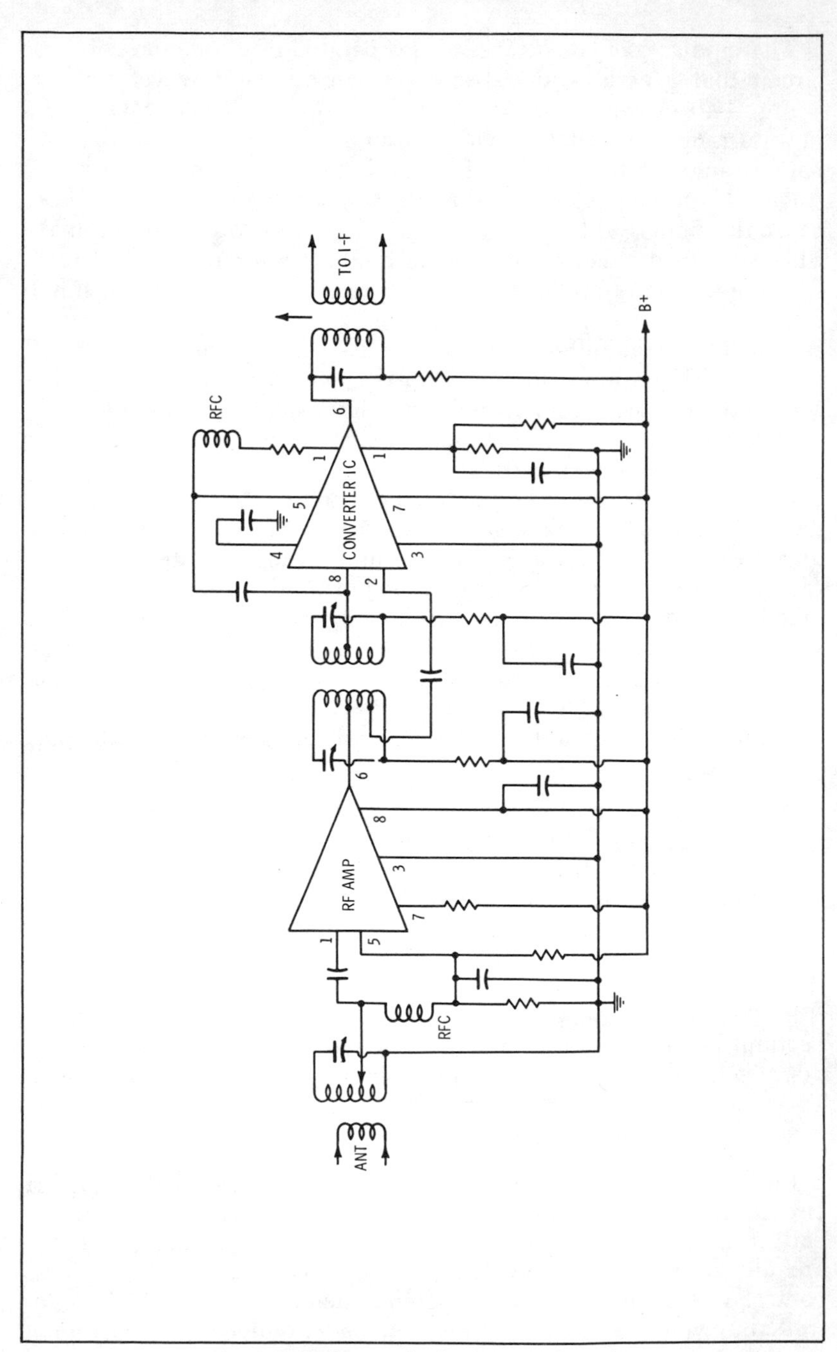

Fig. 4-41. Fm front end using integrated circuit.

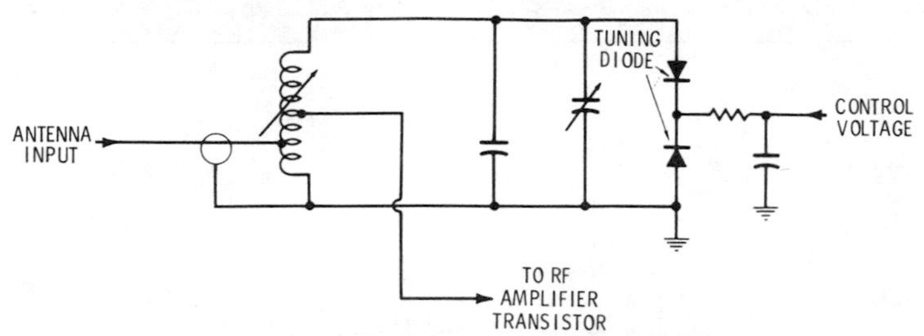

Fig. 4-42. Tuned circuit containing tuning diodes.

I-F Amplifiers—To a major extent, the i-f amplifiers used in an fm receiver determine the overall gain and selectivity of the receiver. Two stages of amplification, in which three double-tuned transformers are employed, are generally used. To secure the required broad-band response, the Q of the i-f transformer windings is made comparatively low. In addition, one or more of the transformers is often overcoupled to broaden its response. The i-f frequency used in most modern fm broadcast receivers is 10.7 MHz. This high intermediate frequency gives excellent rejection of image-frequency interference.

Integrated-circuit units provide excellent results in small space for modern i-f amplifier assemblies. Integrated circuits include a number of types that are essentially subsystems designed to replace several discrete-component stages in specific types of applications. Such circuits are designed to provide multiple functions in specialized hi-fi applications. Certain integrated circuits are designed esecially for use in the i-f sections of broadcast receivers. These units are basically wide-band amplifier-limiter circuits intended for use with external fm detectors. Others can provide high-gain i-f amplification, noise limiting, fm detection, and low-level audio amplification in fm receivers without the use of external components other than tuned coupling networks and bypass elements.

Fig. 4-43 shows a schematic diagram of an integrated-circuit i-f section for an fm stereo receiver. Each IC consists of three direct-coupled cascaded differential amplifier stages and a built-in regulated power supply. Each of the cascaded stages consists of an emitter follower. The operating conditions are selected so that the dc voltage at the output of each stage is identical to that at the input to the stage. This condition is achieved by operation of the bases of the emitter-coupled differential pair of transistors at one-half the supply voltage and selection of the value of the common emitter load resistor to be one-half that of the collector load resistor. As a result, the voltage drops across the emitter and collector load resistors are equal, and the collector of the emitter-coupled stage operates at a voltage equal to V_{BE} plus the common-base potential.

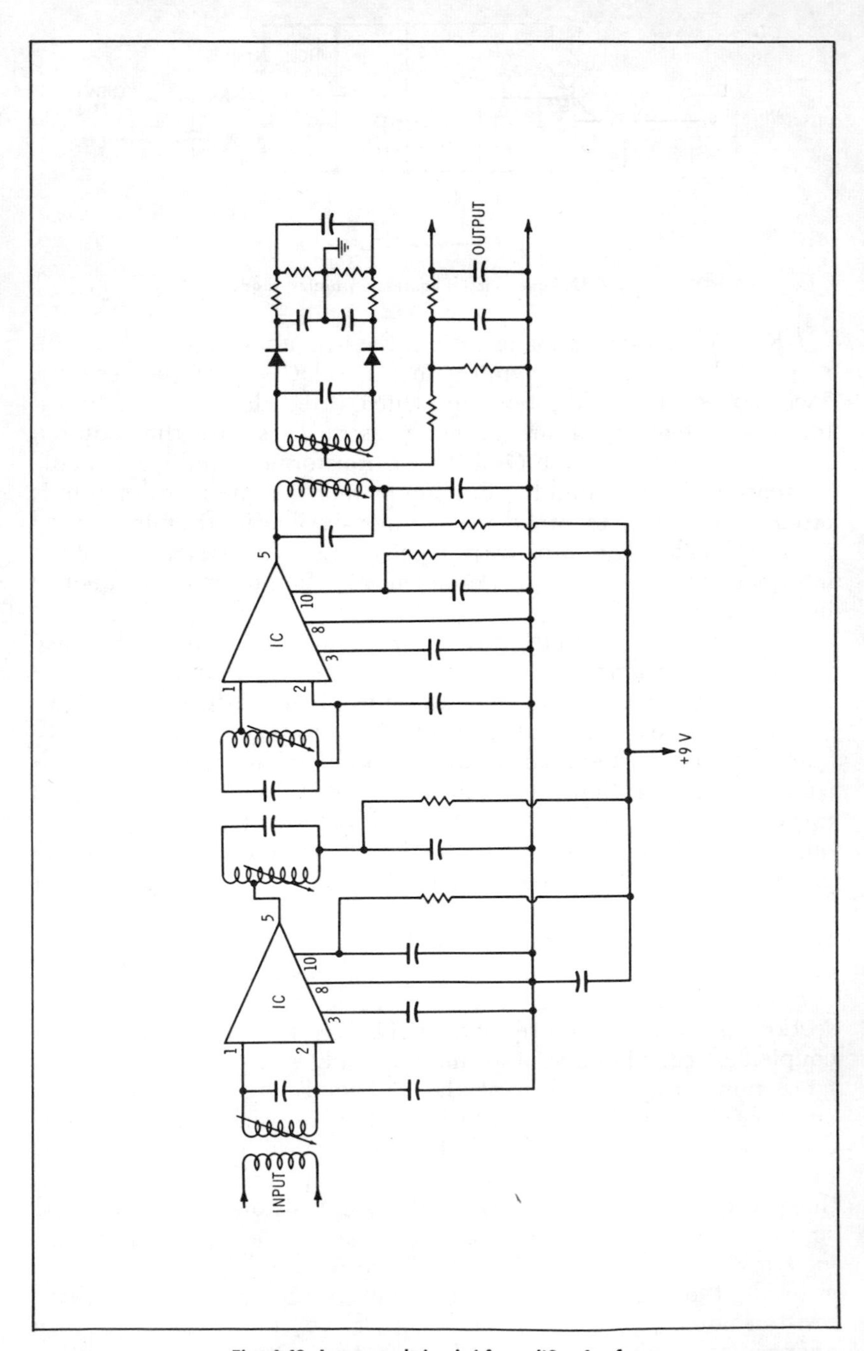

Fig. 4-43. Integrated-circuit i-f amplifier for fm.

The potential at the output of the emitter follower, therefore, is the same as the common-base potential.

The integrated circuits in Fig. 4-43 are designed to operate at various levels of dc supply voltage up to 7.5 volts. Others, which have higher supply-voltage and dissipation ratings, may be operated at dc supply voltages up to 10 volts. For each circuit, the external dc voltage is applied to terminals 10 and 5; dc voltages required at other terminals are derived from the internal power supply. When the circuits are operated at the same dc levels, the characteristics of their amplifier limiter stages are identical. For operation at 7.5 volts with an ac resistive load impedance of 3000 ohms from terminal 5 to ground, the output voltage at terminal 5 with respect to ground is typically 3 volts peak-to-peak.

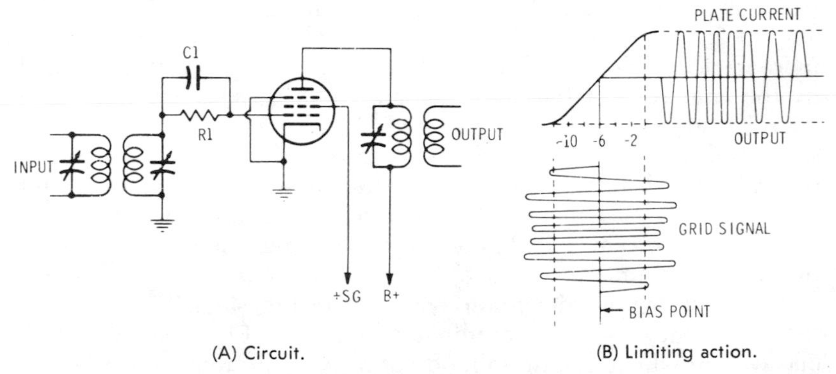

(A) Circuit.

(B) Limiting action.

Fig. 4-44. A typical limiter circuit.

The performance of these integrated circuits in the i-f amplifier-limiter section of an fm radio receiver is at least equal to that of conventional circuits in every characteristic, and is superior in many of them. In particular, the a-m rejection ratio (more than 50 dB) of the integrated circuits is so large that it cannot be measured with commercial fm-am signal generators because of incidental phase modulation of the generators.

Limiters—In receivers using discriminator-type detectors, some means must be provided to remove amplitude variations from the received signal before it is applied to the detector stage. The function of the limiter is to remove noise pulses and to restore uniformity to the signal over the passband. A simplified limiter circuit is shown in Fig. 4-44A. A sharp-cutoff tube is used. The plate and screen voltages applied to the tube are much lower than those normally applied to an amplifier, and no fixed bias is provided. Bias is obtained by placing a capacitor and resistor in the grid return of the tube. Under these operating conditions, a grid signal of comparatively low amplitude will drive the tube to saturation on positive peaks, and to cutoff on negative peaks.

This action is illustrated in Fig. 4-44B. When the input signal to the limiter has sufficient amplitude, all the negative and positive peaks are clipped, and the signal at the plate of the limiter has a constant amplitude. If the signal is not of sufficient amplitude, only partial limiting will take place. Resistor R1 and capacitor C1 play an important part in the operation of the limiter. During the positive half cycles of input voltage, the grid of the tube draws current, loading the input tuned circuit and providing a diode clipping action. During this period, there is current through resistor R1, and capacitor C1 is charged. During the negative portion of the input-signal cycle, C1 discharges through R1, developing a negative grid bias which is proportional to the amplitude of the input signal. When the amplitude of the input signal increases, the bias on the tube becomes more negative. Thus the bias on the tube is automatically controlled by the amplitude of the input signal. The time constant of R1 and C1 is chosen so that it is long enough to maintain substantially constant grid bias during the negative portions of the i-f signal applied to the stage. However, the time constant is short enough to permit an increase in bias when sudden amplitude impulses occur.

In hi-fi receivers, two limiter stages are recommeded to secure proper limiter action.

An integrated-circuit limiter is shown in Fig. 4-45.

Discriminators—The discriminator circuit of Fig. 4-46 illustrates one way in which the audio modulation may be removed from the frequency-modulated carrier. The i-f signal is coupled to the discriminator plates by means of transformer T1. The secondary of the transformer consists of two windings, L2 and L3, tuned by means of capacitors C2 and C3. Tuned circuits L2-C2 and L3-C3 are resonated at different frequencies, one above the frequency of the received signal and the other below it. The frequencies to which the resonant circuits are tuned are equal to the carrier center fre-

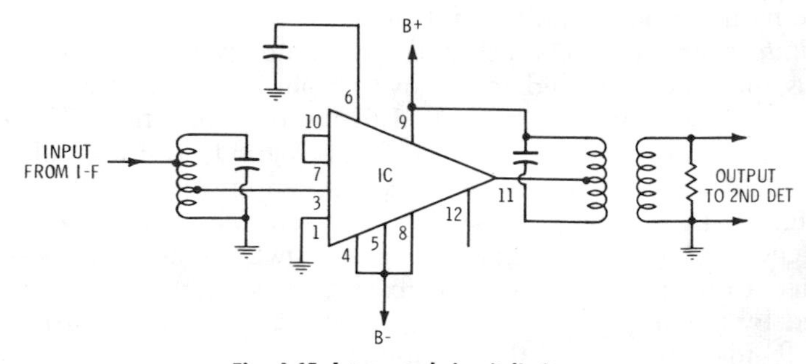

Fig. 4-45. Integrated-circuit limiter.

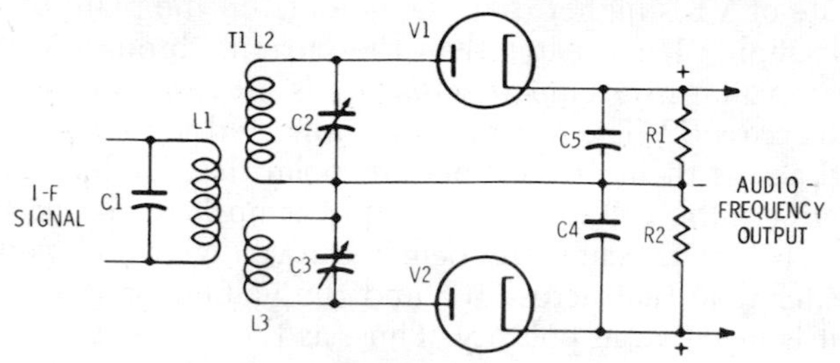

Fig. 4-46. A double-tuned discriminator circuit.

quency plus the maximum rf carrier deviation, and the carrier center frequency minus the maximum rf carrier deviation. For fm broadcast reception, one tuned circuit is tuned to the carrier center frequency plus 75 kHz, and the other to the carrier center frequency minus 75 kHz.

Fig. 4-47 illustrates the response curves of the tuned circuits as indicated by the voltages at the plates of the diodes. When a frequency-modulated signal is applied to the input of the circuit, the instantaneous voltages on the diode plates vary as the signal swings each side of the center frequency. If L2 and C2 are resonant above the carrier frequency, the voltage on the plate of V1 will be higher than the voltage on the plate of V2 when the carrier swings higher in frequency than its center frequency. When the carrier swings lower in frequency than its center frequency, the voltage on the plate of V2 will be higher than the voltage on the plate of V1. When the carrier frequency is above its center frequency and the voltage

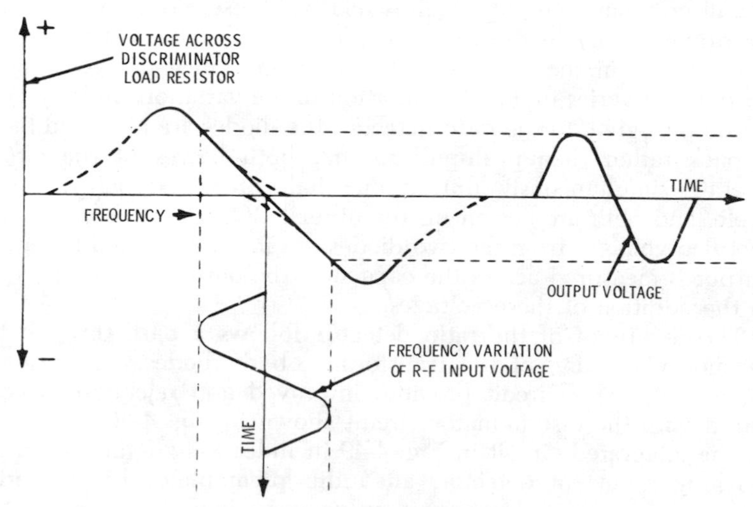

Fig. 4-47. Discriminator response curves.

on the plate of V1 is higher than the voltage on the plate of V2, the current through V1 is greater than the current through V2; consequently, the voltage developed across R1 is greater than the voltage developed across R2. The resistors are connected so that the voltages developed across them are of opposite polarities. Under the conditions described, the voltage at the output is positive. When the carrier swings below its center frequency, the voltage developed across R2 is greater than that across R1, and the voltage at the output of the circuit is negative in polarity. Thus, as the carrier swings above and below its center frequency, it produces a voltage in the output of the discriminator which varies in accordance with the modulation of the fm carrier.

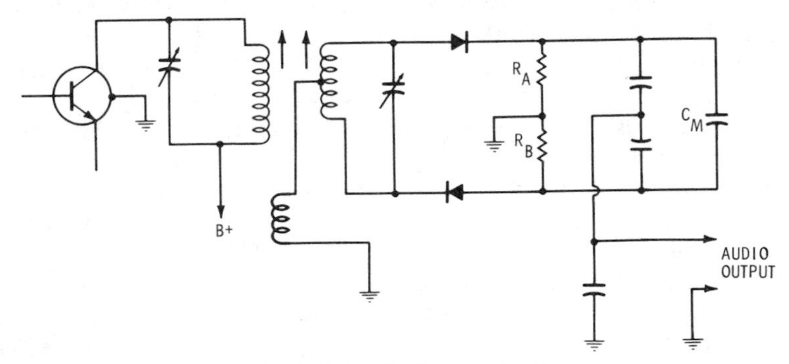

Fig. 4-48. A ratio detector circuit.

Other FM Detectors—The ratio-detector circuit shown in Fig. 4-48 is a discriminator circuit which is relatively insensitive to amplitude variations (a-m) in the fm signal. The phase-shift relationship of the voltages in the tuned transformer results in a detected signal output that varies in direct proportion to the variations in frequency of the fm signal. In the ratio detector, the diodes are arranged back-to-back rather than push-pull so that both halves of the circuit operate simultaneously during one half of the signal-frequency cycle, and both are cut off on the other half cycle. As a result, the rectified voltages from the two diodes are in series. When the audio output is measured across the capacitor, the output voltage is equal to the addition of these voltages.

The dc circuit of the ratio detector follows a path through the secondary winding of the transformer, both diodes, and resistors R_A and R_B. This circuit provides improved a-m rejection as compared with the discriminator circuit shown in Fig. 4-46.

The integrated circuit in Fig. 4-49 includes a high-gain i-f amplifier-limiter, an fm detector, an audio preamplifier-driver, and a zener-diode-regulated power supply on a single monolithic chip.

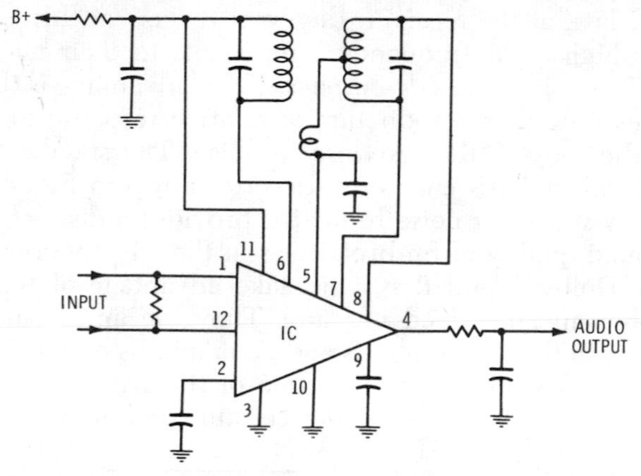

Fig. 4-49. Integrated circuit used as limiter, fm detector, and audio amplifier in receiver.

This circuit is designed for use as a major subsystem for the i-f sections of high-fidelity fm receivers. This unit is supplied in a 12-terminal TO-5 type package and operates over a temperature range of −55° C to +125° C.

The four-stage emitter-coupled i-f amplifier-limiter section of Fig. 4-49 provides a voltage gain of 80 dB at 10.7 MHz. The output stage of this section provides exceptional limiting characteristics, which can be attributed to its use of a transistor constant-current source. The fm detector section is distinguished by circuitry which provides forward bias to the detector diodes and also provides a reference voltage for automatic frequency control (afc). The audio preamplifier-driver provides a low-impedance drive for subsequent audio amplifiers. The power-supply section provides zener-diode-regulated, decoupled voltages for the i-f amplifier, detector, and audio-amplifier sections.

This integrated circuit is designed to operate from a dc supply voltage of +30 volts applied to terminal 11 through a 750-ohm resistance. Terminal 11 may be connected to any positive voltage source through a suitable resistor, provided the maximum dissipation limit or any of the maximum voltage or current limits for the circuit is not exceeded.

Tuner Compensation

Pre-emphasis and de-emphasis are used in standard fm broadcasting and reception to minimize noise at the receiver. The high-frequency portion of the audio modulating signal is purposely boosted at the transmitting station to a relatively higher level than the remainder of the audio-frequency range. This is known as *pre-*

emphasis. Then, at the receiver, the exact reverse of this is done, to restore the high audio-frequency components to their normal relative level. This is known as *de-emphasis*. The advantage is that in the de-emphasis process, the reduction of relative response to high frequencies also reduces the response to noise. The standard pre-emphasis and de-emphasis characteristics are shown in Fig. 4-50.

The Dolby system of noise reduction provides a distinct improvement in sound quality of fm broadcasts in the effective range of the fm station. Dolby A and B systems take advantage of the psychoacoustic phenomenon called *masking*. They use an advanced technique similar to the pre-emphasis and de-emphasis operations shown in Fig. 4-50, except that only a portion of the signal is emphasized and de-emphasized, and only under certain conditions.

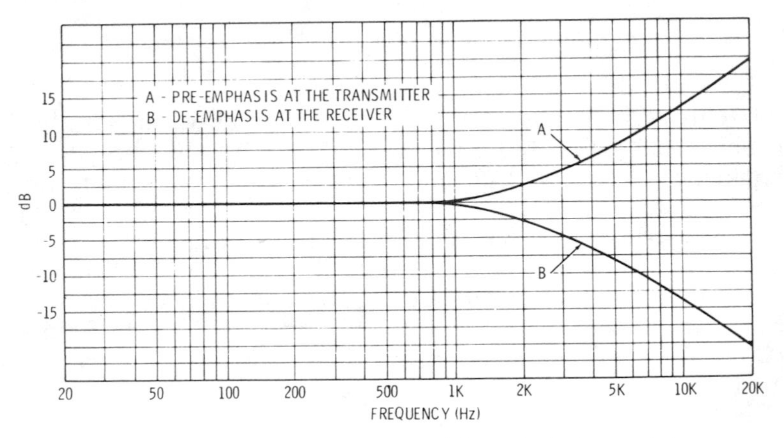

Fig. 4-50. Standard fm pre-emphasis and de-emphasis characteristics.

To the listener, the louder sounds cause the ear to respond mainly to them and to conceal the presence of softer sound. To produce a Dolbyized signal, the station processes the signal by analyzing the music and automatically increasing the strength of quiet musical passages, mostly in the upper frequency ranges. This produces a signal level of these lower-strength passages of music that is in proportion to the combined undesired background noises of the equipment and the atmospherics so that the desired signals stand out and are no longer obscured.

The masking effect does not occur when the sounds are of different pitch. For example, the sound of a piano will not mask the sound of a violin.

The de-emphasis circuitry in the receiver must contain Dolby-licensed circuits to reverse the process, by analyzing the sounds that have been emphasized and automatically de-emphasizing them in amounts equal to the emphasis made to each frequency at the broadcast station. Receivers designed to process these kinds of signals are now available (Chapter 5).

For a further discussion of the Dolby system, see Fig. 4-69 and related text.

Stereo Tuners

As explained in Chapter 3, tuners for stereo are similar to those for monophonic reception, but with differences as follows:

1. The original method of stereo broadcast and reception used separate a-m and fm channels. Two separate tuners were used for stereo, one for amplitude modulation and one for frequency modulation. Both a-m and fm facilities were available in hi-fi tuners before the advent of stereo and were employed as a means of carrying the two channels without new techniques or equipment. For stereo, the tuners were completely separate, so that both were played at the same time for am/fm two-station stereo reception.
2. Provisions are now made for reception of multiplex signals with fm tuners. This consists of providing a separate "multiplex" output for feeding to a multiplex adapter, which is integrated in the receiver or may be bought separately.

The specifications important in the selection of tuners for monophonic reception as discussed up to this point apply also to tuners for stereo reception. In the use of stereo-type tuners for two-station reception, the output of the a-m section was one stereo signal, and the output of the fm tuner was the second stereo signal. These outputs were simply connected to separate channels in the stereo preamplifier. For multiplex reception, the regular fm tuner output and the multiplex (mx) fm tuner output are both connected to the multiplex adapter; the adapter then delivers the left and right signals to the preamplifier.

Some examples of combination tuner units are illustrated in Fig. 4-51.

Combination tuners and dual preamplifiers allow such functions as volume, bass and treble boost, and stereo balance to be controlled on the same panel as the tuner dial. Only a two-channel amplifier and speaker system need be added with such an arrangement. Sometimes the combination process is carried one step further and the tuner, preamplifier, and amplifier are all combined. Such units include everything needed for reception of stereo broadcasts except the speaker system. Examples of these combinations are shown in Chapter 5.

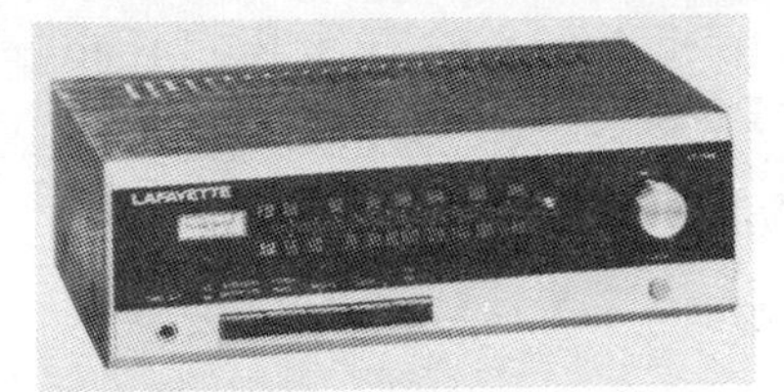

Courtesy Lafayette Radio Electronics Corp.

(A) Lafayette LT-725 am/fm tuner.

Courtesy McIntosh Laboratory, Inc.

(B) McIntosh MR 73 am/fm tuner.

Courtesy Acoustic Research, Inc.

(C) Acoustic Research fm tuner.

Courtesy Electro-Voice, Inc.

(D) Electro-Voice E-V 1255 fm tuner.

Courtesy Heath Co.

(E) Heathkit AJ-1510 digital fm tuner.

Courtesy Marantz Co., Inc.

(F) Marantz Model 20 fm tuner.

Fig. 4-51. Some fm and am/fm tuners.

Stereo Multiplex Adapters

A typical multiplex adapter unit is illustrated in Fig. 4-52. This kind of unit can be connected to any tuner to provide fm multiplex reception. Fig. 4-53 shows an adapter that mounts directly on the receiver chassis, making it an integral unit suitable for fm and fm multiplex outputs.

A block diagram of a typical fm tuner with multiplex is shown in Fig. 4-54. From the antenna through the main detector and de-emphasis network, it is the same as a nonstereo tuner.

The high-pass filter, demodulator, and matrix portion of Fig. 4-54 is often referred to as the *stereo converter* or *stereo translator*. It can be connected to a nonstereo tuner to provide stereo output, if the received signal is transmitted as multiplexed stereo. The multiplex input of the converter is connected to the "multiplex" output of the tuner detector, before the de-emphasis network. The shielded

Courtesy Heath Co.

Fig. 4-52. An adapter to convert early tuners for reception of fm multiplex.

lead between the multiplex jacks on the tuner and the converter should be kept as short as possible, to avoid undue attenuation of the subcarrier.

Stereo Multiplex Operation

As explained in Chapter 3, stereo multiplex is the system whereby the left- and right-channel signals of a stereo program are both sent

Courtesy Fisher Radio

Fig. 4-53. A plug-in stereo multiplex adapter.

out on one fm broadcast channel. In the method standardized in the United States by the Federal Communications Commission, the following objectives are attained:

1. The left (L) and right (R) signals are sent, in separable form, within the 200-kHz fm broadcast channel bandwidth.
2. Both signals are modulated onto the same carrier with negligible loss of signal-to-noise performance compared with that of monophonic transmission. This is accomplished by means of the "interleaving" provided in the system.

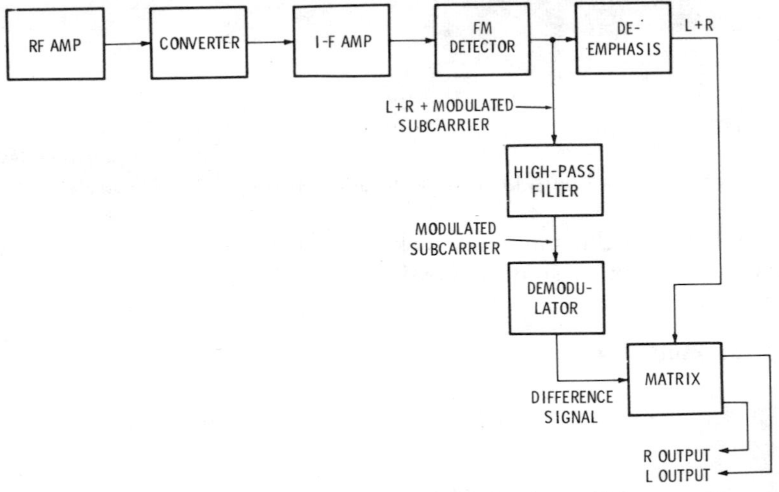

Fig. 4-54. Block diagram of a typical tuner equipped with a multiplex converter.

3. The system is compatible in that the owner of an fm receiver not equipped for stereo reception still receives an L + R signal equivalent to that received from nonstereo broadcasts.

Sum and Difference Signals—Although the L + R signal is all the listener to a monophonic program needs, the proper reception and reproduction of stereo programs requires the use of an additional (L − R) signal. This signal is produced at the transmitter by inverting (phase changing by 180 degrees) the R signal (making it negative, or −R) and adding it in its negative form to the L signal. The resulting L − R signal, as well as the L + R signal, is transmitted on the fm broadcast carrier. The two signals are received and separated at the receiver. Then, in the stereo receiver, they are added together to form the L signal alone, and the L − R signal is subtracted from the L + R signal to form the R signal alone. This is shown mathematically as follows:

$$L + R + (L - R) = L + L + R - R = 2L$$
$$R + R - (L - R) = L - L + R + R = 2R$$

Stereo Multiplex Modulation Signal—The L + R and L − R signals are both sent as part of the signal which frequency modulates the main (radio-frequency) carrier. Before we proceed with details of the modulating signal, it will minimize confusion to emphasize a very important fact: As far as the main carrier of the fm signal is concerned, there is only one composite modulating signal, all of whose components modulate the carrier. The composition of this modulating signal includes the L + R signal, a signal conveying L − R information, and certain other components.

The spectrum of the modulating signal is shown in Fig. 4-55. Its bandwidth is from 50 Hz to 75 kHz. The lowest portion of the modulating-signal spectrum extends from 50 Hz to 15 kHz, and this range includes the components of the L + R signal, or the signal that corresponds to or may be used as a normal monophonic modulation signal. In other words, if we remove all the signal components of frequencies above 15 kHz, we have a signal similar in effect to those signals used before stereo multiplex was developed.

Next in the spectrum is a 19-kHz pilot signal. However, let us ignore this for the moment and proceed to the L − R portion of the signal, extending from 23 kHz to 53 kHz except for a 100-Hz quiet spot in the middle. This region consists of the sidebands resulting when the L − R signal modulates the 38-kHz subcarrier. At the station, after the modulation has taken place, the 38-kHz subcarrier is suppressed, and only the sidebands are used. Since the carrier component *was* at 38 kHz, and since the L − R signal contains

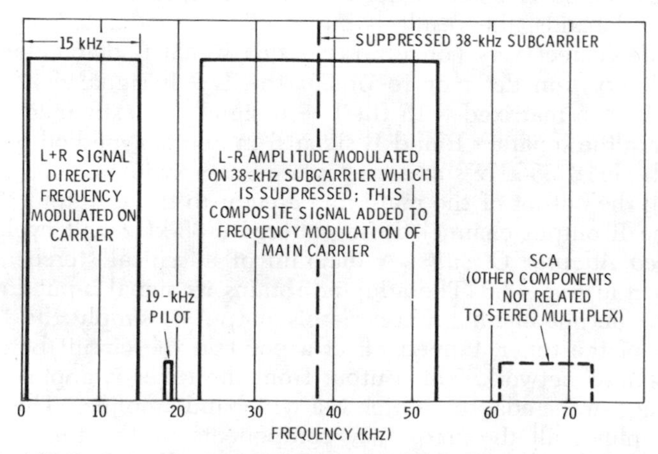

Fig. 4-55. Frequency spectrum of frequency-modulation signal on main carrier.

frequencies from 50 Hz to 15 kHz, the L − R signal sidebands extend from 23,000 Hz to 37,950 Hz, and from 38,050 Hz to 53,000 Hz. These sidebands are added to the modulating signal. In relation to the main carrier, all frequencies in this composite modulation of the main carrier are above audibility, so they do not interfere with transmission and reception of the main-carrier L + R signal for the listener using a monophonic receiver.

The L − R sidebands as an independent signal are not intelligible in the form in which they appear in the modulation signal. They must be: (1) recombined with a 38-kHz carrier, (2) demodulated from that carrier, and (3) decoded.

To process the signal at the receiver, the composite frequency-modulation signal depicted by Fig. 4-55 must be demodulated from the main carrier, and the L + R signal and the L − R sidebands must be separated from the other parts of the signal by filters.

Demodulating the L − R Signal—The carrier with which the sidebands are combined at the receiver must have the desired 38-kHz frequency within very close limits and remain synchronized to prevent distortion when these sidebands are demodulated. The desired synchronization is provided by the 19-kHz pilot carrier added to the modulation signal. This steady 19-kHz unmodulated signal is demodulated from the main carrier at the receiver and passed through a frequency doubler, whose output at 38 kHz is combined with the L − R sidebands to produce the L − R signal. The output from the doubler may be used directly for combining with the sidebands, or, as in some designs, this output can be used to synchronize a 38-kHz oscillator whose output is in turn used to combine with the sidebands.

When the 38-kHz subcarrier has been combined with the L − R signal sidebands, the result is an amplitude-modulated signal. In some stereo receivers (or adapters) this signal is demodulated by an ordinary a-m detector to obtain the L − R signal. The L − R signal is then matrixed with the L + R signal (as explained earlier) to obtain the separate L and R signals. In another method of detection, the local 38-kHz subcarrier is used as a switching signal, connecting the output of the main fm detector to the L output channel and the R output channel during alternate 38-kHz half cycles.

Stereo Adapter Circuits—A diagram of a typical stereo adapter is shown in Fig. 4-56. The adapter obtains its signal input from the "stereo" output of an fm tuner; this output is simply the "audio" output of the tuner, tapped off at a point in the circuit before the de-emphasis network. This output from the tuner is applied to the first stage in the adapter, which is a wideband amplifier. This amplifier amplifies all the frequency components in the tuner output, which include the L + R signal (already in audio form), the 19-kHz

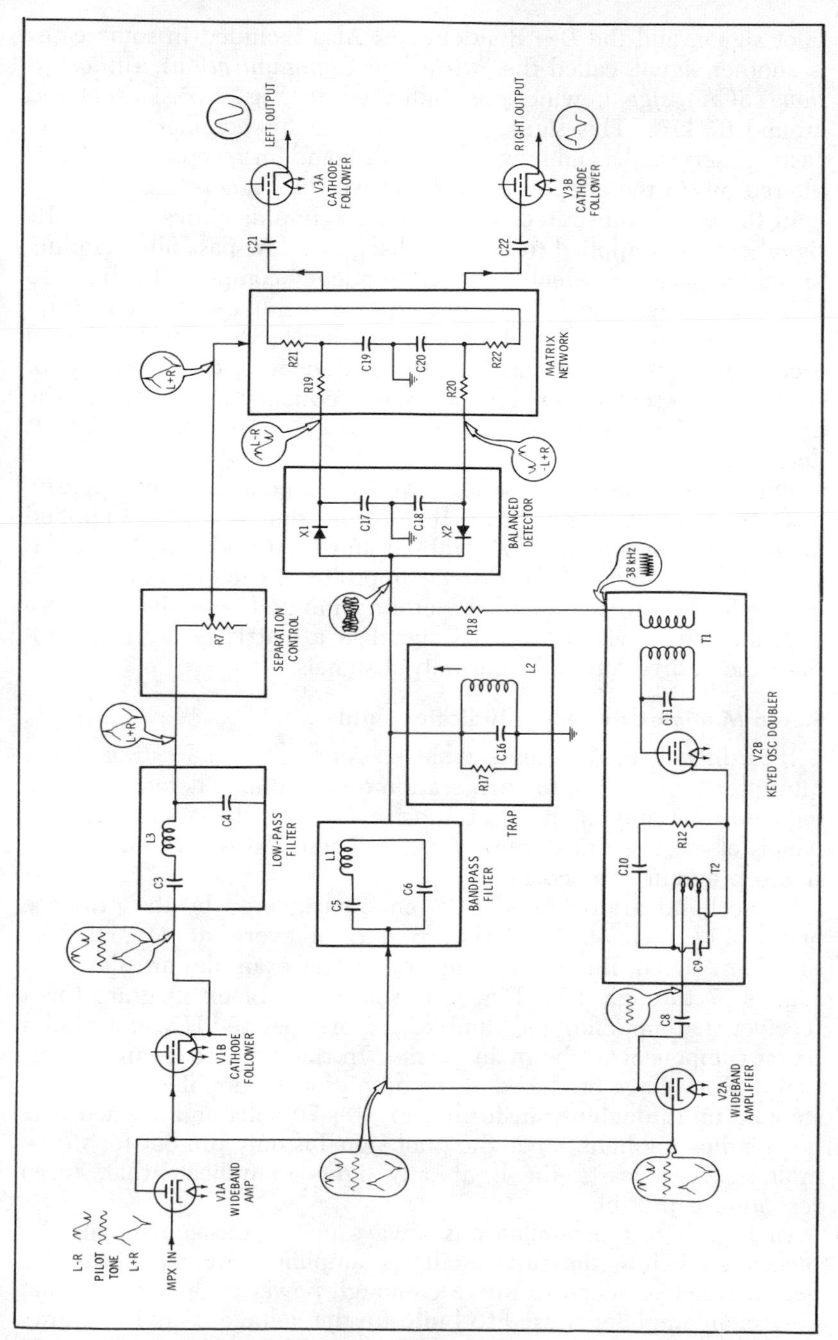

Fig. 4-56. Block diagram of Heath multiplex adapter.

pilot signal, and the L − R sidebands. Also included in some cases is another signal, called the *Subsidiary Communications Authorization* (SCA) *signal,* which, as indicated in Fig. 4-55 is centered around 67 kHz. This signal, which is used for commercial "storecasting" service, has nothing to do with home fm reception, and it is filtered out in the adapter circuits—so we shall ignore it.

In the circuit illustrated, the wideband amplifier uses a tube. Its signal output is applied to three paths: (1) a low-pass filter (inductor and capacitors) which removes frequency components above 15 kHz, leaving just the L + R signal, (2) a bandpass filter and trap, which pass only the L − R sidebands, and (3) a 19-kHz keyed oscillator-doubler which doubles the frequency of the 19-kHz pilot carrier. The oscillator-doubler locks in frequency and phase to the amplified pilot signal, and provides a 38-kHz signal for the L − R detector.

The detector uses two solid-state diodes connected in opposite polarities. The 38-kHz and L − R sideband signals are both applied to the detector, where they combine and then undergo demodulation. Since the diode polarities are opposite to each other, one puts out a plus L − R signal and the other a minus L − R signal. These then are mixed with the L + R signal in a matrix network to produce the desired left and right output signals.

Stereo Multiplex Reception Indicator Lights

It is difficult to determine what stations are broadcasting stereo multiplex by listening to each station on the dial. Therefore, stereo indicators—usually lights or tuning eyes—have been arranged on panels of most recent stereo receivers. These lights turn on when a stereo program is tuned in.

These lights are operated by a circuit triggered by the presence of the 19-kHz pilot signal that is part of every stereo multiplex broadcast signal. Block diagrams and circuit examples are shown in Figs. 4-57 through 4-61. Fig. 4-57 shows the block diagram for a receiver that amplifies and doubles the original 19-kHz pilot modulation component of the main carrier. In this type of circuit, a portion of the voltage produced at the output of the doubler amplifier is stepped up through a transformer to over 70 volts and applied to a neon indicating light. Since the pilot signal is only present for stereo multiplex broadcasts, the light only lights to indicate that stereo reception is possible.

In Fig. 4-58, the oscillator is always on, so the signal must be picked off before the pilot-oscillator amplifier circuits. Since this output is not sufficient to provide enough power to operate a signal device, an amplifier is used to build up the voltage. Fig. 4-59 shows the circuit detail of a stereo multiplex indicator circuit. Fig. 4-60

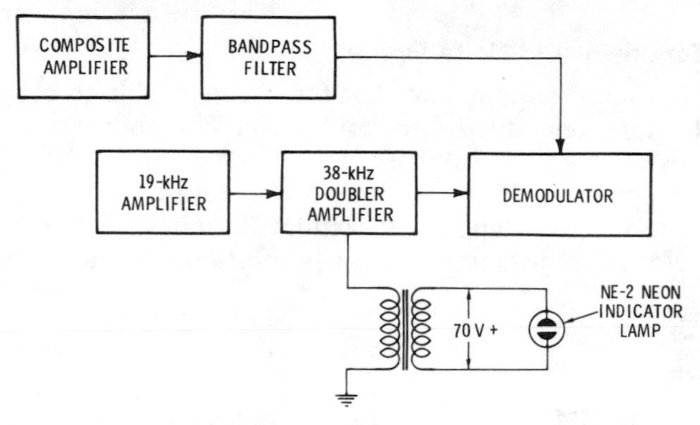

Fig. 4-57. Stereo multiplex indicator light (no oscillator circuit).

shows a Scott multiplex circuit with stereo indicator output (terminal 10 on jack).

Automatic FM/FM Stereo Switching

Some higher-priced fm multiplex receivers have automatic switching circuits which operate to change the receiving mode when fm broadcast material changes from mono to stereo, and vice versa. Fig. 4-61 shows a method of accomplishing this objective and at the same time providing for a stereo beacon light. This circuit is dependent on the 19-kHz pilot signal, and it operates with the agc circuits of the receiver as shown.

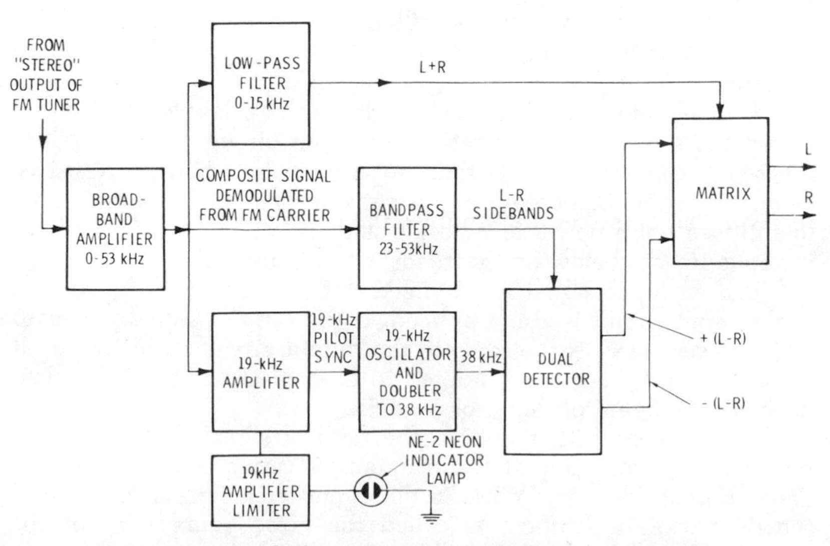

Fig. 4-58. Stereo multiplex with indicator-light takeoff before oscillator.

Tuner Sensitivity and Noise Figure

It is a primary requirement that the tuner be able to bring in the desired stations clearly and without noise. The ability of a receiver to do this is related to its sensitivity. A sensitivity of 10 microvolts on fm reception or 75 microvolts on a-m reception will give good results for receivers operating within 15 miles of the broadcast station. For more distance, look for proportionately more sensitivity.

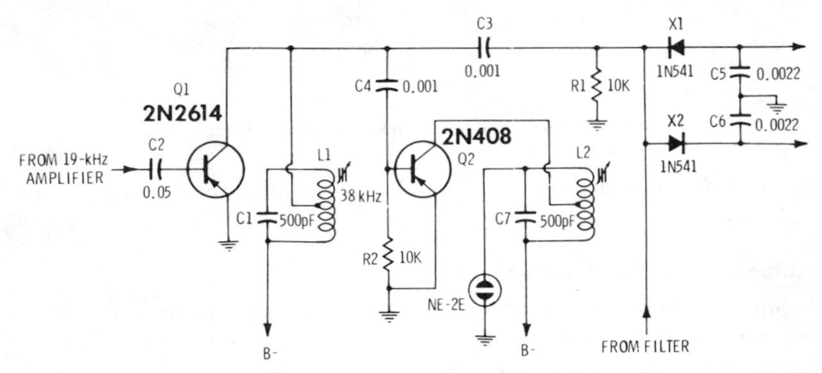

Fig. 4-59. Stereo indicator circuit in Knight Model KN-265.

If you live over 50 miles from the desired station, you should obtain the best tuner available with (probably) less than 3 microvolts sensitivity on fm reception and 10 microvolts on a-m reception. At greater distances, a-m stations will probably be the more reliable source of reception, and your choice of equipment should be so governed.

One of the most important characteristics of a low-level amplifier circuit is its signal-to-noise ratio. The input circuit of an amplifier inherently contains some thermal noise contributed by the resistive elements in the input device. All resistors generate a predictable quantity of noise power as a result of thermal activity. This power is about 160 dB below one watt for a bandwidth of 10 kHz.

When an input signal is amplified, the thermal noise generated in the input circuit is also amplified. If the ratio of signal power to noise power (s/n) is the same in the output circuit as in the input circuit, the amplifier is considered to be "noiseless," and is said to have a noise figure of unity, or zero dB.

In practical circuits, however, the ratio of signal power to noise power is inevitably impaired during amplification as a result of the generation of additional noise in the circuit elements. A measure of the degree of impairment is called the noise figure (nf) of the amplifier, and is expressed as the ratio of signal power to noise

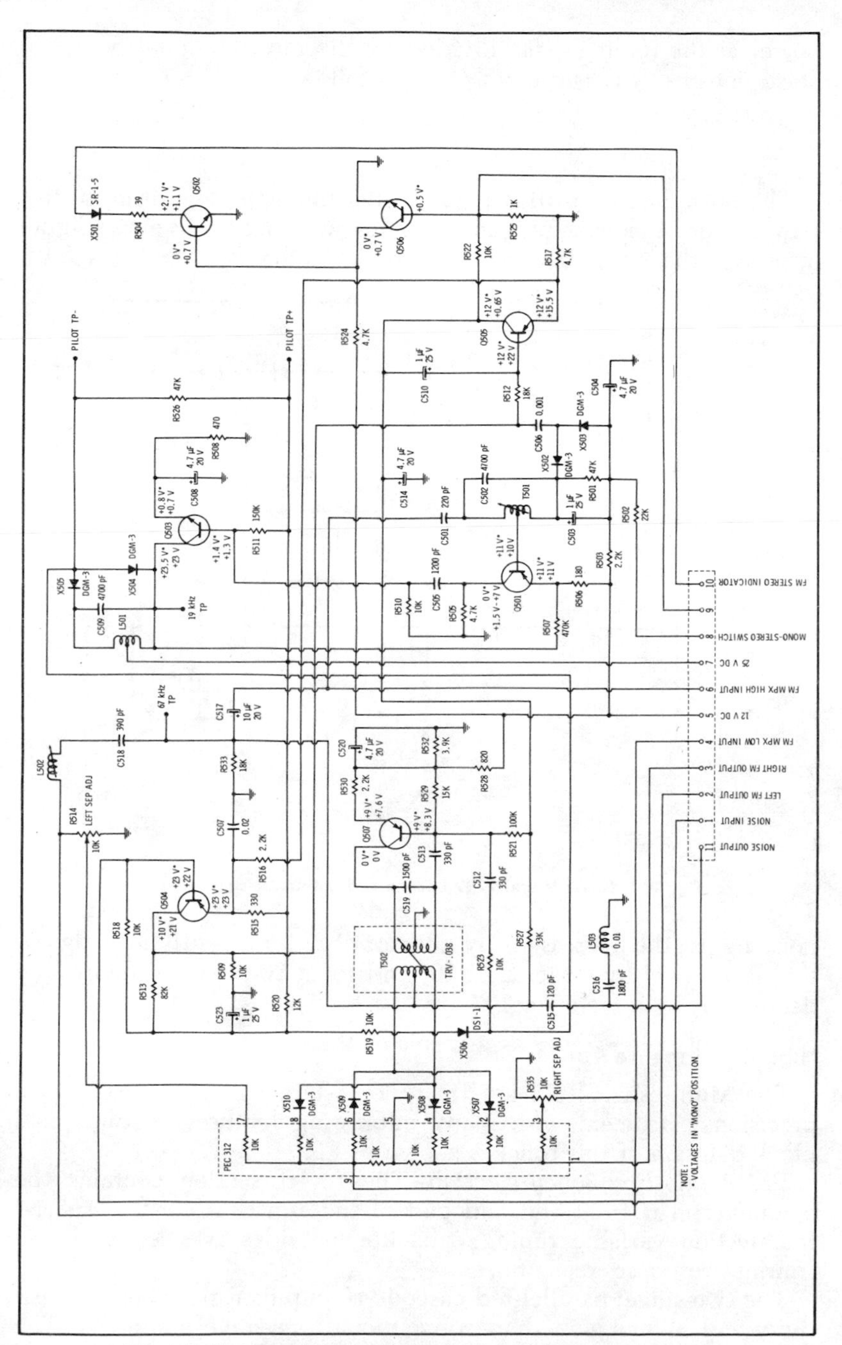

Fig. 4-60. Scott transistor multiplex circuit with indicator output.

power at the input (s_i/n_i) divided by the ratio of signal power to noise power at the output (s_o/n_o), as follows:

$$nf = \frac{s_i/n_i}{s_o/n_o}$$

The noise figure in dB is equal to ten times the logarithm of this power ratio. For example, an amplifier with a one-dB noise figure decreases the signal-to-noise ratio by a factor of 1.26; for a 3-dB

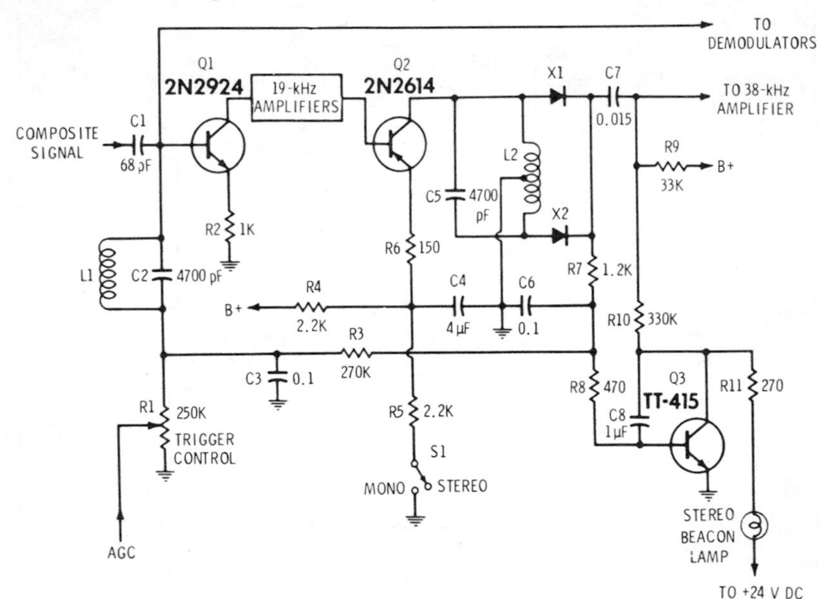

Fig. 4-61. Fisher Model 1249 beacon and automatic stereo circuit.

noise figure, the decrease is by a factor of 2; for a 10-dB noise figure, the decrease is by a factor of 10; and for a 20-dB noise figure, the decrease is by a factor of 100.

High-Performance Tuner

The McIntosh MR 73 am/fm tuner (shown in Fig. 4-51B) is a precision instrument with many interesting features. A functional block diagram of this tuner is shown in Fig. 4-62.

FM Radio-Frequency Section—The fm rf section contains the complete fm rf front end and part of the a-m rf circuits. A special four-section variable tuning capacitor provides rf selectivity and spurious-response rejection.

The two-stage, parallel-fed cascode rf amplifier gives better sensitivity and higher gain than conventional one-stage amplifiers. The use of junction field-effect transistors (JFET's) in this amplifier

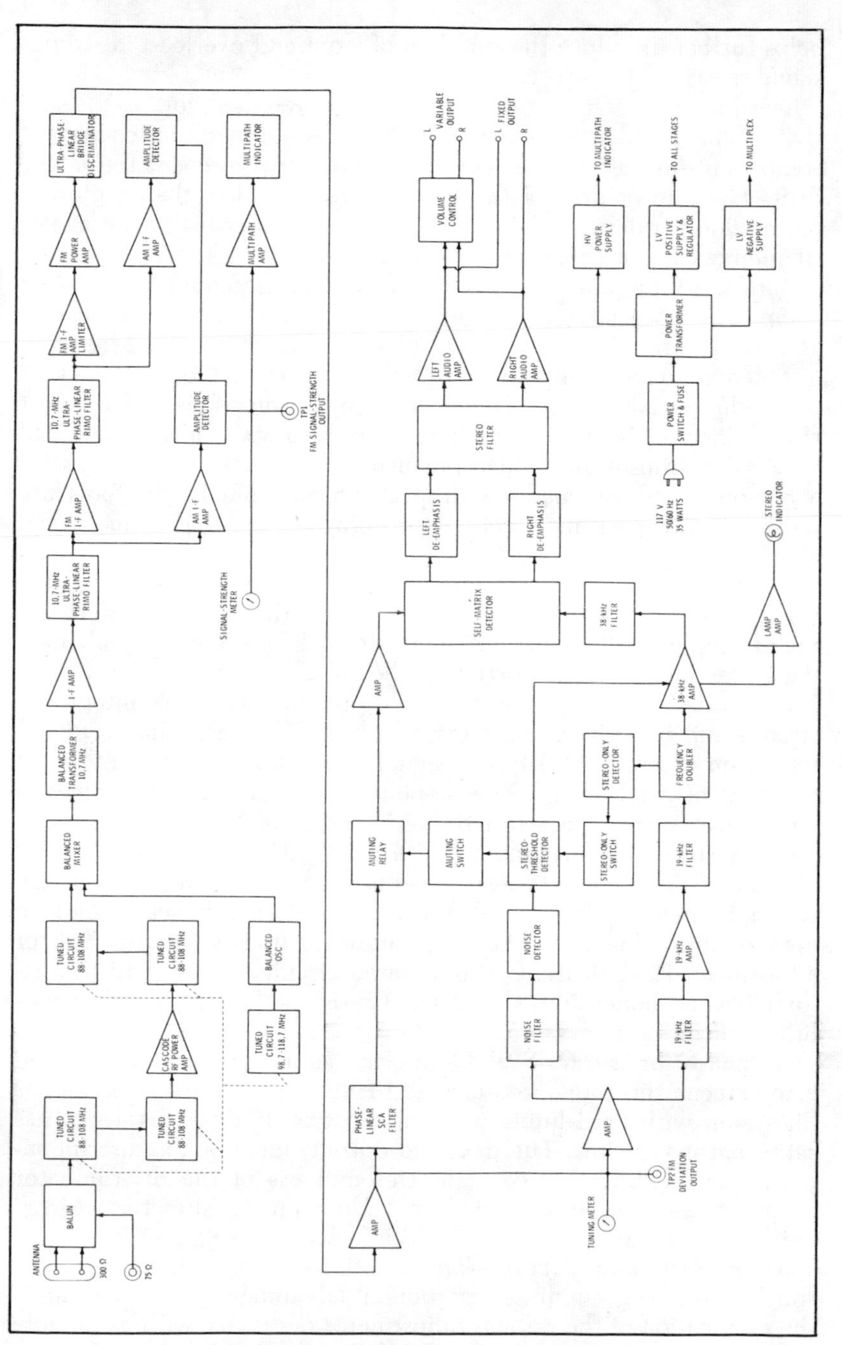

Fig. 4-62. Block diagram of McIntosh MR 73 tuner.

helps further to reduce the problem of front-end overload in strong-signal areas.

The mixer, which also uses a JFET, is designed for high sensitivity and freedom from overload. Low–temperature-coefficient components are used in the fm local oscillator to prevent frequency drift. The rate of drift of the local oscillator is less than ten parts per million per degree centigrade, and this stability makes automatic frequency control (afc) unnecessary.

Both the fm and a-m front ends have been designed in the same completely encased metal modules. This design gives protection against radiation or interference.

Antenna connections are provided for either 300-ohm twin lead or 75-ohm coaxial cable. The normal input impedance of the first rf amplifier is 75 ohms. Impedance match to 300 ohms is provided by a balun transformer which has negligible losses. Connections for a 300-ohm line are made with push-type terminals; no tools are required. A Type F male connector is furnished for 75-ohm coaxial cable.

For greater signal transfer and lower distortion, a special matching transformer has been designed to couple the fm rf section to the fm i-f amplifier. This matching transformer considerably enhances the linear phase characteristics of the i-f amplifier.

Fm i-f and Detector Section—The fm i-f consists of two integrated circuits and two phase-linear crystal filters. They combine to give a total gain of over 120 dB (the signal is amplified to over 1,000,000 times its original level). The response curve has a nearly flat top with linear phase characteristics. The skirts of the response curve are very steep. The maximum width is 240 kHz at −3.0 dB and 400 kHz at −60 dB. The response curve is symmetrical each side of the center frequency. The crystal filters are permanently sealed and do not require adjusting. The i-f cannot drift nor vibrate out of adjustment. Each of the two integrated circuits contains 16 transistors, 3 zener diodes, 5 diodes, and 23 resistors, all on a single monolithic silicon chip.

A "phase" or "Foster-Seeley" discriminator has been designed to complement the integrated-circuit i-f section. The i-f section has high gain with hard limiting characteristics. It develops a capture ratio that is very low. The detected output signal of the discriminator is low in distortion content. De-emphasis of the discriminator output restores the frequency-amplitude characteristics that existed before transmission.

Stereo Multiplex Section—The multiplex section incorporates a special detecting circuit. A particular advantage of this circuit is the elimination of the critical adjustments necessary with commonly used matrixing circuits. The circuit detects the L − R sidebands, then automatically matrixes the recovered information with the L + R main-carrier signal. This yields the left and right program outputs with maximum separation.

The 19-kHz pilot signal is filtered from the composite stereo input signal, amplified by a special limiting amplifier, doubled to the 38-kHz carrier frequency, and then amplified again by a limiting amplifier. The composite signal minus the 19-kHz pilot is combined with the 38-kHz carrier signal. The new combination of signals is fed to the special detector circuit mentioned above. Balanced full-wave detectors are used to cancel the 38-kHz components in the output.

The SCA (Subsidiary Communication Authorization) signal must be removed from the composite output. This is accomplished by the use of a computer-designed "image parameter" band-elimination filter. The filter rejects SCA signals without impairing stereo performance. In the MR 73, fm muting operates by detecting ultrasonic noise which is present between stations or when a weak station is being received. The muting circuit can be activated or defeated by the use of the muting switch on the front panel. The level of muting desired can be adjusted by the muting level control on the top panel. Varying the muting control adjusts the threshold at which the muting takes effect.

When the 19-kHz carrier of a stereo signal is received, the automatic fm stereo switching circuit activates the multiplex decoding circuit. This lights the stereo indicator. The circuit switching is all done electronically with no clicks. The automatic stereo switching can be defeated by turning the mode selector switch to fm mono. (In this position, the stereo indicator will still light to indicate the presence of a stereo signal.) On monophonic transmissions, the stereo switching is inactive at all times, assuring a greater signal-to-noise ratio. The stereo switching circuit has been designed so that noise will not activate it.

A-M Section—The a-m section uses MOS field-effect transistors. The a-m rf amplifier circuit includes a three-section variable tuning capacitor in the metal-enclosed shielded module that also houses the fm rf front end. The a-m rf amplifier uses a dual-insulated-gate MOSFET to obtain more overload protection against strong local stations than can be obtained with conventional bipolar transistors.

The a-m mixer is also a dual-insulated-gate MOSFET. The use of MOSFET's in the rf amplifier and mixer stages makes possible low cross modulation (spurious response) and good image rejection.

Three double-tuned i-f transformers are used to obtain selectivity while still allowing good a-m fidelity. A 10-kHz whistle filter has been incorporated at the output of the a-m detector. Its purpose is to suppress whistles that result from heterodyning between adjacent a-m carriers.

An external a-m antenna may be connected at the rear apron of the unit by using a push connector. An internal transformer matches the external antenna to the input impedance of the a-m front end. An optional ferrite-core antenna is also provided for local or strong stations. A switch on the back panel selects either this antenna or the external antenna.

An a-m sensitivity switch is provided to decrease both the gain of the rf amplifier and the noise between stations.

Audio Preamplifier Section—The audio amplifier increases the level of the program so that it is adequate to drive a preamplifier or other accessory equipment. There is a three-transistor amplifier for each channel. The design uses considerable negative feedback to help achieve low distortion, wide frequency response, and excellent stability. Each audio amplifier delivers 2.5 volts to the FIXED OUTPUT jacks at 600 ohms impedance. A second pair of outputs, the level at which can be varied by the volume control, is available.

The stereo filter is connected in the audio amplifier to reduce noise when a weak stereo station is being received. This filter is designed to permit a good compromise between channel separation and noise rejection.

Power Supply—All signal stages are powered from a 16-volt regulated supply. The 16-volt regulator is elaborate in design, using a specially selected transistor and associated circuit. The regulator uses electronic filtering to maintain a low background hum level, good stability, and good regulation.

A half-wave rectifier and filter supply the dc high voltage needed for the anode of the multipath indicator. A full-wave rectifier supplies dc to the multiplex indicator and to the voltage regulator.

TAPE RECORDERS

Tape recorders consist of tape-transport devices, motors, erase-bias head, recording head, playback head, amplifiers, and controls. Most lower-priced models combine the recording and playback operations in one head, requiring one less unit. The tape is generally transported from a reel through the erase, recording, and playback heads, drawn by a capstan and pressure roller driven by a constant-speed motor.

The tape is finally wound on a second reel, usually operated by a separate motor (see Fig. 4-63). As the tape is pulled through the recording or playback head, variations in flux density in the gap surrounding the tape are produced both in recording and playback. During recording, the signal to be recorded is fed to the recording head, causing a changing flux field which magnetizes the particles in the tape. On playback, the flux from the moving magnetized

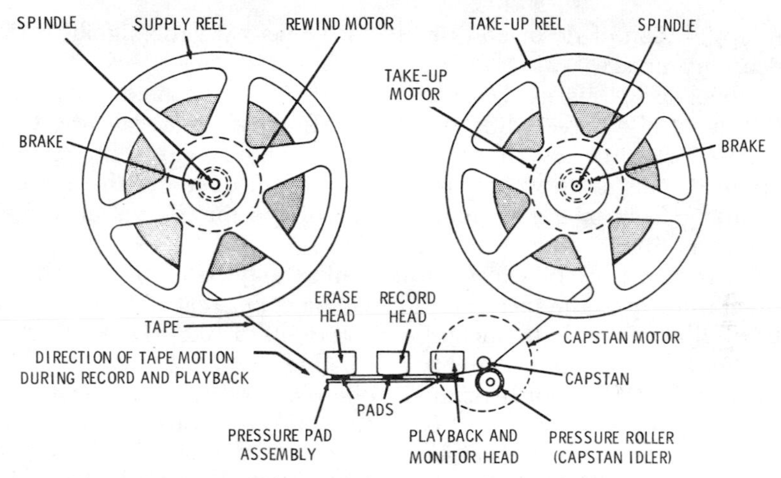

Fig. 4-63. Major parts of a tape-transport mechanism.

particles in the tape induces a signal voltage in the playback head; this voltage is proportional to the residual magnetism of the particles at any point, and therefore is a reproduction of the recorded signal.

The problem of producing a linear recording on magnetic material is similar to that of producing linear output from a class-A amplifier. The magnetization (hysteresis) curve shown in Fig. 4-64 has a changing slope with only a portion straight enough to use for hi-fi

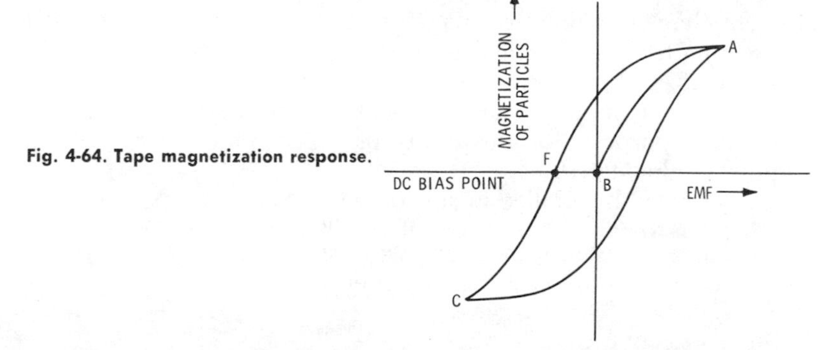

Fig. 4-64. Tape magnetization response.

reproduction. The residual curve shows the relative magnetization left after the medium (tape) has left the magnetizing field. Just as in the operation of a class-A amplifier, only the straight portion of the curve can be used for hi-fi. Thus, the recording must be limited to the straight portion of the residual magnetization curve. This is accomplished by use of bias to change the relative operating level of the recording signal fed to the recording head. This may be done by application of dc or ac (supersonic) bias with the signal to the recording head.

When dc bias is used, a sufficiently positive dc field is applied just ahead of the recording head to magnetically saturate the tape medium. Then negative dc bias is fed with the signal to the recording head. This negative bias is made sufficient to center the average recording level in the straight portion of the magnetization curve, as shown around point F in Fig. 4-64.

Supersonic (ac) bias instead of dc bias may be applied with the signal directly to the recording head. The supersonic signal is above the audible range, in the neighborhood of 40 to 100 kHz—the higher the frequency, the more fidelity is possible.

The mixing effect of the signal to be recorded and the supersonic bias signal is shown in Fig. 4-65. The total signal variation is within the area of linear operation of curve *AB*, as described. The supersonic bias signal is recorded and reproduced along with the audio signal but cannot be heard by the human hearing system. The supersonic bias arrangement is very popular because it has quieter effects at low volume and silent areas of the program, and it is easy to produce and adjust.

Erase

Demagnetization principles, which have been used in other fields for years, are applied in a precise manner to erase a recorded signal from a tape. By applying an ac field to a tape in sufficient strength to produce complete magnetic saturation of all articles on the tape, thereby homogenizing its magnetic pattern, and then gradually decreasing the same ac field to zero strength, the tape will become demagnetized. For tape recorders, the ac field is obtained from the supersonic bias supply. In addition to the conveniences of this source for such a purpose, the possibility of a beat note from a separate supply is eliminated.

Erasure may be applied in several ways. When a tape is recorded on, it is customary to erase continuously all information and noise

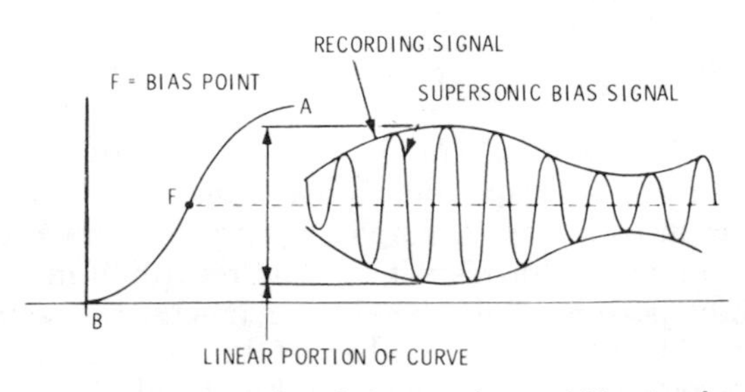

Fig. 4-65. Mixing and applying recording and bias signals.

just prior to the recording operation, as is provided for in tape recorders (see position of erase head in Fig. 4-63). This removes any former programs or noises that may have been on the tape. *Bulk erasing* may be applied to the whole tape all at once. The same principle is used, but the tape is immersed in an ac magnetic field large enough to receive the reel. Bulk erase equipment is more powerful than recorder erase units and is used to erase more quickly and to restore tapes overloaded to near a permanent magnetization condition beyond the erase power of ordinary recorder equipment.

Tape-Drive Motors

Tape-drive motors must have properties similar to those of the very best motors used for phono turntables. Tape drives, however, do not have the advantage of the flywheel effect of a turntable to damp flutter and other changes of speed of the motor; therefore, perfection of motor design is even more desirable for tape. The best tape machines use hysteresis motors, which increase the cost of the machine considerably. For quality reproduction, motor flutter should have a rating of not higher than 0.5 percent, which is the audible level. For high-fidelity operation, the motor should be able to hold all variations in speed to less than 0.5 percent, and the better units are rated at less than 0.2 percent.

The total number of motors in a recorder may vary from one to three. The better machines usually have three—one for feed and rewind, one for take-up, and one for capstan drive.

Heads

The better machines provide three heads, as has been described: one each for erase, record, and playback. Optimum operation requires a separate playback head not only to play back but also to monitor the signal as it is being recorded. This provides a simultaneous check on every link in the recording chain so that if anything is wrong, ones does not have to do the recording over or lose the program. The erase, record, and playback heads are similarly constructed. However, separate design of heads is desirable because optimum operation of each requires different gap spacing and inductance characteristics. Heads contain a ring-shaped, high-permeability core with a fine gap formed to allow a ¼-inch wide by 0.002-inch thick tape to pass with minimum clearance. The core has a winding to provide or pick up signals. Tape contacts the core at the gap either to be magnetized or to induce a field, according to the operation desired. Low-cost commercial tape-recorder heads provide reasonable recording and reproduction at voice frequencies, but most professional units require better heads. Hi-fi heads, such as Dynamu heads, are available to replace standard heads on lower-cost units. These can record and play back with full response ± 3 dB up to 14,000 Hz, if the rest of the elements of the recorder are suitable.

Mechanical Features of Recorders

The transport mechanisms and controls of tape machines can be very elaborate. A tape-transport mechanism draws the tape past the heads at a constant rate of speed, that is, the speed at which the tape is to be played back. The tape is unwound from a reel, passed through the heads, and wound onto a take-up reel. A capstan with a roller device to press the tape against it pulls the tape at a constant speed. This may be reviewed in Fig. 4-63.

The feed reel has a torque applied in a direction opposite to the tape travel. The take-up reel has a similar torque in the direction of the tape travel to relieve the capstan of any difficulty in pulling the tape. On rewind, the pressure on the tape is freed to release it from the capstan, and the feed reel is reversed in direction and speeded up to rewind at high speed.

Miscellaneous

In the better equipment, brakes are usually provided to stop the reels quickly; various speeds are provided for rewind; interlocks are provided to protect tapes from fast rewind speeds during record or playback; electrical cutouts are provided to prevent erase during rewind or playback; release of tape pressure from heads during rewind is arranged to reduce wear; pressure is released on the capstan roller when not in use to prevent a flat spot on the rubber idler wheel; and automatic protection switches operate if the tape breaks or tangles.

Tape decks as separate machines or complete with amplifiers are available. Most equipment can be operated as a separate unit with an outside amplifier feeding the recording head, or the playback head feeding a separate audio system.

Audio amplifiers and controls to feed a recorder of the constant-current type must provide from one-tenth to several milliamperes output. To provide for constant-current output, the output circuit must contain a majority of the impedance of the total coupling network. The amplifier should also contain equalization networks to compensate for the magnetic recording characteristics. This equalization is built in most recorders, but if separate audio systems are used, similar compensation must be used. Compensation for recorders varies for different machines, but essentially each system only has to be arranged to equalize in playback the emphasis and de-emphasis applied to record any tape. Generally, a recording amplifier is compensated to NAB tape-recording standards, and regular preamplifiers (for reproduction) with NAB tape-standard equalization are used.

Recording

Better stereo amplifier and control-center combinations have provisions for two-channel output for stereophonic (or monophonic) tape recording. These outputs can be fed directly to stereo tape recorders, and live stereo programs can be recorded for permanent storage.

This material can be obtained from actual two-microphone pickup (live) or live stereo broadcasts, or from stereo records. The program material may be fed from a tuner or a record player to the amplifier control center, monitored, balanced, emphasized or de-emphasized according to personal taste, and fed to the stereo tape recorder. These tapes provide an optimum medium for permanent storage.

Prerecorded tapes are also available and have advantages over hi-fi records. An important advantage is permanence of fidelity. The quality of program material applied to prerecorded tape is practically permanent. With normal caution, prerecorded tape will not acquire surface noise over hundreds of plays. There is constant retention of frequency range and stereo effects in its program material. Records do not have these characteristics. Discs always develop surface noise from needle wear and handling. Lint, dust, and scratches on records also contribute other undesirable noises and loss of quality and stereo effect on records.

The disadvantage of higher cost per unit playing time of prerecorded tapes is being gradually overcome by reduction of speed in relation to fidelity of response. Cassette tapes with suitable characteristics and equipment operated at 1⅞ inches per second can provide high-fidelity output at a price per minute competitive with standard LP records.

Playback

Many other recent improvements in techniques and equipment for tape recording and playback have made tape more competitive with discs as a home playback method. Other relative advantages and disadvantages of tape for playback were discussed in Chapter 3.

There are virtually no tape recording machines that are not also useful for tape playback. Thus a tape machine offers the advantage of allowing one to make his own recordings as well as to take advantage of permanent playback characteristics of high-quality prerecorded stereo tapes.

However, just "any old tape machine" is not enough to ensure stereo reproduction that satisfies the audiophile. The machine must be of better-than-average quality with respect to frequency response, freedom from flutter and wow, low harmonic distortion, and high signal-to-noise ratio.

All good stereo machines are now designed for "four-track" operation. As explained in Chapter 3, this means that two-channel stereo sound can be recorded on the full tape length in both directions. The four-track arrangement requires special heads, mounted in a single head assembly over which the four-track tape passes.

Most tape-machine manufacturers sell both complete recording and playback systems and separate *tape transports*. A tape transport is the mechanical assembly of drive motors, drive system, and heads and bias-generating equipment to derive a signal from or record a signal on the tape. It does not include the equalizing networks, amplifier(s), and speaker(s) that come with a complete recorder-playback unit. The transport can be used to play a stereo tape, with the outputs of the two playback heads fed to the inputs of the preamplifier of a stereo high-fidelity system. Just about all preamplifiers include a position on the input equalizer switch for tape playback. However, for recording with your tape transport, a separate amplifier may be necessary.

Typical high-quality tape machines for stereo recording and reproduction are illustrated in Figs. 4-66A through 4-66Q. Most machines operate at either of two speeds, with the speed selectable by the operator. A few operate at either 15 inches per second (ips) or 7½ ips, others at either 7½ ips or 3¾ ips. Still others include a third speed, 1⅞ ips; cassette machines operate at this speed exclusively. It is an inherent property of the tape method of recording for frequency response to improve with increasing speed, and a check of manufacturers' specifications will show this. However, it must be remembered that as speed increases, playing time for a given length of tape decreases. Also, a two-track stereo tape which plays back only in one direction provides only half the playing time of a four-track, two-channel tape. Table 4-1 shows various characteristics of tape for several different modes of operation.

Claims as to frequency response at a given speed vary from one machine and manufacturer to another. It should be noted whether the response is claimed with a variation of 2 dB, 3 dB, or just on some "usable response." Machines using a speed of 15 ips obtain better response with less technical difficulty at that speed than machines using lower speeds. Response at 15 ips should be from about 30 or 40 Hz to 20,000 Hz or more with variation in the order of plus or minus 2 dB. The upper limit of response ($\pm$ 2 dB) at 7½ ips is usually about 12,000 to 18,000 Hz, and at 3¾ ips between 7500 and 12,000 Hz. New designs operating at 1⅞ ips are providing comparatively excellent frequency response. Do not let a few hundred hertz of upper-limit frequency response be a final criterion in your choice of a machine, because manufacturers differ in the rigidity and manner of making their tests. Listening is the best test.

Total-harmonic-distortion tolerances are about the same as those for amplifier equipment alone, and about 2 percent should be considered maximum.

Signal-to-noise ratio is important in tape machines because a relatively high noise level is one of the basic problems tape-machine manufacturers have had to overcome. A minimum signal-to-noise ratio acceptable to the high-fidelity enthusiast is at least 40 dB, and 50 or 60 dB is considered highly desirable.

As explained in Chapter 3, one of the disadvantages of tape machines has been the inconvenience of having to thread the head

Table 4-1. Comparison of Tape-Format Characteristics

Characteristic	Open Reel				Cartridge	Cassette
Overall Size	Large				Medium	Small
Continuous Play	Available				Standard	Available
Reverse Play	Available				N/A	Available
Rewind	Standard				N/A	Standard
Record Capability						
Mono	Standard				N/A	Standard
2 Channel	Standard				Available	Standard/ Compatible
4 Channel	Available				Proposed	Proposed
Playback Capability						
Mono	Standard				N/A	Standard
2 Channel	Standard				Standard	Standard/ Compatible
4 Channel	Available				Available	Proposed
Tape Speed (ips)	15	7½	3¾	1⅞	3¾	1⅞
Useful High-End Response (kHz)	20	18	12	10	12	10
High-End Response (kHz) With Crolyn or Equal	25	22	15	12	15	12
Playing Time (*1800-ft Reels)	*	*	*	*		
Mono	45 min	1½ hr	3 hr	6 hr	N/A	2 hr Max
2 Channel	—	1½ hr	3 hr	—	1 hr	2 hr Max
4 Channel	—	45 min	1½ hr	—	30 min	N/A
Signal-to-Noise Ratio	Excellent	Excellent	Very Good	Good	Very Good	Good
Signal-to-Noise Ratio With Dolby	Excellent	Excellent	Excellent	Very Good	Excellent	Very Good
Editing	Excellent	Excellent	Good	Poor	N/A	Poor
Available Prerecorded Library						
2 Channel	None	Poor	Poor		Excellent	Good
4 Channel	None	Poor			Good	N/A

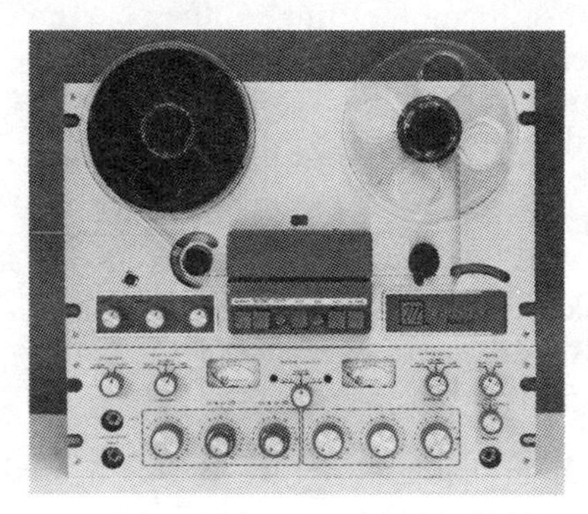

Courtesy Telex Communications Division

(A) Magnecord 1024.

Courtesy Ampex Corp.

(B) Ampex AX300.

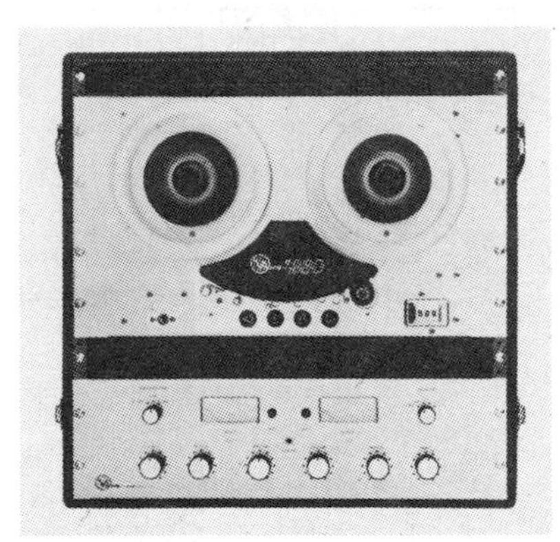

Courtesy Telex Communications Division

(C) Viking Model 230.

Courtesy Roberts Div., Rheem Mfg. Co.

(D) Roberts Model 333X.

Courtesy Lear Jet Corp.

(E) Lear Jet eight-track cartridge tape deck.

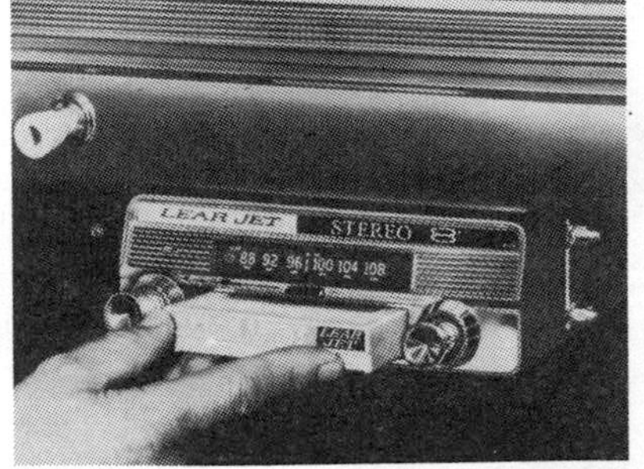

Courtesy Lear Jet Corp.

(F) Lear Jet auto cartridge player/fm radio.

Fig. 4-66. Typical tape machines

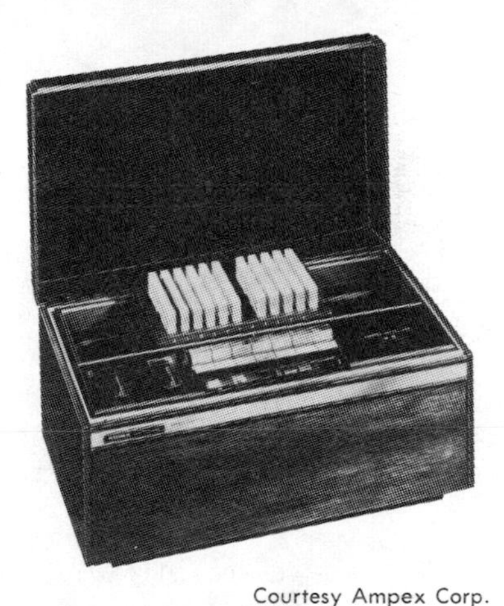

Courtesy Ampex Corp.

(G) Ampex Micro 335 bidirectional, automatic-change cassette deck.

Courtesy Harman-Kardon, Inc.

(H) Harman-Kardon CAD5 cassette deck with Dolby B system.

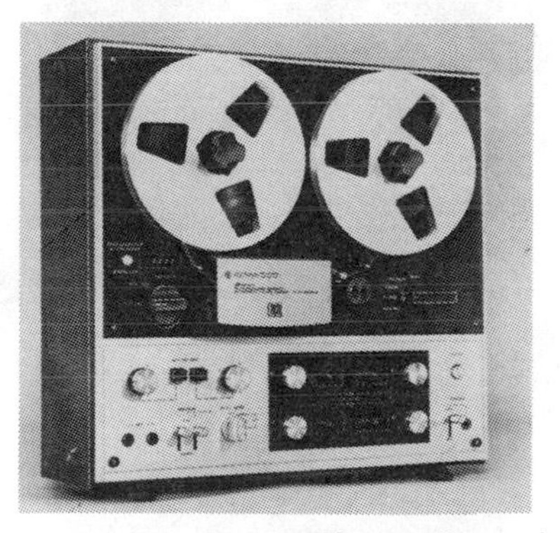

Courtesy Kenwood

(I) Kenwood KW-6044 four-channel deck.

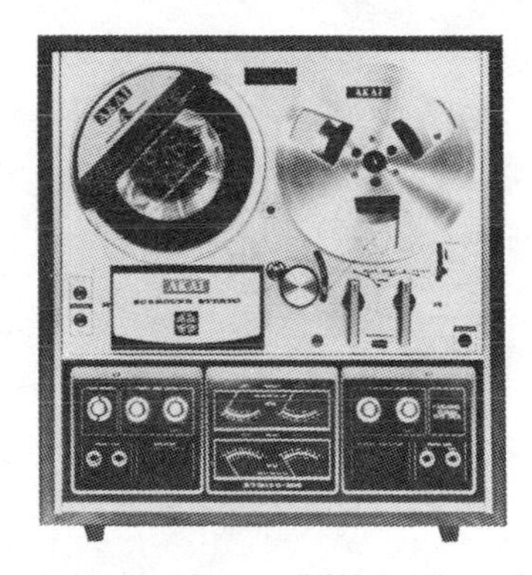

Courtesy AKAI America, Ltd.

(J) AKAI 1730D-SS four-channel tape deck.

Courtesy TEAC Corporation of America

(K) TEAC TCA-42 four-channel tape deck.

(Fig. 4-66 continued on next page.)

for stereo recording and reproduction.

Courtesy JVC America, Inc.

(L) JVC 1400 four-channel/two-channel open-reel tape deck.

Courtesy Fisher Radio

(M) Fisher CP-100 eight-track cartridge player.

Courtesy JVC America, Inc.

(N) JVC 1202 four-channel, eight-track stereo cartridge player.

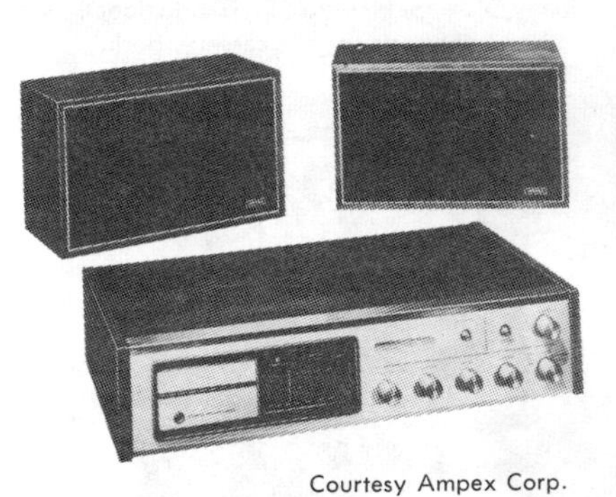

Courtesy Ampex Corp.

(O) Ampex 8400 cartridge stereo music system.

Courtesy JVC America, Inc.

(P) JVC 1350 four-channel, eight-track automobile stereo cartridge player with 24 watts total output.

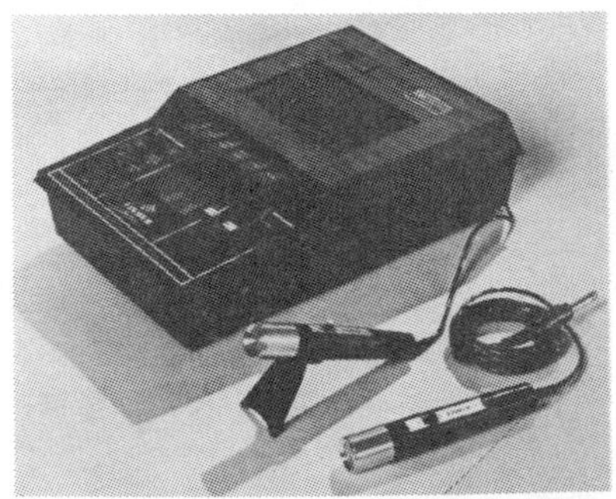

Courtesy Fisher Radio

(Q) Fisher RC-80B stereo cassette recorder with Dolby system and provision for chromium-dioxide tapes.

Fig. 4-66. Typical tape machines for stereo recording and reproduction.—(cont.)

assembly and set up the reels properly. For this reason, cartridge-type tape-machine arrangements have been developed. The cartridge is loaded with standard ¼-inch tape and plays up to an hour of stereo at 3¾ ips. Threading of the tape is not necessary because the tape is already "set up" in the cartridge. The cartridge is merely inserted or removed as desired.

The machine in Fig. 66D accommodates reels, cartridges, and cassettes.

Figs. 4-66E and 4-66F show the Lear Jet 8 eight-track stereo cartridge players, which have a comparatively flat response at frequencies up to 10 kHz. As shown, a model with fm radio included is available for the car, and another model is available as a tape deck for the home. The car model operates on 12 volts dc, and the home model operates on 115 volts ac. These units accept cartridges which can play continuously, with one hour of program material which will repeat until turned off. The tape deck may be indexed (advanced) to any one of four portions of the program material by depressing a button on the panel.

Two-channel cassette stereo recorders are shown in Figs. 4-66G, 4-66H, and 4-66Q. Stereo cassettes come with a range of playing periods including 30, 60, 90, and 120 minutes (total time for both directions). The cassette operates with a tape speed of 1⅞ ips, but with quality equipment and the latest techniques, recording and reproduction with excellent response characteristics up to 12,000 Hz can be achieved.

Figs. 4-66I, 4-66J, and 4-66K show four-channel open-reel recorders which can record two channels and/or play back two or four discrete channels. The Kenwood tape deck shown in Fig. 4-66I has four independent preamplifiers and special provision for selection of correct bias for low-noise, high-density tapes (Crolyn, cobalt, etc.) to provide optimum quality of reproduction from 20 Hz to 20 kHz with a signal-to-noise ratio better than 47 dB (before application of Dolby). This unit also offers three tape speeds—7½, 3¾, and 1⅞ ips—easy rewind, VU meters for four channels, two microphone inputs, two line inputs, and four line outputs, plus headphone jack, four-digit counter, and automatic shutoff.

Fig. 4-66L shows the JVC four-channel/two-channel open-reel stereo tape deck that can record or play back all four discrete channels, as well as two channels on four-track tape. This recorder has specifications that include frequency response of 20 to 25,000 Hz and signal-to-noise ratio of 53 dB.

Figs. 4-66M, 4-66N, 4-66O, and 4-66P show four-channel, eight-track cartridge-tape decks. The players shown have frequency-response ratings of 30 to 12,000 and 15,000 Hz, depending on tape type and associated equipment.

The four-channel, eight-track cartridge player shown in Fig. 4-66P is designed for use in an automobile. This circuit is compatible for playing two-channel cartridges. There is provided fine correction of channel separation, L-R balance control, front-rear balance control, and a program indicator.

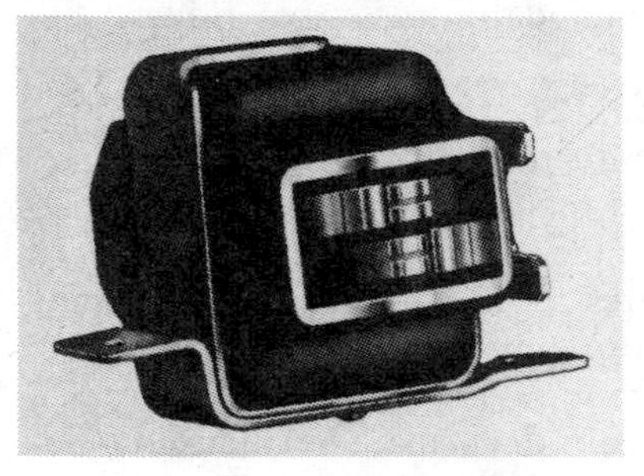

Courtesy Ampex Corp.

Fig. 4-67. Ampex bidirectional head for cassette decks.

The Ampex bidirectional deep-gap cassette head shown in Fig. 4-67 (used in the recorder shown in Fig. 4-66G) combines six separate elements in a single unit to provide bidirectional erase/play/record capability in full two-channel stereo. With all elements in one head, alignment problems are reduced, and recordings made in either direction have identical characteristics. This head has deep-gap design, which is said to improve quality.

For more information on tape formats and capabilities, see Chapter 3.

The Dolby Noise-Reduction System

To enhance the output quality of the slow-speed, narrow-track cassette, the Harmon Kardon CAD5 professional tape deck shown in Fig. 4-66H has incorporated the licensed Dolby B system.

Dolby has developed a sophisticated system for recording and playback of tapes that provides a 10 to 15 dB improvement in signal-to-noise ratio. The original system was designed for master tapes to be used for making prerecorded production tapes and discs, but the system has been modified and applied to recording and playback of prerecorded reels, cassettes, and cartridges for home use.

The Dolby system is a two-step, symmetrical process that operates *before* and *after* recording. Whenever the strength of certain components of lower-level signals being fed to a recorder falls below a predetermined threshold, the system boosts their strength before recording. The signal is then recorded in this encoded form, and during the recording process, the customary noise is added by the electronic circuits of the recorder and by the operation of the mechanical equipment on the tape itself. During playback, the mirror image of the same Dolby process that boosted these signal components before recording returns them to the original level. At the same time, the noise added during the recording process is reduced. The effective reduction in hiss level is approximately 10 dB.

The original system (Dolby A) operates by dividing the audio band into segments as follows:

1. Below 80 Hz
2. 80 to 3000 Hz
3. 3000 to 9000 Hz
4. 9000 Hz and up

Each segment is processed separately. All signals which are of sufficient strength, say 40 dB over the noise level, to provide good performance pass straight through the Dolby system without change. Signals of lesser magnitude are emphasized 10 to 15 dB. The output from the Dolbyized recording therefore contains almost exclusively signals at least 40 dB above the noise level.

To reproduce, the Dolby system processes the boosted signals back to their initial relative level and at the same time reduces the noise by the same amount, leaving the signals unaltered in recording still unaltered in playback. The output is therefore a low-noise reproduction without disturbances such as hum, hiss, cross talk, and other unwanted distortion. The Dolby system provides a 10- to 25-percent improvement in signal-to-noise ratio.

A somewhat simplified system—sometimes called the Dolby B system—has been developed for home use. The B system provides 3 dB of noise reduction at 600 Hz, 8 dB at 2 kHz, and 10 dB at 4 kHz. The type B system works only on the lower-level signals, in the same manner as the A system previously described, but over less of the audio spectrum. This system was designed primarily to eliminate tape hiss.

Fig. 4-68 shows a noise-reduction unit that is a simultaneous record-playback control center incorporating the Dolby audio noise-reduction system. It consists of two separate sections, a complete Dolby record preamplifier and playback Dolby circuits. Among the functions offered by the unit illustrated are the following:

1. Separate input-level controls on both stereo channels for both microphone and line inputs. These maintain input mixing capabilities for any recorder, and add these capabilities to recorders presently lacking them.
2. A master recording-level control that governs both stereo

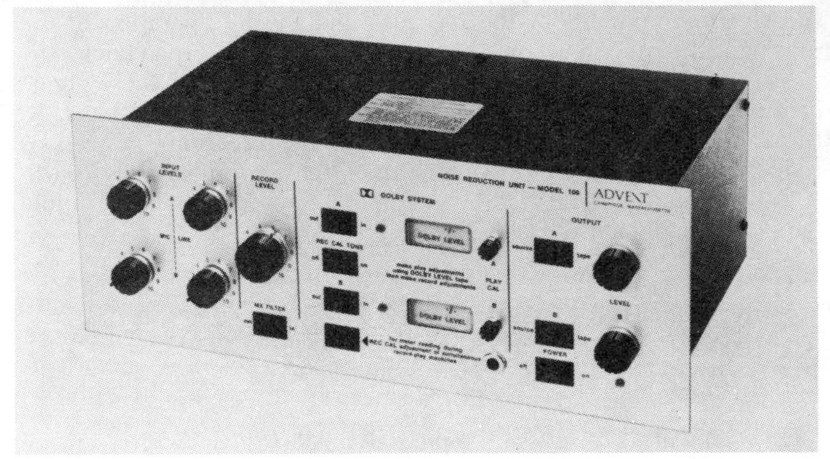

Courtesy Advent Corp.

Fig. 4-68. Advent Model 100 noise-reduction unit.

channels and allows the recording level to be set without disturbing the balance of stereo channels or individual inputs.

3. Output-level controls for each channel that permit matching the requirements of any preamplifier, amplifier, or receiver.
4. A multiplex-filter switch that prevents recording interference from inadequate suppression of multiplex-carrier or pilot-tone frequencies by a tuner.
5. Complete calibration facilities for optimum use of the Dolby system with any recorder. These include two calibration meters, an internal test-tone oscillator, and "Dolby level" tapes (open-reel and cassette) that make it possible to set the unit to a standard characteristic for all Dolbyized tapes, including prerecorded commercial releases.
6. A headphone output.
7. Source-tape monitor switches.

Fig. 4-69 shows graphically how the Dolby system works. In Fig. 4-69A, the original signal is fed into the Dolby record circuit of the CAD5 tape deck (shown in Fig. 4-66H). High-level signals are unchanged, but low-level signals are boosted before recording. Since tape hiss is added to the boosted signal, the hiss is reduced at the same time the boosted signal is brought back to normal by the Dolby playback circuit. The top curve of Fig. 4-69B shows the boosting characteristic of the Dolby record circuit at extremely low signal levels. If the signal level is higher, the boosting is cut back in a precise way. The bottom curve shows the complementary, mirror-image characteristic of the Dolby playback circuit. The play-

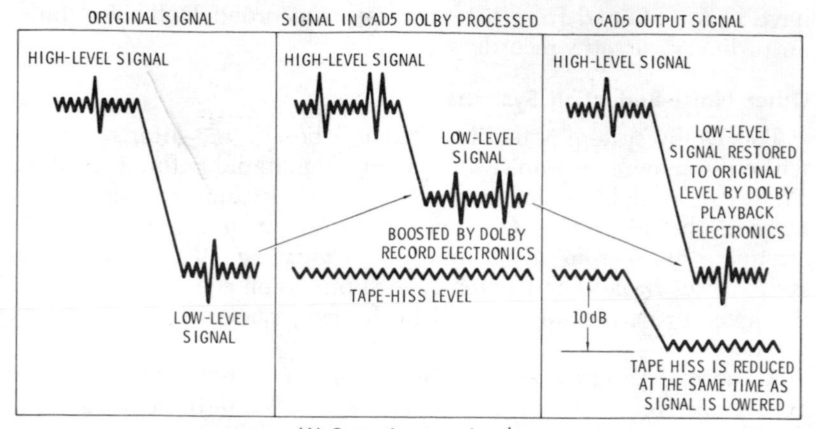

(A) Operations on signal.

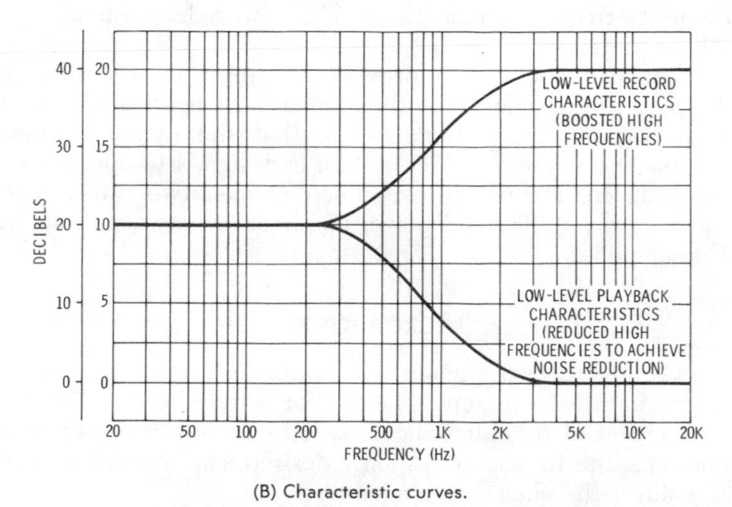

(B) Characteristic curves.

Fig. 4-69. Dolby B processing in Harman-Kardon CAD5 tape deck.

back circuit compensates exactly for the record circuit, resulting in overall flat frequency response and excellent transient performance under all signal conditions.

When the Dolby system is coupled with the new chromium dioxide tapes, the improvement in performance is even greater because of added frequency response.

It is important to note that the Dolby B cassette deck shown can be used for recording and reproducing either using or not using Dolby processing. If a prerecorded tape has been Dolbyized, the instrument switch is operated in the Dolby position; if the tape is not Dolbyized, the switch should be off. However, excellent results have been obtained from listening to prerecorded Dolbyized tapes on ordinary cassette recorders.

Other Noise-Reduction Systems

The Dolby system is intended for use before and after recording without removing any noise already in the material before recording. There are available other units, however, that can filter noise at any stage of the processing. One system has a dynamic filter system to attenuate noise from any source by as much as 10 to 20 dB. This system uses broad-band notch-type suppression which is varied by the level of the incoming signal in the frequency range covered by the notch.

At normal program levels, the system does not operate. When the signal level within the notch lowers to 25 to 50 dB below normal, the filter attenuates the high frequencies. The attenuation increases as the notch-frequency signal level falls. This reduces the noise level 10 dB or more.

Another available system provides compression-expansion capabilities in a similar manner to the emphasis–de-emphasis techniques used in record equalization, except that the dynamic range is adjustable and is greater. This system is designed to operate in any low-impedance, low-power-level line between two units, such as between a preamplifier and power amplifier or between a tape deck and preamplifier.

MICROPHONES

Many types of microphones are available. Each has certain advantages and disadvantages. The type of material to be reproduced, the placement of the microphone, whether it is to be used indoors or outdoors, the frequency response desired, and a number of other factors affect the choice of a microphone.

The basic types of microphones, grouped according to their principle of operation, are:

1. Carbon
2. Crystal
3. Dynamic
4. Ribbon
5. Capacitor

Each of these microphone types has its own characteristics with respect to output level, frequency response, output impedance, and directivity. These characteristics determine whether or not a microphone is suitable for a given application.

Output Level

The output level of a microphone is important because it governs the amount of amplification that must be available for use with the microphone. The output level of microphones is usually given in dB preceded by a minus sign. The minus sign means that the output level is so many dB below the reference level of 1 milliwatt for a specified sound pressure.

The unit of sound pressure used in rating microphones is referred to as a *bar*. A bar is equal to a sound pressure of 1 dyne per square centimeter. Speech provides sound pressures between 0.4 and 15 bars. For music, the pressure ranges from 0.5 to 1250 bars.

Microphones are rated in a number of different ways, and this often causes confusion. If ratings are given in any manner other than in bars, it is a good idea to convert the output level rating to dB below 1 milliwatt for a sound pressure of 1 bar. Table 4-2 gives correction factors which, when applied to the corresponding method of microphone rating, will convert it to output level in dB below 1 milliwatt for a sound pressure of 1 bar. When a rating has been converted to these terms, it is much simpler to use when calculating amplifier gain requirements and the like.

A microphone with a low output level necessitates the use of an amplifier with greater gain, which, in turn, increases the possibility of noise and hum. The absolute minimum noise level which can be practically attained at the grid of the input tube of an amplifier is about −125 dB. From this, it has been determined that to have a reasonably quiet installation, the microphone level should not be below −85 dB.

When very low-level microphones are used, it is often necessary to provide a direct-current heater supply for the input tube, in order to eliminate hum which results when an ac heater supply is used.

Frequency Response

The frequency response of a microphone is a rating of the fidelity of relative output voltage which results from sound waves of differ-

Table 4-2. Comparison of Microphone Ratings

Rating Given	Correction Factor
dB below 1 mW/1 bar	0 dB
dB below 1 mW/10 bars	−20 dB
dB below 1 volt/1 bar	2 dB
dB below 1 volt/10 bars	−18 dB

ent frequencies. The simplest way to find a complete picture of the frequency-response characteristics of a microphone is to plot a curve of its output voltage versus input frequency. Since good modern microphones are relatively flat over their range, it is often considered sufficient to specify the range over which their output does not vary more than plus or minus 1 or 2 dB.

For ordinary home high-fidelity use, a microphone frequency-response curve should be reasonably flat between 40 and 10,000 Hz. With systems designed specifically for speech reinforcement, a lower limit of 150 Hz and an upper limit of 5000 Hz are entirely satisfactory. Where it is desired to reproduce music with the highest possible fidelity, the frequency response should be flat (within 2 dB) from about 40 to 15,000 Hz. Fig. 4-70 shows the response of several types of microphones.

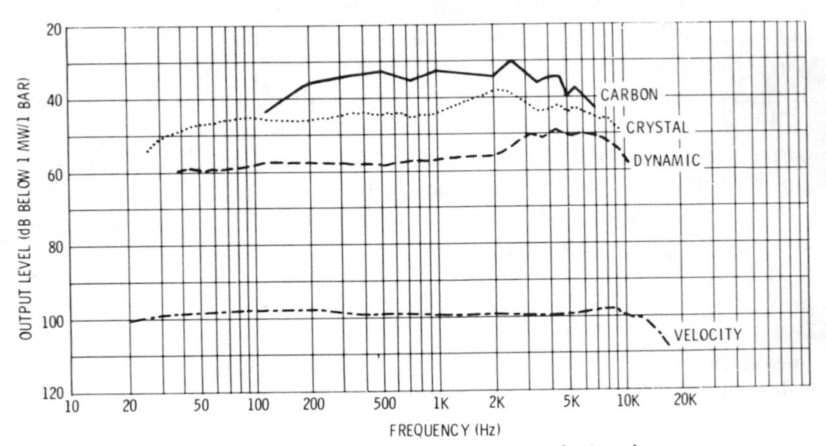

Fig. 4-70. Frequency-response curves for typical microphones.

Output Impedance

The output impedance of a dynamic or velocity microphone requires a transformer or network to match the input impedance of the amplifier. Higher-impedance microphones, such as the crystal types, require only a simple network. Microphones generally employed in public-address systems have impedances of from 20 to 500,000 ohms.

Directivity

Microphones do not respond equally to sounds reaching them from all angles. Their frequency-response characteristics also vary, depending on the angle at which the sound reaches them. A microphone may respond equally to all frequencies between 40 and 10,000 Hz when the sound is originating directly in front of it, while the

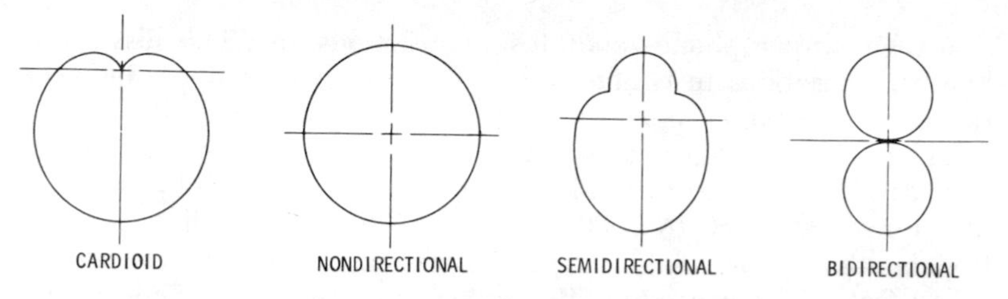

Fig. 4-71. Microphone polar response patterns.

high-frequency response falls off rapidly as the sound originates farther to either side. Where it is necessary to pick up sound from all directions, the directional characteristics of some microphones are not suitable. Fig. 4-71 shows examples of the four important directivity characteristics which can be obtained with the various types of microphones.

The directional characteristics of a microphone can be used to accomplish a number of things. Noise pickup can be reduced by choosing and placing the microphone so that it will not respond to sound originating at the point where the noise is produced. Feedback, which can be very troublesome, can often be completely eliminated by the careful choice and placement of a microphone.

Carbon Microphones

When the maximum output level is required from a microphone, the carbon microphone is often used. While it does have the advantage of high output, the frequency response characteristics of the carbon microphone are poor, and it cannot be used for hi-fi work.

The carbon microphone consists essentially of a diaphragm and a small cup filled with carbon granules. Fig. 4-72 shows the construction of a typical carbon microphone.

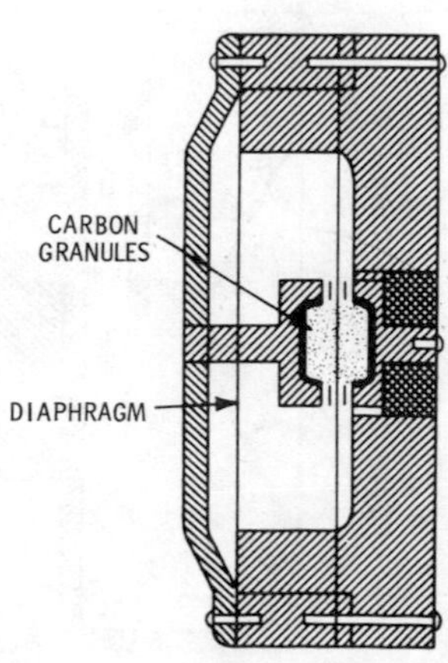

Fig. 4-72. Construction of a carbon microphone.

A carbon microphone generates a continuous hiss. This hiss is due to small variations in contact resistance which take place between the carbon granules.

The average output level of carbon microphones is of the order of −30 dB. The best carbon microphones have a frequency response of approximately 60 to 7000 Hz. They are substantially nondirectional, although their high-frequency response above 300 Hz usually falls off at angles exceeding 40 degrees from the front of the microphone. Although carbon microphones are not used for high-fidelity work, their ruggedness, low cost, and high output make them useful in a few cases.

Crystal Microphones

The crystal microphone is the type most widely used in lower-cost installations. The crystal microphone has a relatively high output level and a high impedance. The impedance of the crystal microphone is high enough that it can be connected through a short cable directly to the grid circuit of a basic amplifier, eliminating the need for an input transformer and preamplifier. A long cable will reduce the output voltage available from a crystal microphone and may effect its high-frequency response.

The most commonly encountered type of crystal microphone employs a diaphragm which moves in accordance with sound waves striking it and exerts pressure on the crystal (Fig. 4-73). This type of construction permits complete enclosure of the crystal and reduces the effects of humidity.

The output level of this type of microphone is usually between −48 and −60 dB. The output impedance is almost always more than 100,000 ohms.

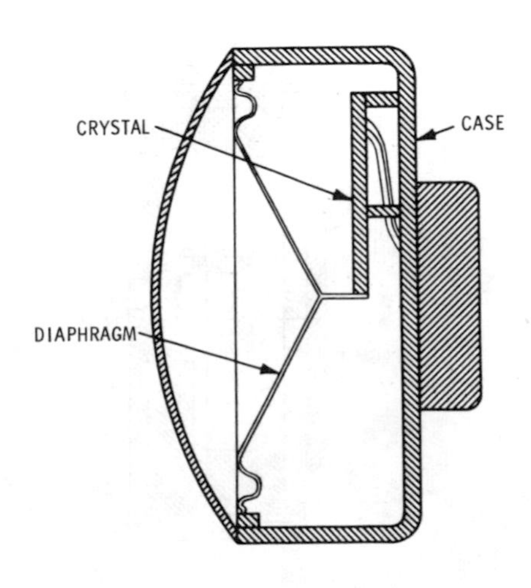

Fig. 4-73. Construction of a crystal microphone.

The crystal microphone is normally nondirectional, although a special pressure-gradient crystal microphone which gives a unidirectional response pattern is now being marketed. This microphone gives excellent results.

Good units may have a frequency response substantially flat between 50 and 10,000 Hz. Units are available with slightly wider frequency-response ranges.

Rochelle-salt crystal microphones should not be used in locations where the humidity is extremely high. They should never be subjected to high temperatures. If such a crystal microphone is subjected to a temperature of 130 degrees, it will be rendered completely useless. Care must always be taken to avoid exposing a crystal microphone to direct sunlight for any length of time.

Dynamic Microphones

The dynamic microphone consists of a metal diaphragm, a coil which is connected to it, and a magnet (Fig. 4-74). In construction and operation, this type of microphone is similar to a dynamic speaker. When sound waves strike the diaphragm, the coil moves. Since the coil is in the field of the permanent magnet, there is induced in the coil a current which is directly proportional to the sound waves striking the diaphragm. This current constitutes the output of the microphone. Dynamic microphones are available with limited or wide-range frequency-response characteristics.

The natural output impedance of a dynamic microphone is between 30 and 50 ohms. Very often, a transformer is incorporated in the microphone, raising its output impedance to a value between 200 and 25,000 ohms. The average dynamic microphone is simple and sturdy. It is not affected by atmospheric changes, has a long life, and is well adapted to all-around hi-fi work.

The output level of most dynamic microphones is about 55 or more dB below 1 milliwatt per bar. The ordinary dynamic micro-

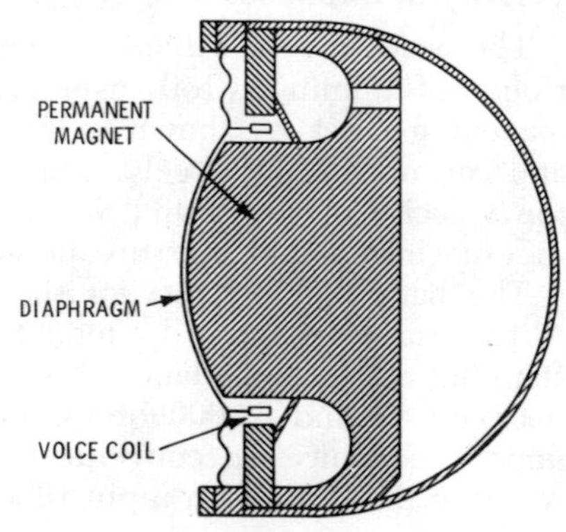

Fig. 4-74. Construction of a dynamic microphone.

phone is essentially nondirectional, although its high-frequency response falls off rapidly on either side, as shown in Fig. 4-75. To make full use of the frequency range of a dynamic microphone, the microphone should face directly toward the source of sound. A special type of dynamic microphone is available for use when high

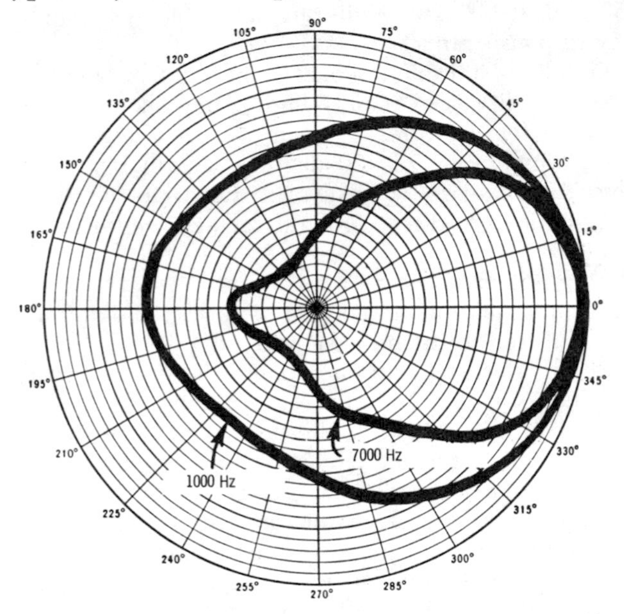

Fig. 4-75. Directivity of a dynamic microphone.

background noise levels are encountered. The response of these units falls off rapidly as the distance between the microphone and the source of the sound increases.

Velocity Microphones

The velocity (or *ribbon*) microphone consists of a very thin ribbon of aluminium foil suspended in the field of a powerful permanent magnet, as shown in Fig. 4-76. The ribbon is corrugated and can move quite freely. The ribbon moves in accordance with the velocity of the sound wave. Response is proportional to the difference in sound pressure between the two sides of the ribbon.

The natural impedance of the ribbon elment is about ¼ ohm. A transformer is usually mounted within the microphone case, stepping up the impedance at the microphone terminals to a value between 25 and 35,000 ohms. For public-address use, the high-impedance units are convenient since they can be connected directly to the grid of an input tube.

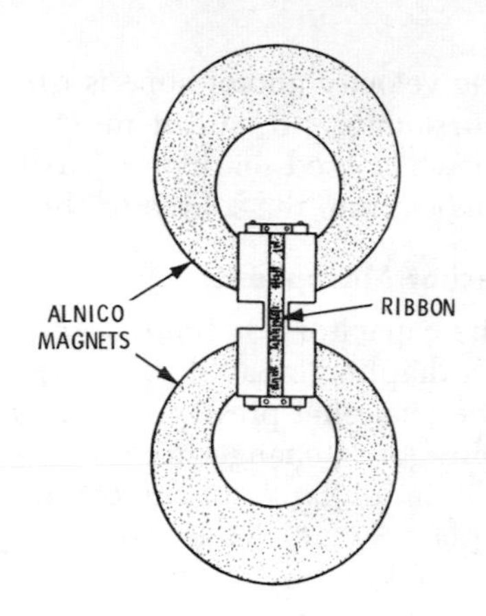

Fig. 4-76. Construction of a velocity microphone.

The output level of velocity microphones is usually 60 dB below 1 milliwatt per bar. Generally, velocity microphones have excellent response characteristics.

The velocity microphone is bidirectional. Maximum response is to sound reaching the front or back of the microphone at a 90-degree angle to the plane of the ribbon faces. This type of microphone is more directional than the crystal and dynamic microphones, and it has an overall response that falls off as the angle of the sound reaching it varies from 90 degrees to the faces of the ribbon (Fig. 4-77).

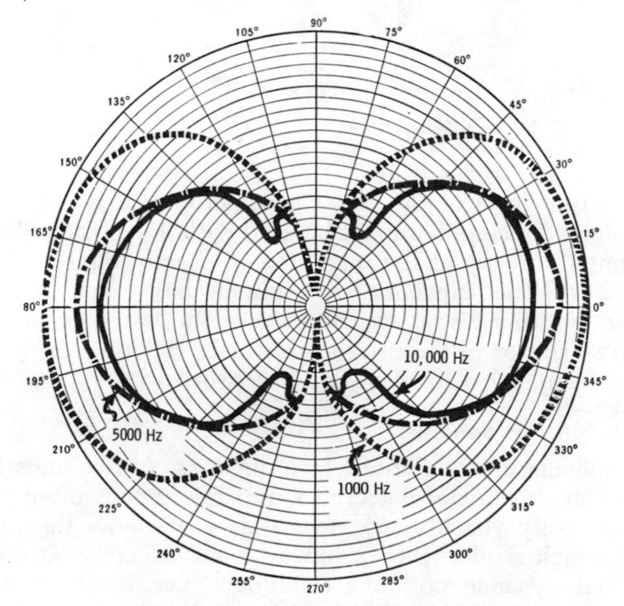

Fig. 4-77. Directivity of a velocity microphone.

The velocity microphone is quite sensitive to the movement of the air surrounding it, and it must be carefully protected from puffs of wind when used outdoors. A ribbon microphone should be at least 18 inches from the source of the sound.

Capacitor Microphones

The capacitor (or *condenser*) microphone consists of a fixed plate and a diaphragm, as shown in Fig. 4-78. The diaphragm is actuated by the changing pressure of the sound waves striking it, causing the diaphragm to change its position in relation to the fixed plate. This results in a change in the capacitance between the diaphragm and the plate, which is utilized to produce a corresponding voltage drop across a resistor connected in series with the microphone and a charging source.

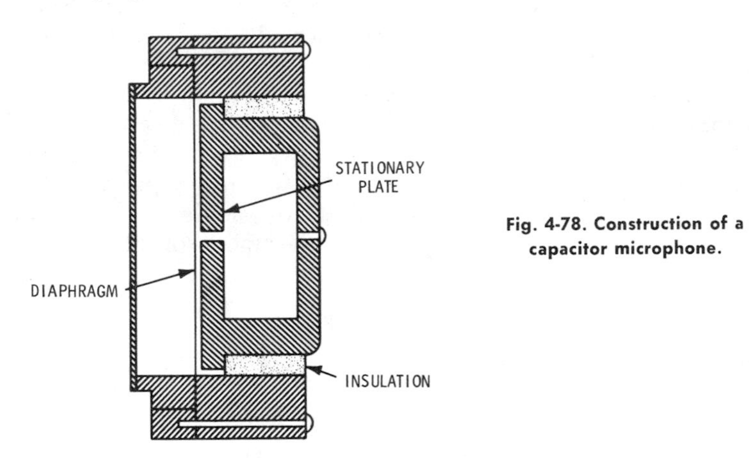

Fig. 4-78. Construction of a capacitor microphone.

The output level of the capacitor microphone is extremely low, and a high-gain amplifier must be used with it. The amplifier should be mounted directly at the microphone, usually right in the microphone case. The capacitor microphone has excellent frequency response and low distortion. Because of the necessity of mounting an amplifier at or in the microphone case, this type of microphone is not recommended for ordinary hi-fi work.

Combination Microphones

Microphones are available which make use of two units to secure a particular directional pattern. A dynamic unit is often combined with a velocity (or ribbon) unit. Fig. 4-79 shows the directional pattern which results when a bidirectional velocity unit and a nondirectional dynamic unit are combined. The resultant directivity pattern is known as a *cardioid,* since it is heart-shaped. Other units

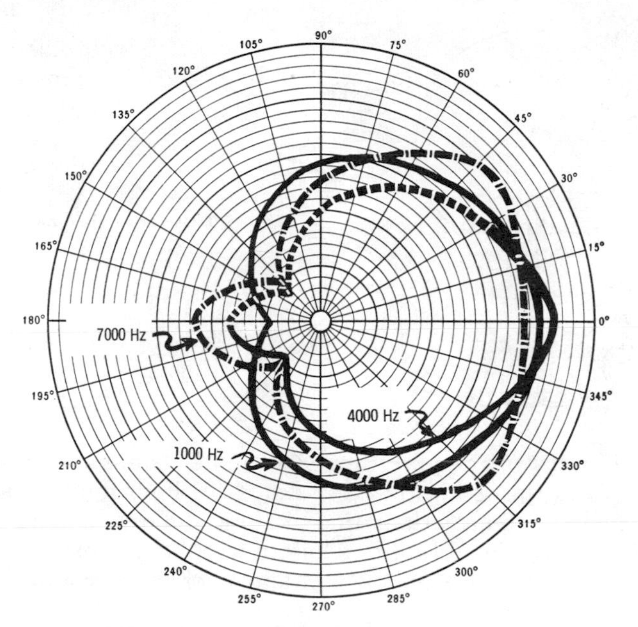

Fig. 4-79. Directivity of a dual-unit microphone.

are also combined to secure similar directivity patterns. Another type of combination microphone, designed especially for stereo recording, is pictured in Fig. 3-4.

Microphone Selection

Microphones should be carefully selected in order to utilize the electrical and physical characteristics of the various types, as described previously. There are no particular rules which can be strictly adhered to in the selection of a microphone. There are, however, a few points which should be kept in mind.

A microphone should be selected with frequency-response characteristics equivalent to those of the other components in the system. A Rochelle-salt crystal microphone should never be used where it is likely to be subjected to a temperature of more than 120 degrees.

Cardioid microphones should be used when "behind-the-mike" pickup must be eliminated. Fig. 4-80 shows the application of a cardioid microphone on a speaker's platform. The back of the microphone faces the audience. Since the microphone is not sensitive to sounds reaching it from this direction, no audience sounds will be picked up and amplified through the system.

In systems where a speaker must move about a great deal, the lapel microphone is very useful. A number of contact microphones are available for use with string instruments.

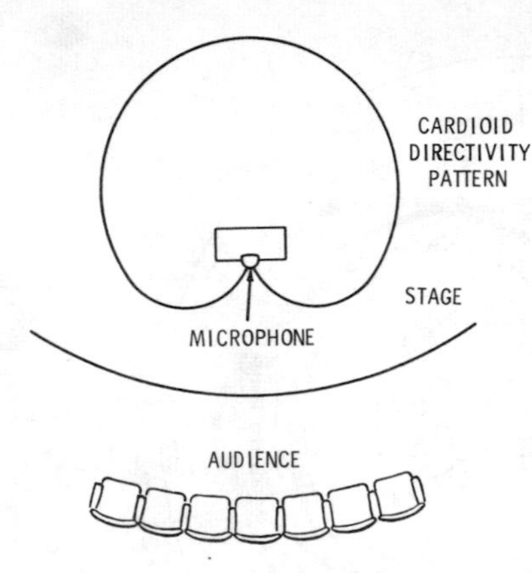

Fig. 4-80. Use of cardioid microphone.

Many installations require the use of more than one microphone. An example of this is illustrated in Fig. 4-81. Four microphones are used to pick up sound originating at three different points. Microphone A picks up sound from a soloist, microphone B picks up sound from a chorus, and microphones C and D pick up sound from an orchestra.

The microphones used have been chosen and placed so that they will pick up the designated sound only. In other words, the microphone in front of the orchestra will pick up sound from the orchestra, but not from the chorus or the soloist. This makes it possible to control the pickup from the three elements—the orchestra, the chorus, and the soloist—individually, so that each may be given the proper degree of reinforcement. A microphone setup such as this is particularly useful in adjusting the level necessary for a vocalist, since a vocalist requires a greater degree of sound reinforcement than does an orchestra.

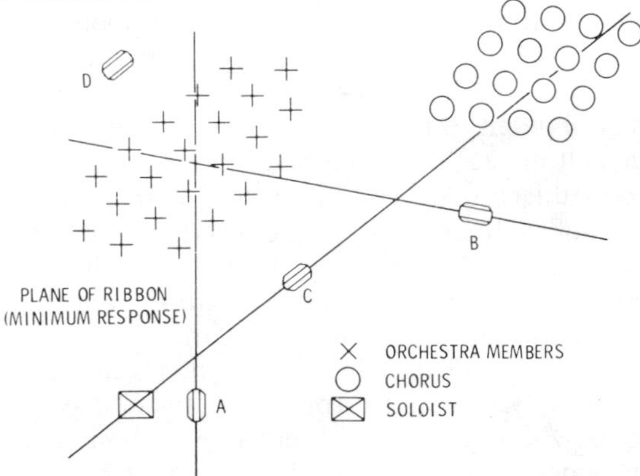

Fig. 4-81. Microphone placement plan.

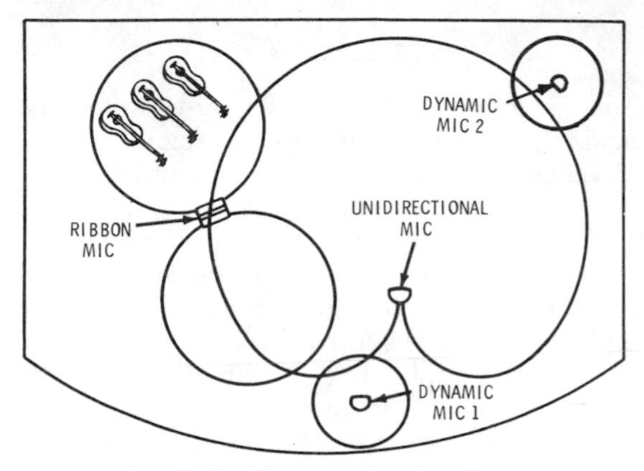

Fig. 4-82. Multiple-microphone installation.

Fig. 4-82 shows another multiple microphone installation. Here, a cardioid microphone is used for overall pickup. It faces the orchestra so that the audience is in the dead area of the microphone. A dynamic microphone is used for pickup from a master of ceremonies or vocalist. A third microphone, this one a velocity microphone, is used to pick up sound from the violin section of the orchestra, and a fourth microphone is used to pick up background sound.

In addition to the differences in pickup patterns, microphone sizes and shapes also vary greatly. In selecting a microphone, consider what uses it will be put to, then select one with the response pattern that will fill the need. Many are quite versatile. For example, the unit pictured in Fig. 4-83 can be held in the hand, worn around the neck on a lavaliere cord and clip, or supported by a desk or floor stand.

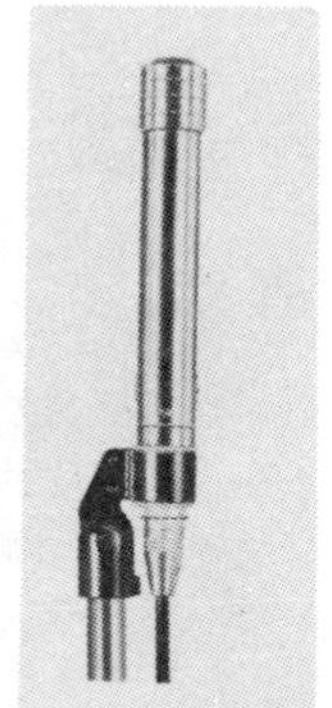

Courtesy Shure Brothers Inc.

Fig. 4-83. Shure Model 535 dynamic microphone.

TELEVISION

Another source of program to feed a hi-fi system is the audio signal accompanying each television picture signal. The audio equipment in the average television receiver is not of high-fidelity nature. If one wishes, arrangements can be made to pick off the audio signal after detection from the television receiver and feed it through a hi-fi system. The pick-off point should be at a point as near the second detector as a proper impedance match will allow. The lead to the preamplifier should be short.

One can obtain a separate high-fidelity fm tuner for television stations. Such tuners have features similar to those of standard fm tuners, except that they cover the television frequencies only, and the channels are calibrated on the dial.

5

Amplification and Control

The amplifier/preamplifier section of a hi-fi system includes circuits and components necessary to provide power output, inversion, voltage amplification, preamplification, compensation, and a full set of operating controls.

Most commercially available amplifiers and preamplifiers for high-fidelity performance are designed for stereo, rather than monophonic operation. This actually means that they can be used for either. However, because there will always be some monophonic amplifiers and because a stereo system is really a coordinated multiple monophonic system with some modifications, we shall discuss these sections first as applied to monophonic form. Then, the methods of combining and modifying for stereo will be given, with examples of actual units.

Preamplifiers are required to bring the level of low- and medium-level signals up sufficiently to provide a useful input to amplifier units. But their most distinctive function is to provide frequency-response equalization and controls for volume, loudness, balance, phase, and bass and treble emphasis or de-emphasis. In other words, the modern preamplifier is also an "audio control center." However, some special-purpose preamplifiers, such as those for amplification and equalization of the outputs of variable-reluctance cartridges, do not have controls.

The suitability of an amplifier for an installation depends on a number of factors, the most important being its power output, gain, input impedances, output impedances, frequency response, distortions, controls, and the characteristics of the power supply from which it operates. The requirements of the installation and limitations of other components should be considered before the amplifier is chosen. These factors and requirements will be discussed in Chapter 8, on systems design, selection, and installation.

Basic amplifiers are units with power output stages, inverters, and limited voltage amplification only. Preamplifiers are necessary to increase medium- and low-level input signals, such as those from reluctance phono pickups and velocity or dynamic microphones, to the minimum input level required for full output of the basic amplifier. Preamplifier-equalizers raise the signal and provide for record equalization.

Audio control centers including preamplifiers, equalizing networks, boost, droop, loudness, compensation, filtering, decoding, and switching circuits are more elaborate and usually are used instead of the plain preamplifier or a simple preamplifier-equalizer. Lower- and medium-cost amplifiers are available which provide in the same unit for all functions just described. However, it is to be expected that the overall quality will have a relation to the price.

The power amplifier unit or section provides amplification and power to actuate the speaker. Average hi-fi speakers for average living rooms require at least 10 watts of audio output power per channel from the amplifier, to handle most conditions. When several rooms, large rooms, or outside areas are to be supplied with the highest fidelity, 60 to 200 watts of power might be desirable in a home installation. This audio output power is provided by the basic amplifier unit or by the amplifier section of a preamplifier and power-amplifier combination unit.

According to the need, from 50 to 100 dB of amplification of very low-level signals from pickups and other inputs, plus compensation and controls to obtain various emphasis, is provided by the preamplifier-equalizer part of the system or by a separate hi-fi control-center unit.

Hi-fi systems may be designed around either of these two types of amplifier arrangements, that is, the combination of preamplifier, control, and power amplifier, all in one unit such as shown in Figs. 5-1A and 5-1B, or the control center plus basic amplifier units used separately, as shown in Figs. 5-2A and 5-2B. In addition, there are complete compacts (see Fig. 5-74).

Circuit arrangements and components may be somewhat similar in both arrangements, but more flexibility and better results can be expected from separate units. It is easier to provide optimum design for separate units, due to differences in conditions of control, amplification, power, vibration, shielding, regulation, and noise problems in each of the separate units, such as the tuner, preamplifier, and the basic amplifier. However, it is a matter of practical system design when a tuner is to be used to decide whether the operating controls are more desirable on the tuner panel, in a separate preamp control panel, or on the amplifier unit itself. The answer to this will depend on the system requirements, the location conditions, and the user's preference.

The preamplifier and control section (either as a separate unit or as part of a complete amplifier unit) may provide controls for power on and off, switching the various possible inputs or combinations of inputs into the system, volume, record equalization or compensation, bass and treble boost or droop, filtering, and loudness. This section should also provide amplification of the very lowest input signals to be used to a level sufficient to drive the power-amplifier section to full output.

Record equalization is provided to compensate for pre-emphasis and de-emphasis applied to records in their manufacturing process. Records are made with all high audio frequencies heavily pre-emphasized; i.e., the recorded volume of the highs is proportionately much higher than the normally recorded mid-range frequencies, for the purpose of reducing noise during playback. On playback, the emphasized signal overrides or masks the undesirable random noise and needle-scratch noises in the high audio-frequency spectrum. When the recording is made, the volume level of the high frequen-

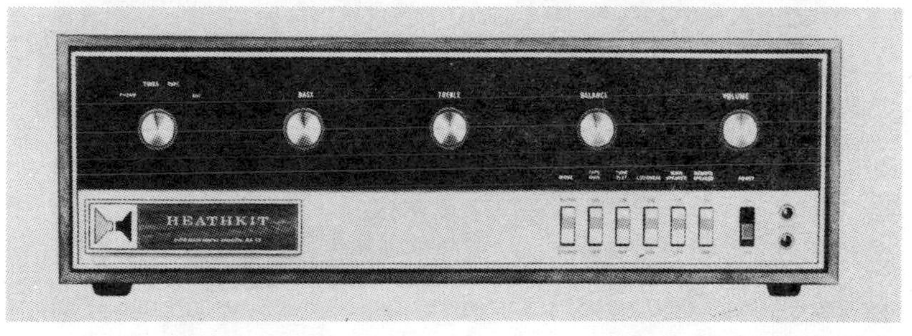

Courtesy Heath Co.

(A) Heathkit Model AA-15 stereo amplifier.

Courtesy AKAI America, Ltd.

(B) AKAI Model AA-6100 four-channel amplifier.

Fig. 5-1. Stereo master audio controls.

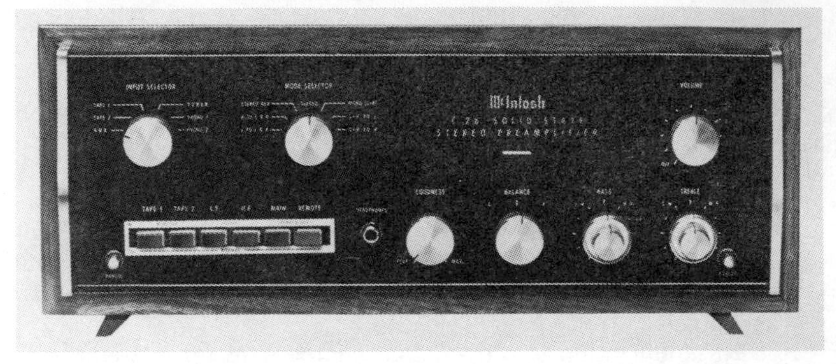

Courtesy McIntosh Laboratory Inc.

(A) McIntosh C-26 preamplifier.

Courtesy Harman-Kardon, Inc.

(B) Harman-Kardon Citation Twelve amplifier.

Fig. 5-2. Separate stereo amplifiers.

cies above 1000 Hz (RIAA, AES, NAB, ORTHO) is increased as shown by the curve in Fig. 5-3. When the record is played back, the preamplifier can be adjusted to proportionately de-emphasize these signals in an equal but opposite manner, as shown by the curve in Fig. 5-4, developing an audio output very close to the original recorder input.

The same technique is used in fm broadcasting and accounts for much of the superiority of frequency modulation over amplitude modulation when proper compensation is applied in the reproducing amplifier.

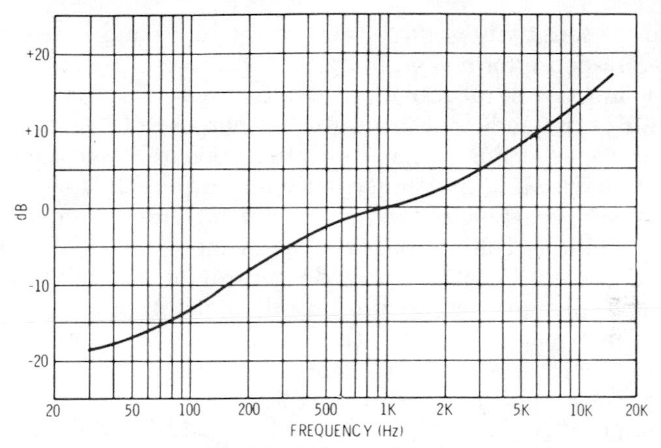

Fig. 5-3. RIAA recording curve.

In a similar but opposite manner, the very low frequencies are de-emphasized as shown in Fig. 5-3. This is done to compensate for the wide swings of the record cutter caused by the very low notes. Such wide cuts in the record groove must be reduced, or the grooves would have to be made with more spacing, causing a reduction of possible recorded time. Large movements of the stylus also introduce uncontrollable distortions in the recording head. Emphasis in playback must be provided to compensate for this de-emphasis on the records. Fig. 5-4 shows how this is provided for.

Loudness controls are provided to compensate for the normal variation in response linearity of the ear at different volume levels. It was described in Chapter 1, Fig. 1-1, that as the volume (intensity) of sound is reduced, reality supposedly becomes more distant

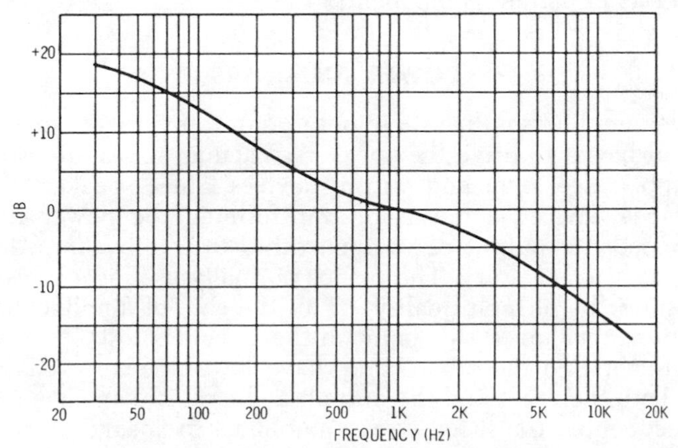

Fig. 5-4. RIAA playback curve.

because the sensitivity of the human ear to highs and lows drops off from the response to the "midrange" frequencies.

The loudness control provides circuits to emphasize the low and high ends of the audio spectrum for the purpose of bringing reality closer as reproduction volume is proportionately reduced. Some amplifiers have multiple ganged controls to provide for automatic loudness compensation as volume is varied. Other equipment employs switching circuits to provide one or more values of emphasis.

Bass and treble controls provide control of emphasis and deemphasis similar to that provided by the equalization controls, but in a variable manner so that they can be adjusted according to individual taste. These controls, one for treble and one for bass notes, are usually adjustable to provide plus or minus 15 or 20 dB variation of either the highs or lows at the ends of the audio spectrum. Advanced equipment provides for rolloff frequency adjustment with constant attenuation slopes and special filters for disturbances such as turntable rumble and objectionable high frequencies.

A balance control provides adjustment of the relative volume output of each of two channels of a stereo amplifier. A phase switch provides for reversal of the input of one of two stereo channels with respect to the other. This is used to correct the phasing of outputs of pickup cartridges and fm multiplex outputs, and to correct differences in amplifier output phase if two power amplifiers of different design are used.

There are other controls that may be included for three- or four-speaker stereo. These will be discussed later.

How the preamplifier and control sections function will be described in detail following an explanation of the basic power amplifier and its important components.

POWER AMPLIFIERS

Power amplifiers utilize vacuum tubes or transistors to develop the required power to drive the speakers. Vacuum tubes are essentially high-impedance input and output devices. Since speakers are low-impedance devices, a means of transforming the power from the high-impedance output of vacuum tubes to the low impedance of the speaker is necessary. The device normally used is a transformer, which must be of high quality, or all the care of amplification in a quality manner up to this point in the circuit is lost.

Transistors, on the other hand, have low-impedance characteristics at both the input and the output, and circuits have been devised to direct-couple transistor power amplifiers to speakers without an output transformer. In many cases the output circuits can be designed to match closely a fixed-ohmage load within the output load range that will provide uniform output with extremely low distortion. However, there are applications of transistor audio circuits in which transformers are the best means of matching because of variable load conditions that cannot be directly matched properly with transistors.

The most costly and major limiting design factor in power amplifiers is the output transformer. The output transformer is considered to be the most important part of the amplifier, and because of its close relationship to the speaker system, we will consider it first. It is one of the most critical components in the high-fidelity system. Modern resistance-coupled amplifier design is such that excellent frequency response and low distortion are inexpensively obtained in the low-level stages preceding the output stage. But at the relatively high power level of the output circuit there are more chances for distortion, and failure to use a high-fidelity type of transformer can nullify all the advantages developed in the careful design of the preceding portions of the amplifier.

In the selection of an amplifier, an amplifier kit, or a circuit for construction, consideration and investigation of the quality of the output transformer is of primary importance. Also, high-fidelity enthusiasts frequently replace output transformers and experiment with their characteristics. Such experimenting often leads to substantial improvement in performance and often allows response adjustments to suit individual tastes. Although the design and construction of these components must of course be left to specialists, the high-fidelity enthusiast will have a keen interest in the factors which make the output transformer so critical and cause its cost to represent a large proportion of his high-fidelity budget. For this reason, the following brief review of the factors most important in the selection of output transformers is included.

Symbolically, and in its basic principle, the output transformer is the same as any transformer designed to couple power to a load from a source. However, unlike ordinary power transformers, output transformers must maintain a high degree of efficiency over a range of frequencies, rather than just at the power frequency, and they must not distort the original signal waveform.

The effects of various factors in the transformer on performance can be most easily visualized in reference to an equivalent circuit, illustrated along with the actual circuit in Fig. 5-5. The equivalent circuit, being in the form of a simple series-parallel connection of inductors and resistors, allows analysis of the effects of any one factor. This equivalent circuit is not complete, because there are also effects from distributed capacitances of the windings, capacitance between windings, and capacitances to ground. Because the output transformer is heavily loaded (low impedance), the capacitances can be neglected unless the design is very poor. The equivalent circuit is made up of two main parts: (1) the primary portion, including effects derived from the primary circuit, and (2) the reflected portion, corresponding to effects reflected from the secondary into the primary. Any impedance so coupled is "transformed" to a new

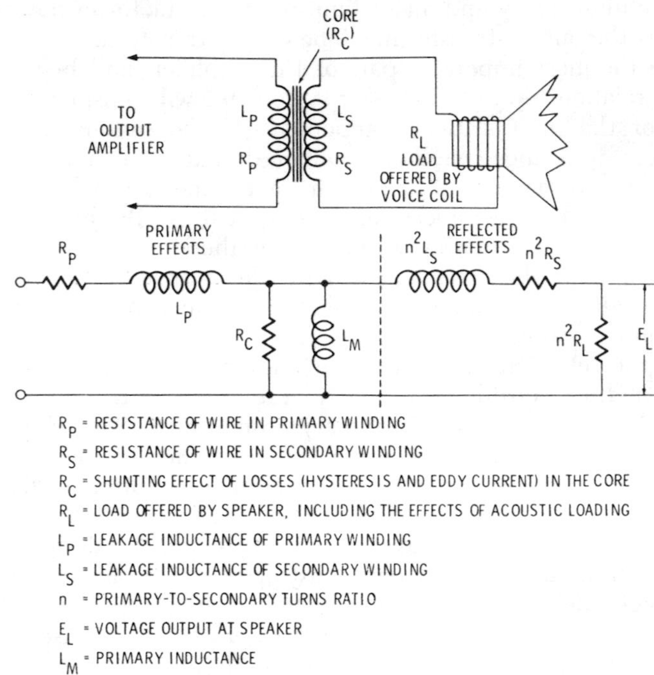

Fig. 5-5. Actual and equivalent circuits of an output transformer.

value by the turns ratio; the impedance transformation ratio is the square of the turns ratio. This is why the reflected inductance and resistance from the secondary are each multiplied by n^2.

Winding Resistances R_p and R_s

As shown by the equivalent circuit, these resistances act in series with the source, adding to its internal impedance. The higher the resistances, the greater is the voltage drop across them at high currents (high output power), and thus the poorer the voltage regulation. Since current through a resistance dissipates power, these resistances represent a power loss in the transformer, reducing its efficiency. The ill effects of excessive resistance dictate that wire as large as possible, consistent with size, weight, and coupling coefficient, be used.

Leakage Inductances L_p and L_s

Leakage inductances also act in series with the source. The reactance they offer to impede the signal depends on frequency, being equal to $2\pi fL$. At low frequencies the leakage reactances are negligible, but at high frequencies they must be minimized; otherwise they interfere with frequency response. Leakage inductance is the inductive effect resulting from the flux of one winding which does not link with the other winding. In good high-fidelity transformers, leakage reactances are minimized by careful attention to the physical shape, mounting, and orientation of the windings on the core. In some cases, the primary and secondary windings are interwound; that is, a few turns of the primary are wound, then a few turns of the secondary, then some more of the primary, and so on. This is obviously an expensive procedure, and it is one of the reasons for the relatively high cost of a good output transformer.

Core Losses

There are two types of core losses. *Eddy currents* are electric currents generated in the core material due to the fact that it simulates a conductor being cut by lines of magnetic force. The core material has appreciable resistance and thus dissipates power wasted in heating the core. It is because of eddy currents that cores are made of many laminations (thin sheets) instead of solid metal. The flux threading through the core is thereby divided into a small portion for each lamination; this, plus the fact that the resistance of the core is increased by the laminations, greatly reduces eddy-current losses as compared with those in solid material. The thinner the laminations (maintaining the same total volume of material), the lower are the eddy current losses, but also the more expensive is the construction. The other kind of core loss is that due to *hysteresis.* This is the tendency of the core material to retain residual magnetism and thus resist the positive and negative flux alternations necessary for transformer action. Hysteresis is a function of core material. Much research has been done on determining and developing high-grade core materials. Core laminations employed in output transformers for high-fidelity use are of carefully selected materials with minimum hysteresis loss and of as high resistivity as possible to minimize eddy currents.

Both eddy currents and hysteresis losses have the same effect on transformer performance as would be produced by the connection of a shunt resistor across the input to the transformer, as illustrated by R_c in Fig. 5-5. The greater the core losses, the lower is the equivalent resistance of R_c and, therefore, the greater is its shunting effect.

Primary Inductance

Before a transformer can operate properly, magnetic flux must be set up in the core and be maintained there. This is the job of the primary inductance. It is the inductive reactance which limits no-load current to a reasonable value. If primary inductance is too low, it shunts the input signal. Since inductive reactance is proportional to frequency ($X_L = 2\pi fL$), its shunting effect is worst at low frequencies, being one of the limiting factors for low-frequency response.

Practical Performance Factors in the Output Transformer

The preceding discussion covers the important factors involved in the performance of the output transformer in general. For high fidelity we are interested in how these factors affect performance in practical, commercially available types. Such effects can be made clearer by redrawing the equivalent circuit twice, once including only those factors important at high audio frequencies and once for factors important at low audio frequencies. Such equivalent circuits are shown in Fig. 5-6. In transformers of reasonably good quality, core losses become negligible, so they are not shown here. Also, as previously mentioned, distributed and other capacitances, important in the high-impedance interstage transformer, are negligible here because of the relatively low impedance involved.

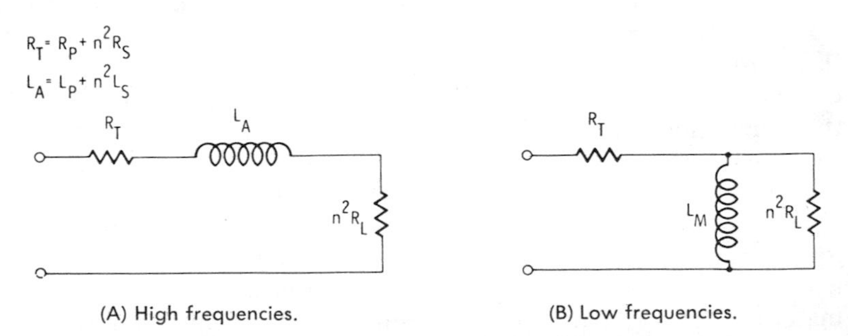

(A) High frequencies. (B) Low frequencies.

Fig. 5-6. Effective equivalent circuits at high and low frequencies.

In Fig. 5-6A, it can be seen that the combined leakage reactances form the important high-frequency factor. If equivalent combined leakage inductance L_A becomes excessive, most of the input voltage at high frequencies appears across it, rather than across the load. At the same time, a large L_A causes phase shift and phase distortion of the output signal. The closer the coupling coefficient between the primary and secondary is to unity, the lower is the leakage inductance. Manufacturers of high-fidelity transformers employ special methods of interwinding the turns to minimize leakage inductance.

In Fig. 5-6B, it is apparent that the important low-frequency factor is the primary inductance, L_M. It was not included in the high-frequency circuit of Fig. 5-6A because its reactance is so high at high frequencies that it appears as an open circuit. At some low frequency, however, the reactance of L_M becomes low enough to cause an important shunting effect across the reflected load. This means a loss of low frequencies and a phase shift at low frequencies. For this reason, manufacturers of high-fidelity output transformers try to use as many turns as possible. But in doing so, they make it harder to avoid more leakage inductance, so a careful compromise between the two must be made. By careful use of interwinding and choice of physical dimensions and shapes, both of these factors are made much more favorable in high-fidelity transformers than in the run-of-the-mill types.

Choosing and Testing Output Transformers

The high-fidelity enthusiast who wishes to purchase an output transformer is faced with an array of commercially available units, ranging in cost from about a dollar to as much as one-hundred dollars or more. Tiny, cheap transformers made to a price for low-cost radio receivers can be ruled out immediately, since they can hardly be considered as high-fidelity equipment. Also, all high-fidelity systems worthy of the name use push-pull output; therefore, we will not consider single-ended transformers.

From here on, it depends on the person himself and the limits he has set on his system as a whole. The speaker system is of first consideration in this respect, since most amplifiers preceding the output transformer are so high in quality that they are not a factor. Using a high-grade output transformer with a relatively poor speaker system does not allow full realization of the capabilties of the transformer. A transformer with performance comparable to that of the speaker is cheaper and sounds no worse. On the other hand, if you intend to change to a speaker of better quality in the future, it is often wise to get the best possible transformer in the beginning, and thus eliminate the transformer as a serious limiting factor in later changes and tests.

If a top-notch transformer is in order, the purchaser is quite safe in buying one of two or three special types known by high-fidelity men everywhere. Companies such as those whose wide reputation rests primarily on such a unit produce a uniform and dependable product.

Few systems warrant the ultimate in transformer specifications, and few pocketbooks can stand making the whole system the best available. Among more moderately priced output transformers, manufacturers are not always definite about performance. For example, sometimes the drop-off in response over the specified frequency range is not stated. If it is 2 dB or less, then the range means something. Examples of good and poor response curves are shown in Fig. 5-7. The curves actually show relative response with respect to response at 400 Hz. In other words, to compare only frequency response, and not total output, the response at 400 Hz (and thus over the entire middle portion of the range) is plotted along the same line, even though output level under particular actual operating circumstances may be different. Note how much greater a frequency-response range could be claimed if the drop-off is taken at 5 dB or 10 dB.

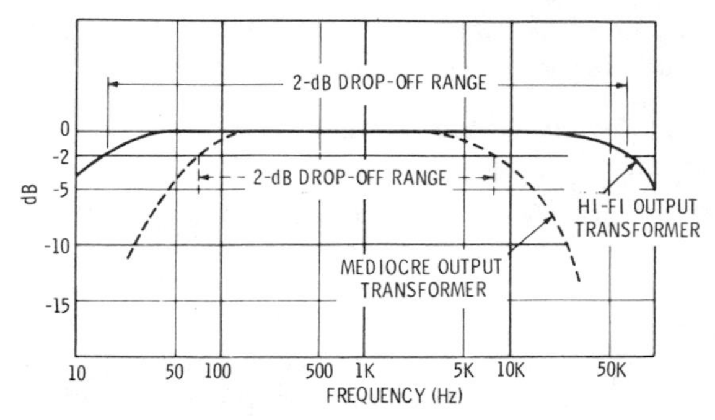

Fig. 5-7. Output-transformer response characteristics.

Some specifications do not take into account the tolerance in the balance between the two halves of the primary. If there is unbalance, distortion is introduced in proportion to the degree of unbalance. Also, unbalance will cause the magnetic flux from the two halves of the primary to fail to balance in the core; unbalance in the fluxes causes possible overload and core saturation, with magnetic overload distortion, hum, and overheating.

Because these things are not apparent in over-the-counter buying, the high-fidelity enthusiast may wish to be prepared to make certain basic tests on a transformer. For this purpose, the following checking and testing procedures are given:

1. *Check the relative weight.* Heavy transformers are not necessarily good ones, but good ones for high-fidelity performance are heavy. The low-frequency response depends on core volume. While it is true that the efficiency of core materials has been greatly improved of late, it still takes a husky chunk of iron to do a good job down to 50 Hz. If you have examined and handled top-notch trans-

formers, you have developed an idea of what the weight must be and can make comparisons.

2. *Try it in an amplifier.* Listening is, of course, the ultimate functional test, providing conditions are so controlled that any distortion heard can be attributed to the transformer. However, testing equipment is required to make other than an aural test. To make laboratory-type tests, you should know the characteristics of the amplifier used, because such things as unbalance in the output tubes or other inherent amplifier imperfections can introduce distortion which might mistakenly be blamed on the transformer. If these characteristics of the amplifier are known, then you can temporarily hook the transformer into it and make tests as shown in Fig. 5-8. Run tests on amplitude distortion and frequency response. These tests check for unbalance and other deficiencies automatically.

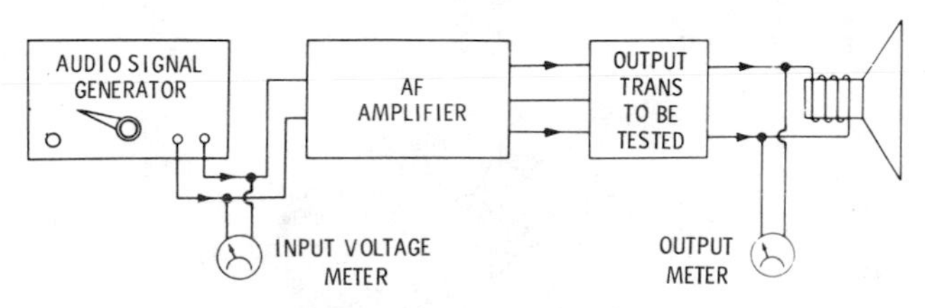

Fig. 5-8. Setup for output-transformer tests.

3. *More advanced tests with an impedance bridge.* It is not always handy to install the prospective transformer in an amplifier, or perhaps you are building your first amplifier and do not have a standard amplifier, but you still want to buy and check your transformer before installation. In this case, there are a few tests which, though far from a complete analysis, give a good idea of whether you are buying a well-designed component. These tests call for an impedance bridge. The method of testing is illustrated in Fig. 5-9. It is as follows:

(A) First check for unbalance, because if there is too much unbalance the transformer is unsuitable and further tests are not necessary. Unbalance is checked as shown in Fig. 5-9A. The resistance and inductance of each half of the primary winding, with the secondary winding open, are measured. The percentage difference between them is the percentage of unbalance. Unbalance of 1 percent is considered low enough to be good, 5 percent is only fair, and more is poor.

(B) Now check the total primary inductance, as illustrated in Fig. 5-9B. The reason this test is appropriate will probably be more clear on re-examination of Fig. 5-5. If, in this equivalent circuit, the load is removed (to simulate an open secondary), it can be seen that the only inductances in the circuit are the leakage inductance of the primary and the primary inductance. The leakage inductance, even in relatively poor transformers, is very much less than the primary inductance, and it is therefore negligible in this test. Thus, the measurement of the inductance across the primary leads with the secondary open is a reasonable check on primary inductance. However, because the bridge measurement is at low signal level, it makes no check on the inductance at high audio levels. How the inductance holds up under increases of signal amplitude depends on

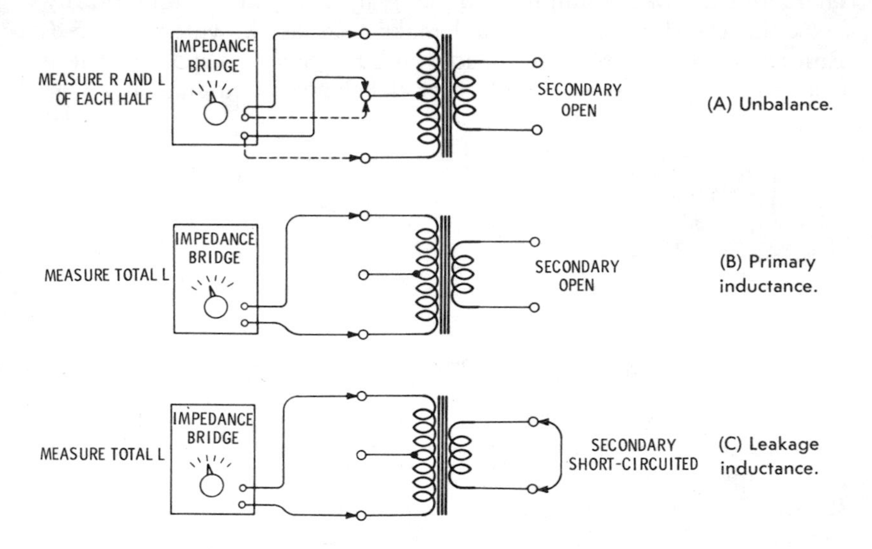

Fig. 5-9. Simple tests on output transformers.

the adequacy of the core. If the weight, volume, and material of the core are sufficient, the bridge measurement of inductance should be a good indication of primary inductance. The minimum acceptable value is about 15 henrys; the method of interpreting the measurement is explained in the following paragraphs.

(C) *Check leakage inductance.* This is done by short-circuiting the secondary winding and again measuring the inductance across the total primary winding, as illustrated in Fig. 5-9C. By reference to Fig. 5-5, it can be seen that short-circuiting the load shunts the primary inductance with the relatively low value n^2L_S, so effectively there is just the leakage inductance in the circuit. Now our bridge measurement indicates a relatively low value. Maximum leakage inductance allowable in a good transformer is of the order of about 75 millihenrys.

Frequency response is usually stated in terms of the range between the frequencies at which the output falls off 3 dB. These limiting frequencies are: (1) that low frequency at which the primary inductance has a reactance equal to the load impedance (the load impedance is that offered to the output tubes for proper operation and is listed as such in the tube manuals), and (2) that high frequency at which the leakage reactance becomes equal to the load impedance. For example, if the load impedance is 6000 ohms, a primary inductance of at least 19 henrys is required to reach 50 Hz at low-frequency drop-off ($X_L = 2\pi fL = 6.28 \times 50 \times 19 = 5970$ ohms). For the same impedance, 95 millihenrys leakage inductance would allow

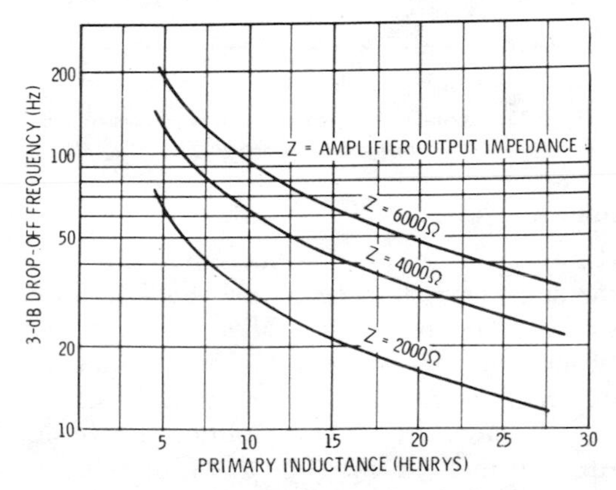

Fig. 5-10. Primary inductance required versus minimum response frequency.

response to 10,000 Hz, whereas 63 millihenrys allows response to 15,000 Hz. For the convenience of the reader, graphs showing lower and upper response limitations, respectively, for typical values of load resistance and a range of inductances are included in Figs. 5-10 and 5-11.

It must again be emphasized that the foregoing relations hold only if the inductances are correct for at least normal operating levels. This is where the power rating of the transformer becomes important. For instance, if a transformer is rated at 5 watts, and is to be operated at 10 watts, distortion is likely to result because the core is overloaded. The core will saturate, and the peaks of the audio waveform will be flattened, as illustrated in Fig. 5-12. It is good policy to operate well below the power rating of the transformer to allow for good dynamic range in the system. In other words, although normal operating average power may be about 5 watts in a typical case, the peak power on momentary loud passages may go to several times that.

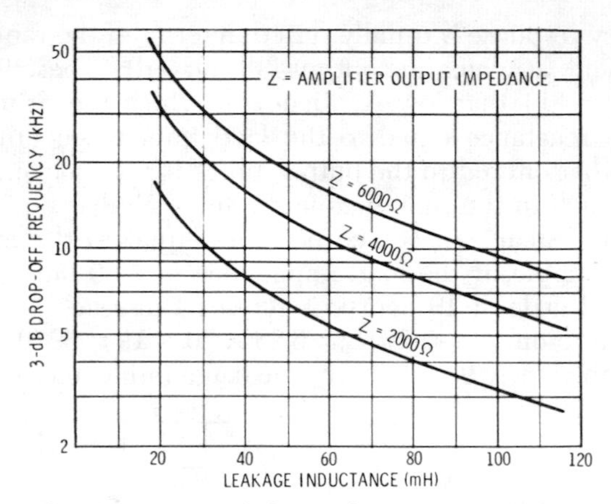

Fig. 5-11. Maximum leakage inductance versus highest response frequency.

Output Impedance

The output impedance of an amplifier should equal the input impedance of the speaker system to be driven. For universal application, it is desirable to choose an amplifier having a transformer with a range of output impedances including one to match a 500- or 600-ohm line. This facility is useful when it is desired to operate a speaker at some distance from the amplifier, as considerable loss is experienced in long lines unless a matched-impedance feed line of 250 ohms or more is used. If a 500-ohm line impedance is used, it is necessary to use a 300- to 500-ohm transmission line feeding a 500-ohm output-impedance transformer to the speaker. This transformer should be placed at the speaker and should have outputs matching the speaker system. This transformer should be equal in overall quality to the output transformer. Standard impedance output values desirable are: 4, 8, 16, 32, and 500 ohms.

Characteristics of Tube Versus Transistor Power Amplifiers

The ability of an amplifier to deliver its rated power into a speaker load depends on the impedance match between the two. The advent

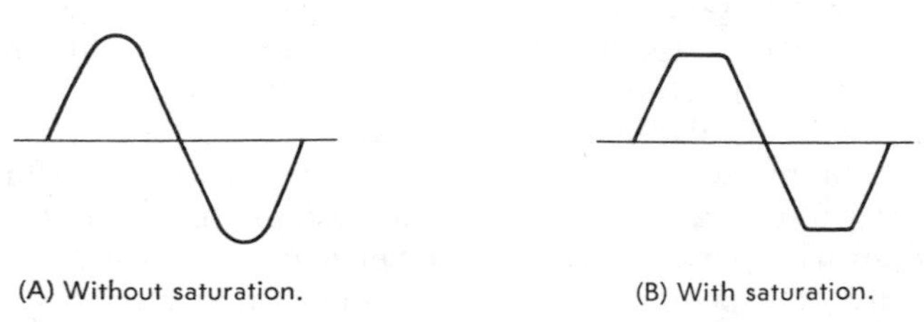

(A) Without saturation.

(B) With saturation.

Fig. 5-12. Waveform of a transformer in which the core is saturating.

of solid-state amplifiers, which usually do not use multiple-impedance output transformers, requires proper selection of speakers in regard to impedance.

Vacuum-tube amplifiers, because of the high impedance of the output tubes, require a transformer to match to the typical speaker, and it is easy to include several "taps" for matching speakers of different impedances. Transistor amplifiers, however, generally do not require an output transformer because the efficient matching impedance of a typical transistor output stage is some value between 4 and 16 ohms, close to the nominal impedance of most speakers. A transistor amplifier has a single optimum load impedance determined by its circuit design, and therefore it is not able to deliver full power into a range of fixed load impedances, as is readily possible with a vacuum-tube amplifier.

Speakers are not fixed-impedance devices. The impedance of any speaker or speaker system varies over its useful frequency range. The Electronic Industries Association specifies that the rated impedance shall be the minimum impedance over the useful frequency response of the speaker. However, even where the manufacturer adheres to this rating procedure, combinations of speakers in systems and the settings of crossover controls for mid-range and treble balance may significantly alter the impedance characteristics. The minimum (rated) impedance may rise to several times the rated value at other frequencies, usually reaching its low-end peak at the resonant frequency of the speaker enclosure system. Since speaker impedance is not purely resistive, other factors also limit speaker ability to utilize the amplifier rated power. Large multiple-speaker systems may appear as largely capacitive loads, just like a full-range electrostatic speaker, and the electrostatic speaker may drop well below its rated impedance at some frequencies.

Because the load is not a constant impedance, no amplifier can deliver equal power into any speaker system at all frequencies. Tube amplifiers with multiple-impedance output transformers can match several specific loads (but not simultaneously), and an impedance mismatch has a similar power limitation on any amplifier, tube or transistor. The Institute of High Fidelity has chosen 8 ohms as the standard impedance for rating power amplifiers—the modal nominal impedance, but speakers vary from 4 to 32 ohms.

A transistor amplifier which delivers 60 watts maximum into an 8-ohm load delivers between 30 and 40 watts into a 16-ohm load, and less into a 32-ohm load. Lower load impedances, such as 4 ohms, cause more current in transistors than do higher impedances. Excessive current can lead to rapid transistor failure unless the power output is deliberately limited to prevent excessive current. There, care should be used in application, matching, and operating.

In order to minimize problems of excessive current, some solid-state amplifiers require the use of series resistors with 4-ohm speakers to prevent damage. This soaks up a high percentage of the available power and greatly diminishes the effective damping of the speaker. The same problem exists with speakers (supposedly specially designed for transistor amplifiers) which include series resistors in order to present a more nearly uniform impedance characteristic. Because the majority of 16-ohm speakers are more efficient than the most popular 4-ohm speakers, it is of greater importance to most users to have the best match between 4 and 8 ohms.

It is worth noting that direct-coupled transistor power amplifiers cannot be paralled to obtain higher power output at usable speaker impedances. Because these amplifiers are essentially current or constant-voltage amplifiers, when they are paralleled the same voltage will be available only at an output impedance which is too low for general speaker use, and there is risk of damage to the amplifier through interconnection of the two outputs. For higher-power applications, the preferred procedure is to drive each channel of the amplifier with identical signals, and connect each output to a separate speaker system.

Diffused silicon transistors permit good circuit performance at high frequencies. Silicon transistors are desirable for power output stages because of their ability to perform at much higher junction temperatures than can germanium transistors. This means smaller heat-radiating fins can be used for the same power dissipation. On the negative side, silicon often has higher saturation resistance, which gives decreased operating efficiency that becomes appreciable in operation from low-voltage supplies.

POWER AMPLIFIER CIRCUITS

Next to the output transformer and the speaker, the power amplifier circuit and the other components used therein are usually the most important factors governing the percentage of distortion present in a system. Since it is usually desired to keep distortion to the lowest value consistent with economy, the power amplifier should be carefully designed. Some form of degenerative feedback should always be used. Feedback reduces harmonic distortion, stabilizes the output impedance, and reduces the effects which resonance in speakers has on the output.

Push-pull arrangements will give a low percentage of distortion. Triodes in push-pull give excellent results. The use of beam tubes in ordinary circuits results in distortion higher than is permissible unless sufficient degenerative feedback is used. From 15 to 20 dB of feedback should be used in beam-power amplifiers.

Beam-power tubes used in push-pull arrangements for higher than average power (100 watts or more per channel) in certain modern circuit designs give negligible distortion and are recommended when highest-quality reproduction and high output of music are desired.

Transistors are used in quasi-complementary-symmetry circuits which are essentially class-B circuits that provide outstanding performance for the most stringent requirements for high-fidelity systems. Silicon transistor arrangements can be used to supply 70 to 100 watts per channel with negligible distortion, high reliability, and comparatively low cost. In addition, these arrangements require a minimum of space and provide a maximum of weight reduction. They can be direct-coupled to eliminate expensive, heavy, and bulky output transformers, and they have become the most widely used output device in equipment with power ratings under 100 watts per channel.

Output Circuits

Amplifiers in general are classed by letters, A, B, and C, according to the method of adjusting the tubes or transistors. Audio power output stages are generally adjusted for class A, class AB, or sometimes class B, but never for class C. Preamplifier stages are usually voltage amplifiers and are adjusted to operate in the class-A region.

A class-A amplifier circuit has a fixed bias, and the applied signals are so arranged that plate or collector current is present at all times.

A class-B amplifier is one in which the operating fixed bias is adjusted approximately to the cutoff value for the tube or transistor so that plate or collector current is quite low with no signal and so that there is plate or collector current for each half-cycle of an applied ac signal.

A class-AB amplifier is adjusted so that the fixed bias and applied alternating signal cause plate or collector current to be present for appreciably more than half but less than the entire signal cycle.

Class-C amplifiers are so arranged that plate or collector current is present for less than half of the signal cycle. Class-C circuit arrangements are not used in audio work.

Class-A and class-AB arrangements are well known for their high-fidelity characteristics. In years past, circuits arranged to operate in class A were the only acceptable amplifier circuits for hi-fi equipment, but today the best amplifiers operate in class AB with special compromise arrangements (untralinear and unity-coupling designs) to bring about audio reproduction with practically unmeasurable distortion and with excellent response and high outputs.

Class-A amplifiers are used for linear audio amplifier service at low power levels. When power amplifiers are used in this class of operation, the amplifier output usually is transformer-coupled to the load circuit, as in Fig. 5-13. At low power levels, the class-A amplifier can also be coupled to the load by resistor, capacitor, or direct coupling techniques.

There is some distortion in a class-A stage because of the nonlinearity of the active device and circuit components. The maximum efficiency is not realized. The class-A transistor amplifier is usually biased so that the quiescent collector current is midway between the maximum and minimum values of the output-current swing. Collector current, therefore, is present at all times and imposes a constant drain on the power supply. This drain is a distinct disadvantage when higher power levels are required or operation from a battery is desired.

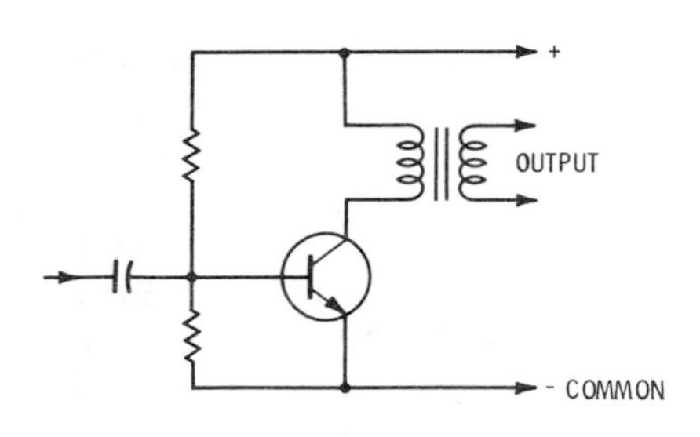

Fig. 5-13. Basic class-A transformer-coupled transistor amplifier.

Class-B and class-AB power amplifiers are usually used in pairs in a push-pull circuit because conduction is not maintained over the complete cycle. A circuit of this type is shown in Fig. 5-14. If conduction in each device occurs during approximately 180 degrees of a cycle and the driving wave is split in phase, the class-B stage can be used as a linear power amplifier. The maximum efficiency of the class-B stage at full power output can be as high as 78.5 percent when two transistors are used. In a class-B amplifier, the maximum power dissipation is 0.203 times the maximum power output and occurs at 42 percent of the maximum output.

Standard single-ended parallel or push-pull parallel circuit arrangements do not produce sufficient fidelity to be used for hi-fi systems. Push-pull arrangements produce less distortion because even harmonics, plate-current effects in the transformer, and hum pickup in the plate circuit normally tend to cancel out.

There are several successful types of push-pull amplifiers in current use in output circuits utilizing triode, tetrode, transistor, and ultralinear (hybrid) circuits. Pentode amplifiers have such nonlinear characteristics that they cannot be considered as high-fidelity equipment. Triode arrangements with triode-connected tetrodes (Fig. 5-15) and silicon transistors offer the simplest approach to high fidelity.

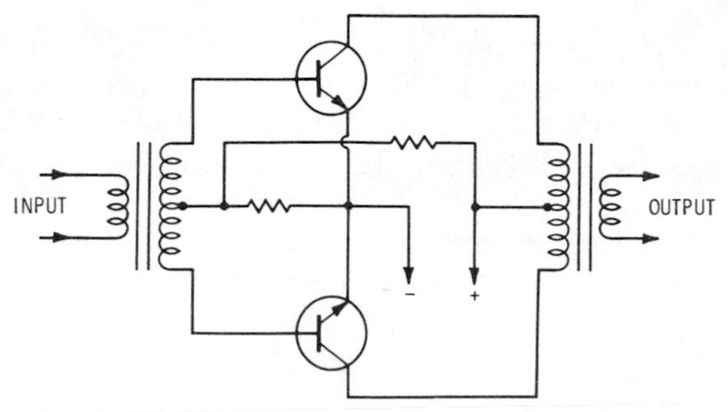

Fig. 5-14. Basic class-AB transistor amplifier.

This approach may be used with or without additional overall systems circuit improvements such as negative feedback (to be discussed later). Compared with straight tetrode-connected operation, transistor and triode-connected output stages using comparable components provide better stability, noncritical performance, less distortion, and better speaker-damping characteristics.

However, triode amplifiers (compared with tetrodes) have the following disadvantages: (1) Triode amplifiers have the lowest efficiency; therefore they require large and more expensive components, tubes, and power supplies to obtain outputs equivalent to

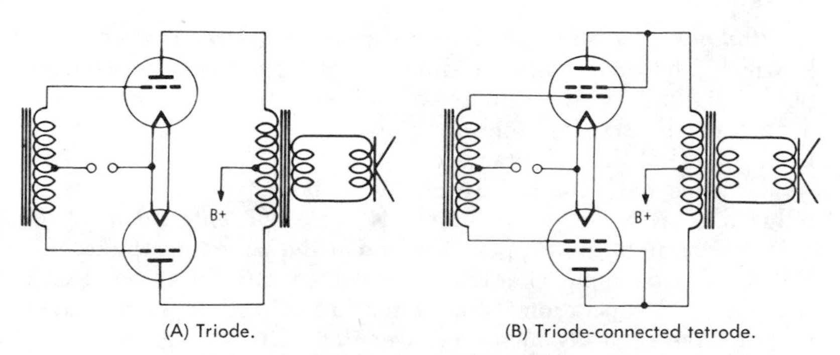

(A) Triode.

(B) Triode-connected tetrode.

Fig. 5-15. Power-amplifier stages.

those obtained from tetrode amplifiers. (2) Triode amplifiers require much more driving power. This condition further increases cost. The overall percentage of distortion will be increased by each additional stage required to raise the power sufficiently to drive the triode amplifiers. The higher the power required from any intermediate stage, the higher the distortion of the output of that stage will be. All these distortions are added to the final triode amplifier stage distortion, partially negating the advantages to be gained in the use of the triode output stage itself.

Tube-type amplifiers presently are used mostly where power-output requirements exceed 100 watts per channel. For power outputs under 100 watts per channel, transistors offer many advantages and, therefore, are most popular. The following discussions of amplifier theory apply to both tubes and transistors. In some cases, tubes and transistors are interchangeable in a circuit arrangement, provided that the components necessary to be matched in impedance and other characteristics are changed accordingly but the circuit principles do not change.

Tetrode Amplifier Circuits

The efficiency of a tetrode amplifier can be as high as 60 percent. Tetrodes operated in certain circuits have very high amplification characteristics, requiring but a fraction of the drive of triode and transistor amplifiers. These advantages make possible the reduction of cost to a practical commercial level for manufacture of very high-power high-fidelity amplifier equipment. The disadvantages of tetrode amplifier design have been overcome by circuit, component, and electron-tube improvement to the extent that highly satisfactory results can be attained. Technically, the output characteristics of tetrode amplifiers have been so improved that the measurable distortion is negligible. A tetrode final amplifier circuit is shown in Fig. 5-16.

Ultralinear operation involves a compromise circuit arrangement in which a tetrode is connected neither as a triode nor as a tetrode, but special taps on the primary of the output transformer are connected to the screens, thereby providing an amplifier operation having some of the advantages of both tetrode and triode operation. At the same time, the disadvantages of both types of circuits are reduced. This circuit arrangement is more efficient and has more gain than straight triode operation, and at the same time the stability and speaker-damping characteristics are practically as good as in straight triode operation. Most important, distortion in the power output stage is hardly increased over that of triode operation. This

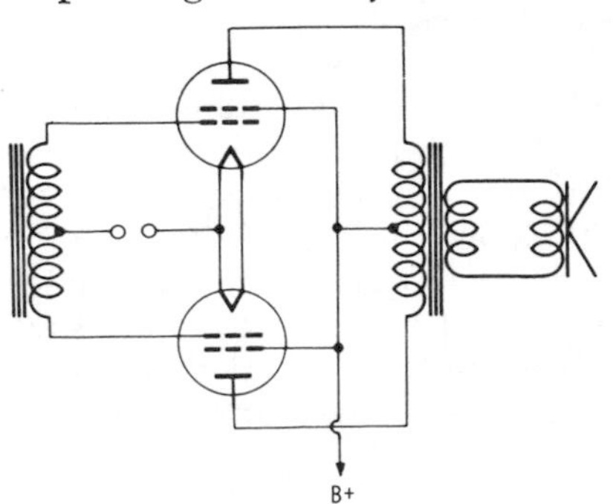

Fig. 5-16. Tetrode power amplifier.

compromise design is generally accepted as one of the better contemporary approaches to high-fidelity power-amplifier circuits. The circuit arrangement to achieve this is shown in Fig. 5-17. The tap point varies from 20 percent to 40 percent of the winding, with considerable difference of opinion between different manufacturers as to the optimum point. Tap-point design is usually determined empirically by listening tests. Always, when the human ear is involved, measurement by a fixed standard is impossible. Therefore, one must listen to each unit to determine which is preferable to him.

Another design considered optimum by current standards is the patented McIntosh unity-coupling circuit (Fig. 5-18). "Unity coupling" is a name which identifies a group of audio amplifier circuits. These circuits are designed to reduce distortion at high frequencies (the treble range) which conventional push-pull circuits may generate.

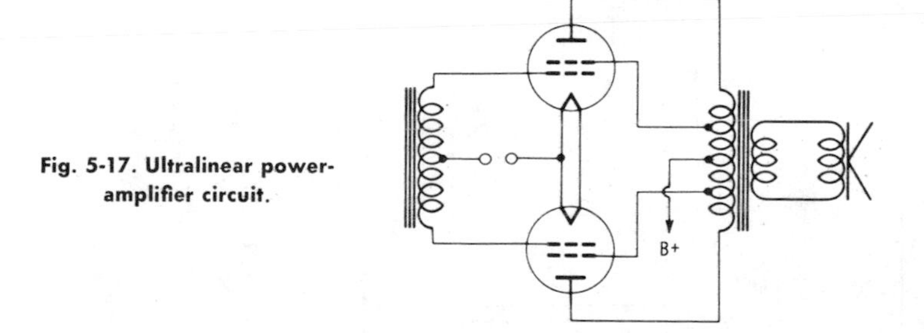

Fig. 5-17. Ultralinear power-amplifier circuit.

To illustrate this problem graphically, consider a tone waveshape as shown in Fig. 5-19A. One push-pull tube operates from A to B, the other from C to D. Both tubes operate from B to C. The small interval BC can be reduced to zero, though it seldom is practical to do so.

One half of a push-pull amplifier ideally produces a wave as shown in Fig. 5-19B. It consists of one-half a sine wave and a flat. The half cycles from the two halves of the amplifier are added together to make the full wave in the output transformer.

The conventional transformer consists of two halves wound as shown in Fig. 5-19C. At high frequencies, the flat between successive half-waves changes shape. Graphically, the change appears as shown in Fig. 5-19D. The new shape from B to C is due to the collapse of the magnetic energy in the output transformer, which is not completely coupled from one half to the other. When two waves of this shape are added together, they appear as shown in Fig. 5-19E. The departure from the original tone shape represents the addition of new tones generated by the amplifier itself.

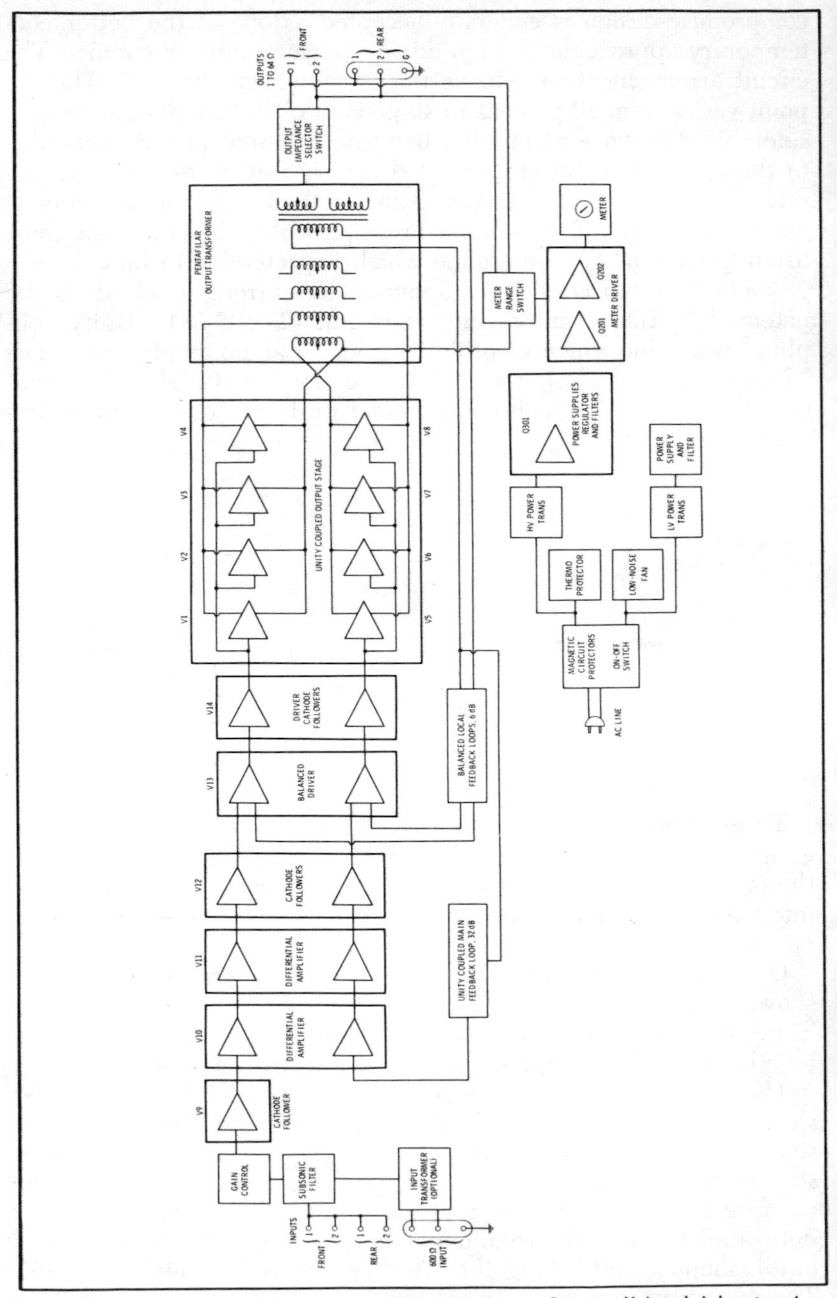

Courtesy McIntosh Laboratory Inc.

Fig. 5-18. Block diagram of McIntosh 350-watt amplifier.

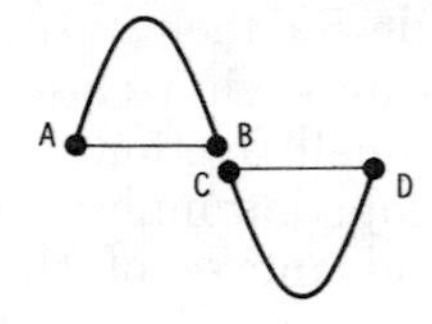

(A) Pure tone waveshape.

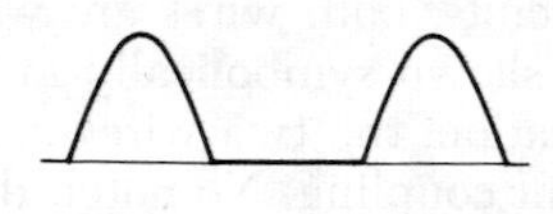

(B) Output of one half of amplifier.

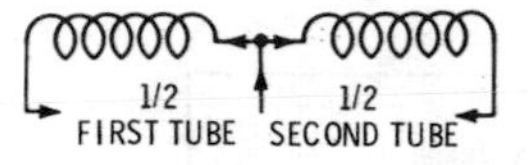

(C) Conventional transformer primary.

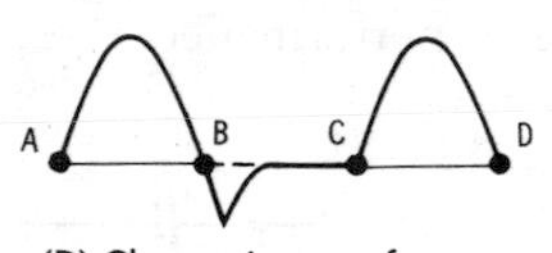

(D) Change in waveform.

(E) Amplifier output with distortion.

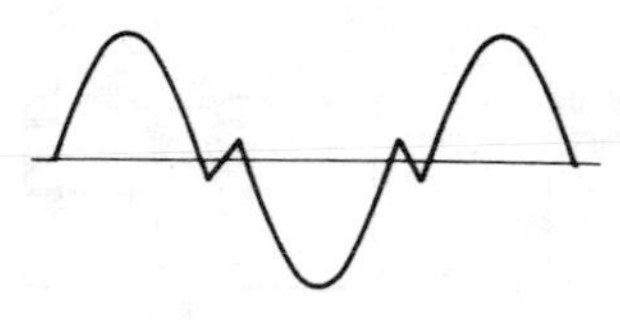

Fig. 5-19. Waveforms in push-pull stage.

The unity-coupled circuits in their simplest form use only one half of a conventional push-pull transformer, as shown in Fig. 5-20A. The coupling under this condition is nearly 100 percent of unity. However, it is still desirable to use two transformer windings so that only one power supply is needed. In McIntosh unity-coupled

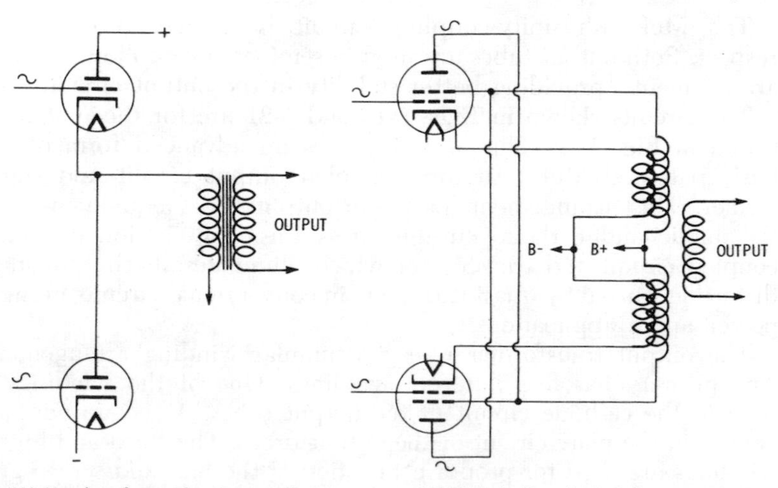

(A) Simplest form of unity coupling.

(B) With half cathode-follower circuit.

Fig. 5-20. Use of unity coupling.

circuits, both wires are wound side by side for their entire length, as shown symbolically in Fig. 5-20B. Having such a close physical relation, the two wires are within a very small fraction of 100 percent coupling. No notch distortion is developed at high frequencies, as shown in Fig. 5-19A. An additional advantage of this circuit improvement is that more transformer turns can be used at low frequencies to reduce distortion there, too. Therefore, unity coupling improves both the high-frequency performance and the low-frequency performance.

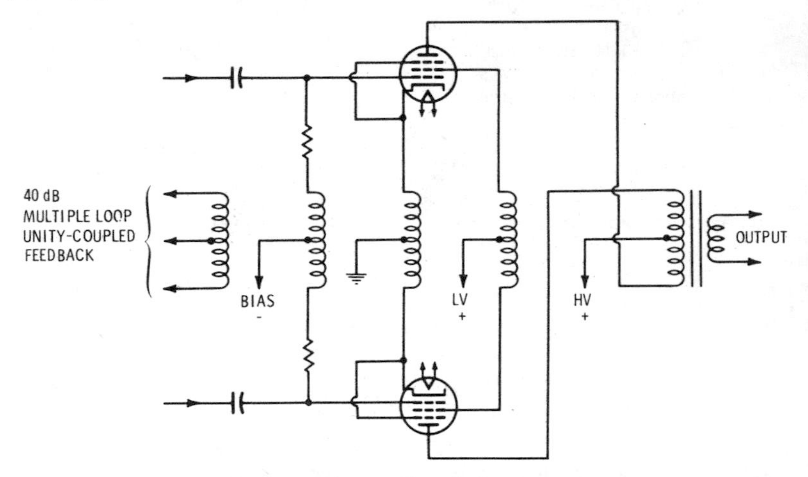

Fig. 5-21. McIntosh Pentafilar winding arrangement.

The McIntosh unity-coupling circuit is unique in one other respect. Both output tubes are used in semi–cathode-follower circuit arrangements, providing better stability in the output circuit.

The circuits shown in Figs. 5-18 and 5-21 are for the unit illustrated in Fig. 5-22. This amplifier uses an advanced form of the basic patented McIntosh unity-coupled output circuit and transformer. This arrangement loads the output tubes equally in both the anode and cathode circuits. It is this combination of unity-coupled circuit and transformer which eliminates the output-stage distortion caused by quasi-transients in conventional circuits in high-power audio applications.

The output transformer uses a Pentafilar winding arrangement. The primary has five different windings. One of the windings is used in the cathode circuit of the output tubes. A second winding is used in the plate circuit of the output tubes. The third and fourth windings are used for proper connection of the bias and screen-grid voltages to the output and driver stages. The fifth winding supplies the feedback signal for two negative-feedback loops. In this unity-

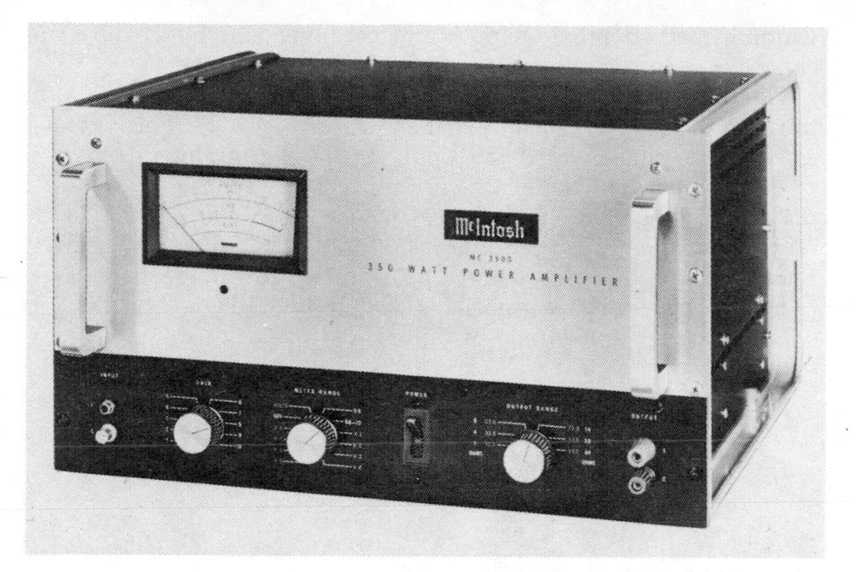

Courtesy McIntosh Laboratory Inc.

Fig. 5-22. McIntosh 350-watt single-channel amplifier using tetrodes and unity coupling.

coupled feedback, one loop is the push-pull coupling to the balanced driver stage. The other, the main feedback loop, couples the feedback winding to the input differential amplifier. The windings are all wound on the transformer at the same time. This winding technique, developed and perfected by McIntosh, results in extremely tight coupling which reduces leakage inductance.

A greater amount of negative feedback can be obtained in an amplifier using primary feedback. The stability of the amplifier is not affected. If the feedback winding is wound with its turns side-by-side with those of the primary and has the same number of turns, the feedback winding will have the same phase and voltage as the primary windings for frequencies up to 500 kHz. Therefore, it is unity coupled with the primary.

By using unity-coupled negative feedback, it is possible to obtain up to 40 dB of feedback with very good stability and extremely low nonlinear distortion. It is relatively easy to maintain a flat frequency response with very low phase shift in the electronic circuits of an amplifier prior to the output transformer.

To achieve flat frequency response well beyond 20 kHz, very close coupling is required between the primary and secondary windings in the output transformer. This is accomplished by dividing the Pentafilar primary into ten different winding sections. The secondary is divided into eight different winding sections. These winding sections are then interleaved. This expensive and difficult winding method provides optimum coupling and holds the shunt capacitance to a minimum.

There are other variations of circuits for connecting the output tubes to the speaker, such as that shown in Fig. 5-23. This circuit is essentially a single-ended push-pull stage. The output tubes are in series in regard to the plate-current swing, and the grids are in push-pull. When the current rises in one tube, the current in the other falls. This gives the same result as the operation of a conventional push-pull output stage. The primary advantage here is the reduction of switching transients and a high degree of linearity without special transformer windings.

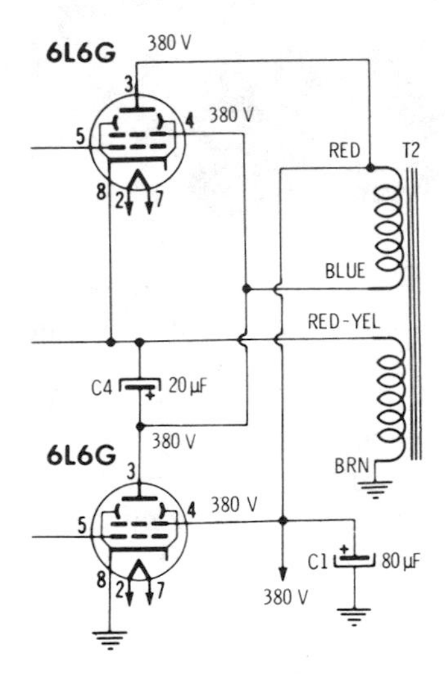

Fig. 5-23. Single-ended push-pull amplifier circuit.

Fig. 5-24A shows the application of power transistors to high-fidelity amplification in a simplified class-B circuit. Class-A amplifier circuits are usually utilized for preamplifiers with transistors, but a disadvantage of class-A operation in transistor power-amplifier circuits is the requirement that there be collector current at all times. When two transistors are connected in class-B push-pull, one transistor amplifies half of the signal, and the other transistor amplifies the other half of the signal. These signals are combined in the output circuit to restore the original signal in an amplified state.

Transistors are not usually used in true class-B operation because of an inherent nonlinearity, called crossover distortion, that produces a high degree of distortion at low power levels. The distortion results from the nonlinearities in the transistor characteristics at very low current levels. For this reason, most power stages operate in a biased condition somewhat between class A and class B. This intermediate class is defined as class AB. Class-AB transistor amplifiers operate with a small forward bias on the transistor to minimize the nonlinearity. The quiescent current level, however, is still low enough that class AB amplifiers provide good efficiency. This advantage makes class-AB amplifiers an almost universal choice for high-power linear amplification, especially in equipment operated from a low-voltage supply.

Fig. 5-24B shows a transistor amplifier with the output transformer eliminated. The advantages of class-B or AB operation utilizing transistors can be obtained without the use of an output transformer. In this circuit, the secondary windings of driver transformer T1 are phased so that when there is a negative signal from base to emitter of transistor Q1, there is a positive signal from base to emitter of transistor Q2, and vice versa. The negative signal to the base of Q1 causes Q1 to draw current, and the positive signal to Q2 cuts Q2 off. When the condition is reversed, transistor Q2 conducts and Q1 is cut off. Resistors R101, R102, R103, and R104 provide dc bias to reduce crossover distortion and to keep the transistors a little above cutoff for no-signal conditions.

Fig. 5-24C shows a simplified complementary-symmetry circuit utilizing pnp and npn transistors together in a class-AB push-pull amplifier with direct coupling to a speaker. These two transistors are connected in a single stage so that the dc current path in the output circuit is completed through the collector-emitter circuits of the transistors.

In the circuits shown in Figs. 5-24B and 5-24C, there is practically no direct current through the speaker voice coil, and direct connection is possible.

Fig. 5-24D shows a direct-coupled transistor power amplifier with excellent low-frequency response. It has the advantage of dc feedback for temperature stabilization of all stages. This feedback system stabilizes the voltage division across the 2N553 power output transistors, which operate in a single-ended class-AB push-pull arrangement. These transistors also operate class B in the Darlington connection to increase the current gain. Using an npn transistor (2N696) for Q3 gives the required phase inversion for driving one of the output transistors and also has the advantage of push-pull emitter-follower operation from the output of 2N1925 input transistor Q1 to the load. Emitter-follower operation has lower inherent distortion and low output impedance because of the high percentage of voltage feedback.

The output transistors have a small forward bias of about 15 mA to minimize crossover distortion and also to operate the output transistors in a more favorable beta range. This bias is set by the voltage drop across R1 and R2, which shunt the input to the intermediate amplifiers. The intermediate amplifiers are biased at about 2 mA (to minimize crossover distortion) with the voltage drop across silicon diode X11. Junction diodes have a temperature characteristic similar to that of the emitter-base junction of a transistor. Therefore this diode also gives compensation for the temperature variation of the emitter base resistance of three of the transistors. These resistances decrease with increasing temperature; thus the decrease in forward voltage drop of approximately 2 millivolts per degree centigrade of the diode provides some temperature compensation.

The input transistor is a class-A driver with an emitter current of about 4 mA. Negative feedback to the base of this stage lowers the input impedance of the stage and thus requires a source impedance that is high enough that the feedback current will flow into the amplifier rather than into the source. Resistor R10 limits the minimum value of source impedance.

About 10 dB of positive feedback is applied by way of C101 and R101. This action helps to compensate for the unsymmetrical output circuit and permits the positive-peak signal swing to approach the amplitude of the negative peak. This positive feedback is offset by about the same magnitude of negative feedback.

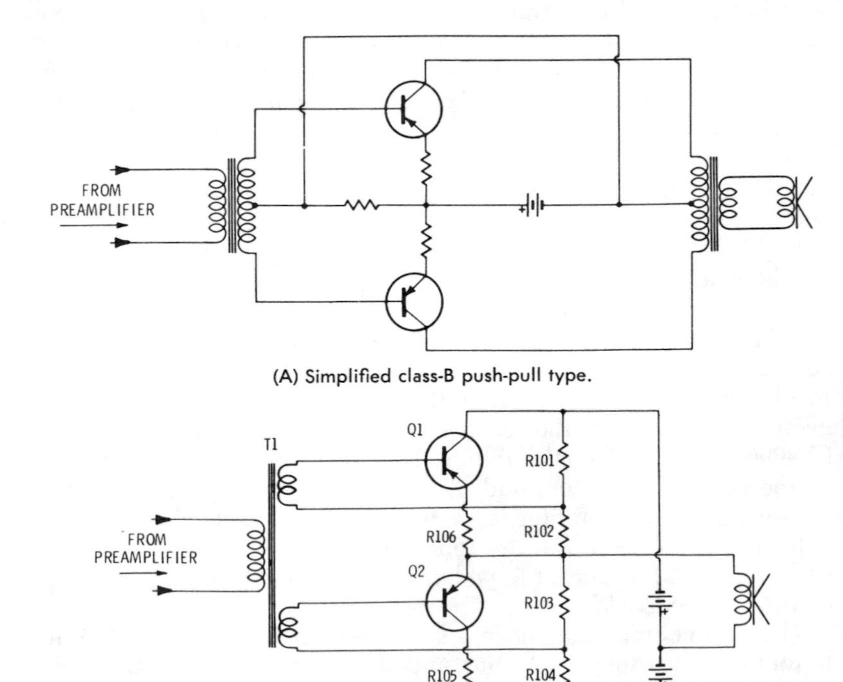

(A) Simplified class-B push-pull type.

(B) Class-B push-pull type without output transformer.

Fig. 5-24. Transistor

A ½-ampere fuse is used in the emitter lead of each output transistor for protective fusing of the output transistors and also to provide local feedback, since the ½-ampere type 3AG fuse has about 1 ohm of dc resistance. This local feedback increases the bias stability of the circuit and also improves the declining frequency response of the output transistors at the upper end of the audio spectrum. Because of possible lower transistor efficiency above 10 kHz, care should be taken when checking the maximum continuous sine-wave output at these frequencies. If continuous power is applied for more than a short period, sufficient heating may result to raise the transistor current enough to blow the ½-ampere fuses. Since there is not sufficient sustained high-frequency power in regular program material to raise the current to this level, actual performance of the amplifier does not suffer, because the power level in music and speech declines as the frequency increases beyond about 1 to 2 kHz.

The speaker system is shunted by 22 ohms in series with 0.22 μF to prevent the continued rise of the amplifier load impedance and its accompanying phase shift beyond the audio spectrum.

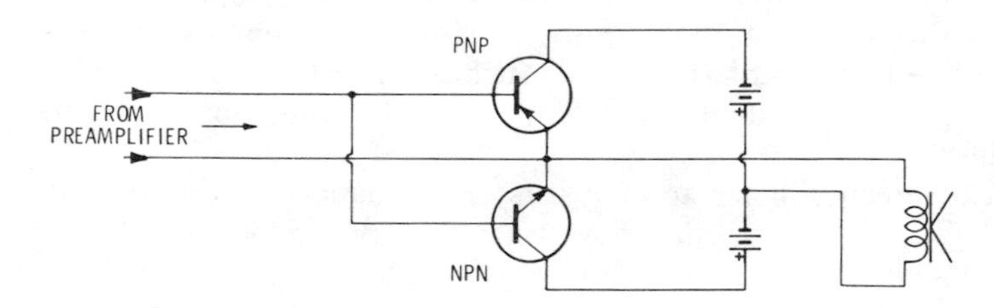

(C) Use of pnp and npn complementary symmetry.

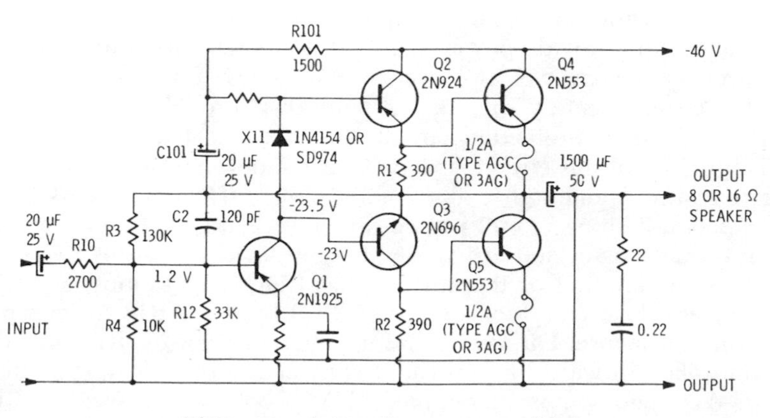

(D) Direct-coupled transistor power amplifier.

power amplifiers.

The overall result, from using direct coupling, no transformers, and ample degeneration, is an amplifier with output impedance of about 1 ohm for good speaker damping, low distortion, and good bandwidth. The power response at 1 watt is flat from 30 Hz to 15 kHz and is down 3 dB at 50 kHz. At this level, the total harmonic and intermodulation distortion are both less than 1 percent. At 8 watts, the intermodulation distortion is less than 2½ percent, and the total harmonic distortion is less than 1 percent, measured at 50 Hz, 1 kHz, and 10 kHz. The performance of the amplifier of Fig. 5-24D is about the same for both 8- and 16-ohm loads.

Fig. 5-25 shows the basic configuration of a quasi-complementary-symmetry circuit that provides outstanding performance in high-fidelity amplifiers. This basic circuit is used for four separate audio-amplifier circuits that can provide continuous sine-wave power outputs of 12, 25, 40, and 70 watts (rms) with only minor changes in components. The input, predriver, and protection-circuit transistors remain the same for all output-power levels.

These universal quasi-complementary-symmetry amplifiers feature rugged hometaxial-base silicon npn output transistors. These transistors and the complementary driver transistors are operated class AB in an arrangement that ensures a small zero-signal current drain.

Other features of the circuit include direct-coupled preamplifier and predriver stages and short-circuit protection or safe-area limiting.

The preamplifier stage consists of a balanced-bridge circuit (Q1 and Q2) that maintains quiescent zero dc voltage at the output. Feedback is coupled through resistor R6, and ground reference is provided through resistor R2 and capacitor C2. The common emitters are returned to the positive supply through resistor R3, diode X1, and resistor R5. Diode X1 and capacitor C4 minimize turn-off transients and provide power-supply decoupling. The bridge circuit is direct-coupled to a class-A predriver stage (Q3), which is coupled to the complementary drivers (Q4 and Q5) through R12. The dissipation-limiting protection circuit is also connected at this point. The purpose of this circuit is to prevent the output stage from being driven into conduction if abnormally high dissipation occurs. The dissipation-limiting circuit provides a shunt path for the drive current from the associated driver and output devices. Resistor R12 provides some limiting of the current that transistor Q9 must support during overload conditions. Capacitor C9 bypasses R12 to improve transient response. Diodes X2, X3, and X4 and resistor R11 provide a controlled forward bias on the drivers and output devices so that class-AB operation is maintained. The "bootstrap" capacitor, C6, supplies the extra voltage swing necessary to saturate the upper output pair (Q4 and Q6) through resistor R12. Resistor R13 and

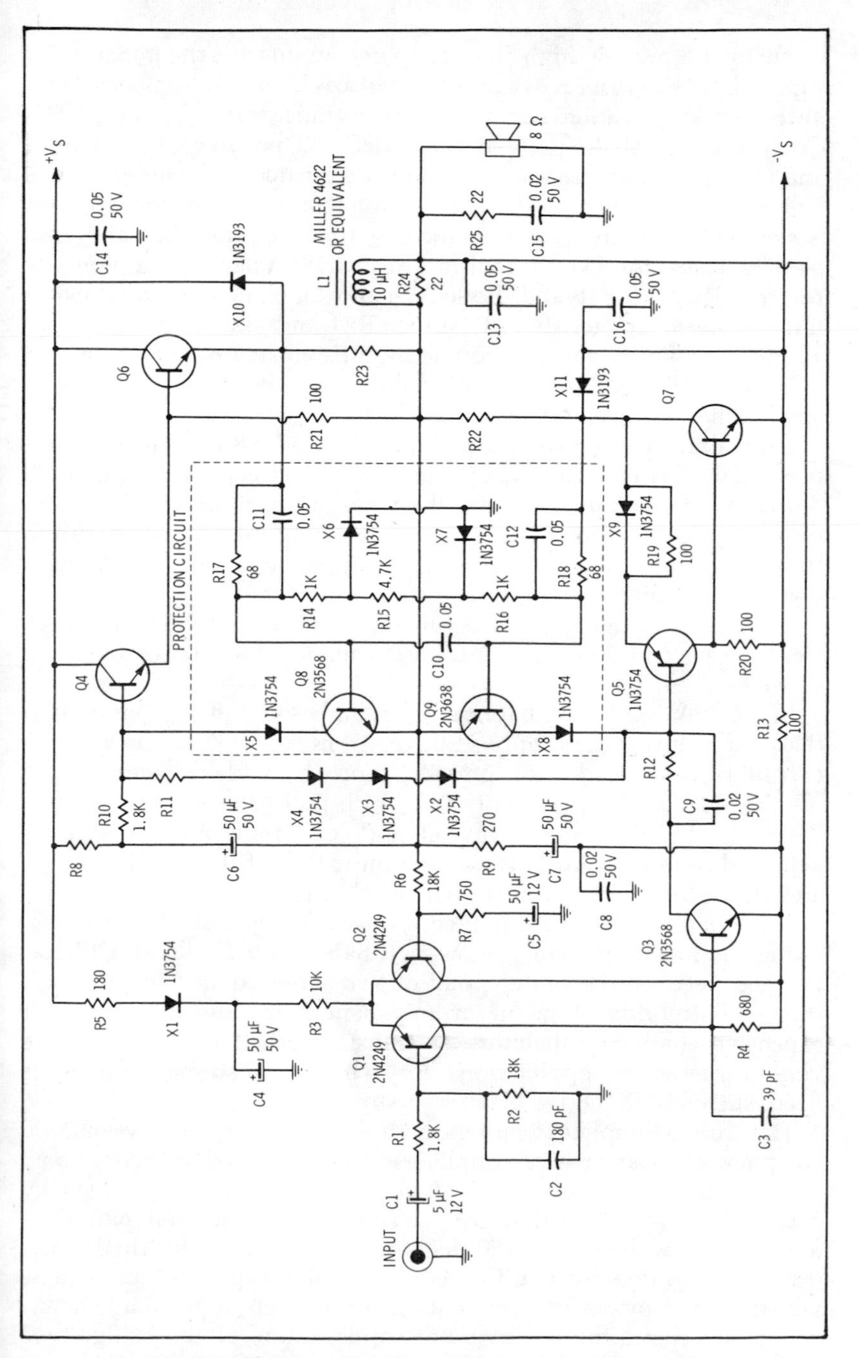

Fig. 5-25. Circuit of a quasi-complementary-symmetry audio amplifier.

capacitor C8 provide high-frequency decoupling for the negative dc supply line. Resistors R20 and R21, with R22 and R23, provide the necessary stabilization for the output transistors (Q6 and Q7). Current is sampled across resistor R23 for positive-cycle sensing and coupled to transistor Q8 through resistor R17. Simultaneous voltage sampling is provided by resistor R14 and diode X6. Current is sampled across resistor R22 for negative-cycle limiting and coupled to transistor Q9 through resistor R18. Voltage sampling by resistors R15 and R16 and diode X7 causes a change in the slope of the limiting characteristics. Resistors R24 and R25, capacitors C13 and C15, and inductor L1 provide high-frequency rolloff (above 50 kHz) so that a good margin of stability can be maintained under any loading conditions. Capacitors C10, C11, and C12 provide additional stability during limiting. Diodes X5 and X8 prevent forward-biasing of the collector-base junctions of transistors Q8 and Q9 during alternate half-cycle signal swings. Capacitors C14 and C16 provide parasitic suppression. Diode X9 and resistor R19 ensure a transconductance match between the upper and lower Darlington pairs to minimize low-level distortion.

Each amplifier provides the full rated power output to frequencies well beyond 20 kHz at a total harmonic distortion of less than 1 percent.

Fig. 5-26A is a block diagram of an audio-amplifier configuration that, for a given dc supply voltage, transistor voltage-breakdown capability, and load, can provide four times the power output obtainable from a conventional push-pull audio-output stage. Alternatively, given power-output and load requirements may be achieved from this circuit configuration with half the supply voltage and transistor voltage-breakdown capabilities required for conventional circuits. This performance is possible because the load can swing the full supply voltage on each half-cycle. The load is direct-coupled between the center point of two series-connected push-pull stages. This bridge type of arrangement eliminates the need for expensive coupling capacitors or transformers. These features are very attractive in applications for which the supply voltage is fixed, such as automotive or aircraft supplies.

The bridge-amplifier configuration consists essentially of two complementary-symmetry amplifiers with the load direct-coupled between the two center points. Each amplifier section is driven by a class-A driver stage that uses a transistor Darlington pair. The amplifiers must be driven 180 degrees out of phase. This dual-phase drive is provided by a differential-amplifier type of input stage, which also provides the advantage of a high input impedance. Fig. 5-26B shows the basic circuit configuration of the bridge-type audio-amplifier circuit.

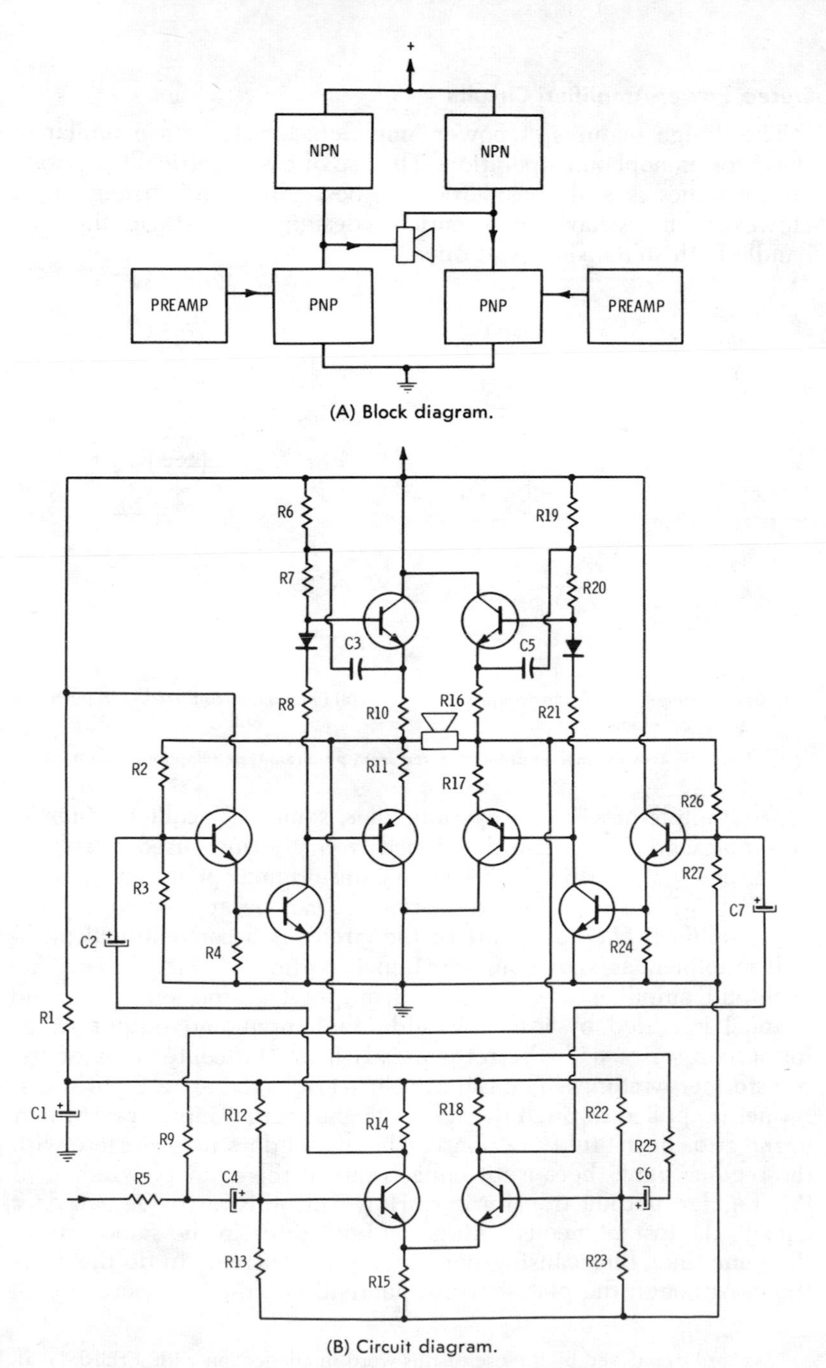

(A) Block diagram.

(B) Circuit diagram.

Fig. 5-26. Bridge-type audio amplifier.

Stereo Power-Amplifier Circuits

The design features of power amplifiers for stereo are similar to those for monophonic operation. The use of two electrically separate power stages is still considered the best all-around arrangement. However, efforts have been made to design single stages that can handle both stereo signals at once.

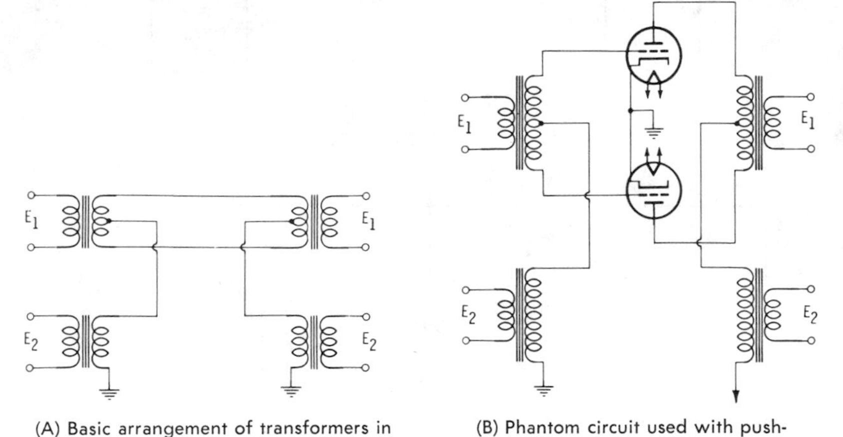

(A) Basic arrangement of transformers in phantom circuit.

(B) Phantom circuit used with push-pull amplifier.

Fig. 5-27. Two-channel amplifier derived from phantom-type telephone circuit.

An example of a "two-way" amplifier, sometimes called a *biortho* or *simplex* system, is one developed from the long-used telephone "phantom" circuit. Fig. 5-27A shows the original phantom* circuit, and Fig. 5-27B shows how the principle can be applied to a push-pull amplifier. The basic part of the circuit is a conventional push-pull amplifier, as shown in the upper portion of Fig. 5-27B. The push-pull amplifier uses input and output transformers. A second channel is added by inserting additional input and output transformers in series with the return leads from the center taps of the transformer windings. The signal voltage from the extra input transformer is applied through the halves of the regular input transformer to the grids of the tubes in phase. This signal does not interfere with the regular signal because it cancels out in the primary winding of the regular output transformer. However, this extra signal does change the instantaneous voltage on both grids in the same way at the same time, thus causing both tube plate currents to do the same thing. Although the plate-current fluctuations from this source can-

*Do not be confused by the use of this word in connection with a third, "middle" speaker and its associated circuit, discussed later.

cel out as far as the regular output transformer is concerned, they do cause the same fluctuations in the total plate current into the center tap. This total current passes through the extra output transformer, whose secondary winding delivers the output from the extra channel.

The circuit of Fig. 5-27 could be used for stereo by applying the L signal to the regular push-pull channel (E_1) and the R signal to the extra circuit (E_2). However, there are serious disadvantages to this. One is that the extra circuit operates as a single-ended (rather than push-pull) amplifier with the tubes in parallel. This type of circuit is much more subject to distortion than is the push-pull amplifier, and its power-handling capability with tolerable distortion is much lower. Since the L and R signals usually do not differ very much in amplitude, the lopsided arrangement would be inefficient.

To overcome this objection, the circuit is rearranged to use sum (L + R) and difference (L − R) signals. The sum signal carries most of the power of the stereo information. Since the L and R signals are usually nearly equal, the difference (L − R) signal is relatively small—it is the difference between two signals which are almost the same in amplitude. This small signal is applied to the extra channel, and does not tax its relatively low distortion-free power-handling capability.

The new arrangement of the circuit is shown in Fig. 5-28. No input transformer is necessary; the −L signal is applied to one grid, and the R signal is applied to the other grid. The negative L signal is obtained simply by passing the L signal through an odd number of conventional grounded-cathode amplifiers, each of which provides a phase reversal.

In the circuit of Fig. 5-28, each tube inverts its signal, so the relative polarities of the signals across the entire transformer primary and each half of the transformer are as indicated. The result is that the signal coupled through the main transformer (T1) is equivalent to L + R with polarities as indicated by the arrows at the secondaries.

Now consider the other channel. The primary winding of transformer T2 carries current from the center tap of the primary of T1. Therefore, this current is the sum of the plate currents of the two tubes, and the voltage has the relative polarity shown. The output of T2 is the L − R signal. This output is combined with the L + R outputs of T1 to provide the overall L and R outputs of the combination circuit. Although the L − R output enters into the derivation of both the L and R outputs, its amplitude is relatively small, so it can be handled as an extra output with little distortion.

Whether the simplified circuit just described or a dual power amplifier is used, there are certain switching functions usually provided in a stereo power amplifier. One of these is to provide for parallel operation for the monophonic mode. A switch is provided simply to connect the L and R outputs in parallel. If you happen to have an amplifier not so equipped, or have two separate amplifiers which you are combining, you can connect the outputs together yourself. Two things are important to remember.

1. The phasing must be correct. This is indicated by the connection that gives the most output.
2. The rules of impedance matching must be observed for most efficient operation. For example, if the speaker has an impedance of 8 ohms, each of the two parallel-connected outputs should be of 16 ohms impedance.

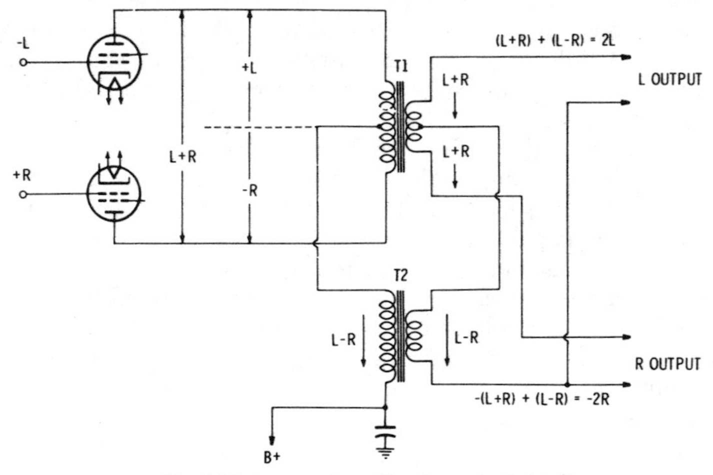

Fig. 5-28. Improved combination output circuit.

Actually, there is some doubt as to real need for such a parallel connection during monophonic operation. Many audiophiles with good stereo systems say that they find playback of monophonic recordings and broadcasts very pleasing through two channels, even though there is no stereo effect involved. Of course, paralleling does provide a single output with the power capabilities of both channels, and this may be useful under some special circumstances.

Another switching function sometimes found in the power-amplifier portion is the "quasi-stereo" switch. This switch permits either the L or the R signal alone to be fed to both channels.

Some output circuits also include a channel-reversing switch, which might more accurately be called a "channel-swapping" switch. Operation of this switch causes the speaker that was receiving the L signal to receive the R signal, and vice versa. This is useful when the listener is checking to see if his L and R connections through the system are correct. For example, if he makes his own tape recordings, he may have an error in the connection of tape-machine outputs, and this can be quickly corrected by the "reversing" switch. Otherwise, once your pickup cartridges and tuner outputs have been properly connected, there is no further use for this switch.

Driver-Amplifier and Phase-Inverter Stages

The amplification of the most sensitive power-amplifier stage is insufficient to provide full output from the signal of the average input device such as a phono pickup, microphone, or tuner. Additional amplifier stages are added to all power-amplifier stage designs to provide the required gain. The low driving power requirements of tetrode tubes operated class AB, as is standard in modern amplifier design, allow use of a voltage amplifier, a driver, and inverter stages with low-power outputs. These can be operated on a linear portion of the tube characteristic and produce sufficient gain per stage in such a manner that usually two to four stages, according to the particular design, can produce more gain than required.

The voltage amplification is usually accomplished in one, two, or three of these stages, and then the signal is fed to inverter and driver stages to provide two components of input signal 180 degrees out of phase with each other to drive the push-pull power-output stage.

There are various arrangements for providing linear, undistorted, low-level power for driving the final amplifier. Block diagrams of various arrangements are shown in Fig. 5-29.

Sensitive, high-gain tetrode power-amplifier stages require comparatively small driving power and are usually connected directly

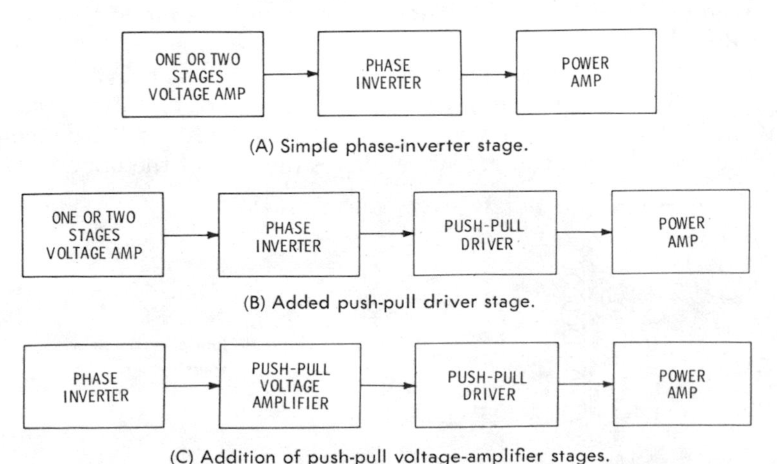

(A) Simple phase-inverter stage.

(B) Added push-pull driver stage.

(C) Addition of push-pull voltage-amplifier stages.

Fig. 5-29. Block diagrams of various amplifier arrangements.

to a simple cathodyne phase-inverter stage as shown in Fig. 5-29A, even though the gain of the cathodyne phase inverter is less than unity. In this case, design requirements are simple, as the cathodyne is fairly foolproof and develops negligible distortion.

Where the power amplifier has a low amplification factor, as when straight class-A operation is used, more power must be provided, and a circuit such as that in Fig. 5-29B may be used. The advantage of this will be found in the power-amplifier operation only because each added push-pull stage must be balanced, stable, and most carefully designed to eliminate distortion to feed the final amplifier stage properly.

The circuit shown in Fig. 5-29C is used in one of the top-brand amplifiers and has been designed with considerable care to achieve balance throughout the push-pull voltage-amplifier stages. Such a design is practical for use in laboratory-type equipment.

Phase Inversion

The signals at the grids of the power stage must be as nearly 180 degrees out of phase and equal in amplitude as possible at all audio frequencies. Since the input and preamplification stages are almost always of single-ended design, it is necessary to split the amplified signal into two equal components, one lagging the other by exactly one half-cycle (180 degrees).

The simplest device for splitting the phase is a transformer with a single input winding and a center-tapped output winding (Fig. 5-30). Such a transformer will provide the required phase split and equality of amplitude but at a considerable expense. An input transformer of this type must be specially designed and constructed. It is large and heavy and, generally speaking, impractical as compared with the simplicity and low cost of resistance-coupled phase-inverter circuits such as those shown in Figs. 5-31 and 5-32.

Fig. 5-31 shows various commonly employed tube-type phase inverters. The circuit shown in Fig. 5-31A utilizes a self-balancing arrangement. The incoming signal drives the grid of the upper tube,

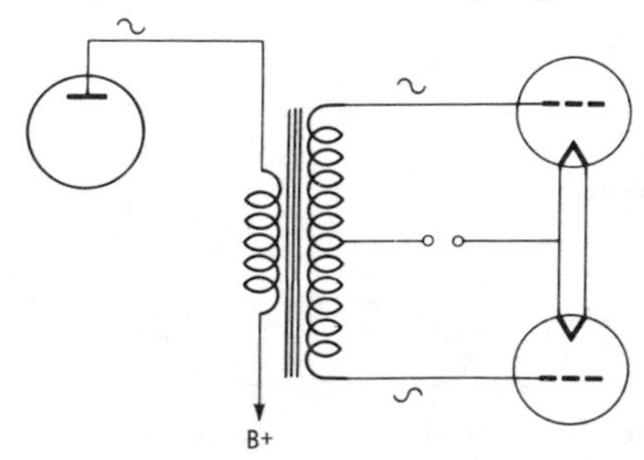

Fig. 5-30. Phase inversion with a transformer.

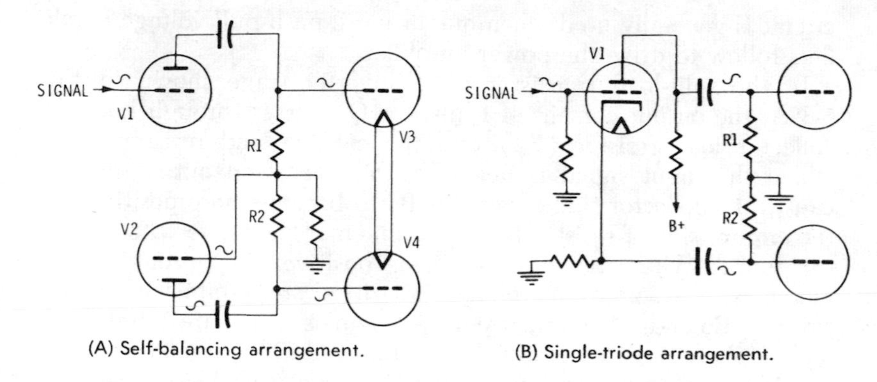

(A) Self-balancing arrangement.

(B) Single-triode arrangement.

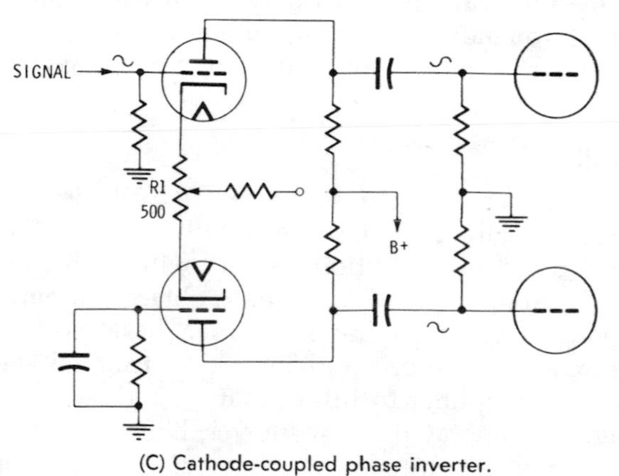

(C) Cathode-coupled phase inverter.

Fig. 5-31. Phase-inverter circuits.

V1. The output of V1 drives the upper tube (V3) in the power amplifier, and a portion of this same output feeds through to drive the grid of V2. The output of V2 is 180 degrees out of phase with the input; therefore, the phase has been inverted as required.

The cathodyne circuit shown in Fig. 5-31B is the most commonly used phase inverter in tube-type equipment because of its simplicity and because it is self-balancing. The signal fed to the grid of V1 is reversed 180 degrees at the plate but is in phase at the cathode, thereby splitting the phase in accordance with the requirements for driving the power-amplifier stage. The same value of current flows through R1 and R2 (equal values of resistance), therefore causing balanced output to each tube in the power amplifier.

The circuit shown in Fig. 5-31C is called a cathode-coupled phase inverter. This circuit has provision for equalizing the outputs from the two sides by adjustment of R1 (with signal applied). This latter circuit is generally used when one or more push-pull voltage amplifiers follow to drive the power amplifier.

In the split-load transistor phase-inverter stage shown in Fig. 5-32A, the output current of transistor Q1 passes through both the collector load resistor (R4) and the emitter load resistor (R3). When the input signal is negative, the decreased output current causes the collector side of resistor R4 to become more positive and the emitter side of resistor R3 to become more negative with respect to ground. When the input signal is positive, the output current increases, and opposite voltage polarities are established across resistors R3 and R4. Thus, two output signals which are 180 degrees out of phase with each other are produced. This circuit provides the 180-degree phase relationship only when each load is resistive and constant throughout the entire signal swing. Other transistor phase-inverter arrangements are shown in Figs. 5-32B, 5-32C, and 5-32D.

Voltage Amplifiers

The gain of power-amplifier stages is relatively low, and phase-inverter stages usually have less output than input. The average amplifier requires several additional stages with high gain per stage to bring the signal up to sufficient level to obtain full output according to the power-stage capabilities. Further, basic power amplifiers usually require a separate preamplifier with output voltages of 1 to 2 volts to drive the amplifier to full output.

These stages are operated as low-power, high-gain voltage amplifiers. Voltage amplifiers are almost always resistance coupled, operating on a small and straight portion of their characteristic curve and providing considerable voltage gain with negligible distortion. Pentodes are used for highest gains, and triodes are used for greatest stability. One to three voltage amplifiers are usually contained in a basic amplifier, according to the sensitivity and gain requirements.

Good voltage-amplifier design can be achieved with negligible distortion of the signal, but it is not so easy to keep down internal noise and hum pickup. Special design arrangements have been made to reduce noise and hum. Hum can be reduced by use of special isolated filament circuits with short and shielded leads carrying alternating currents. Another attack is to use direct-current filament supplies. Low-noise design has also been improved by use of special input tubes and circuits and other components, such as low-noise transistors, resistors, and controls.

Negative Feedback

In the dynamic operation of a modern basic amplifier, there is usually provided feedback of a portion of the output signal to the

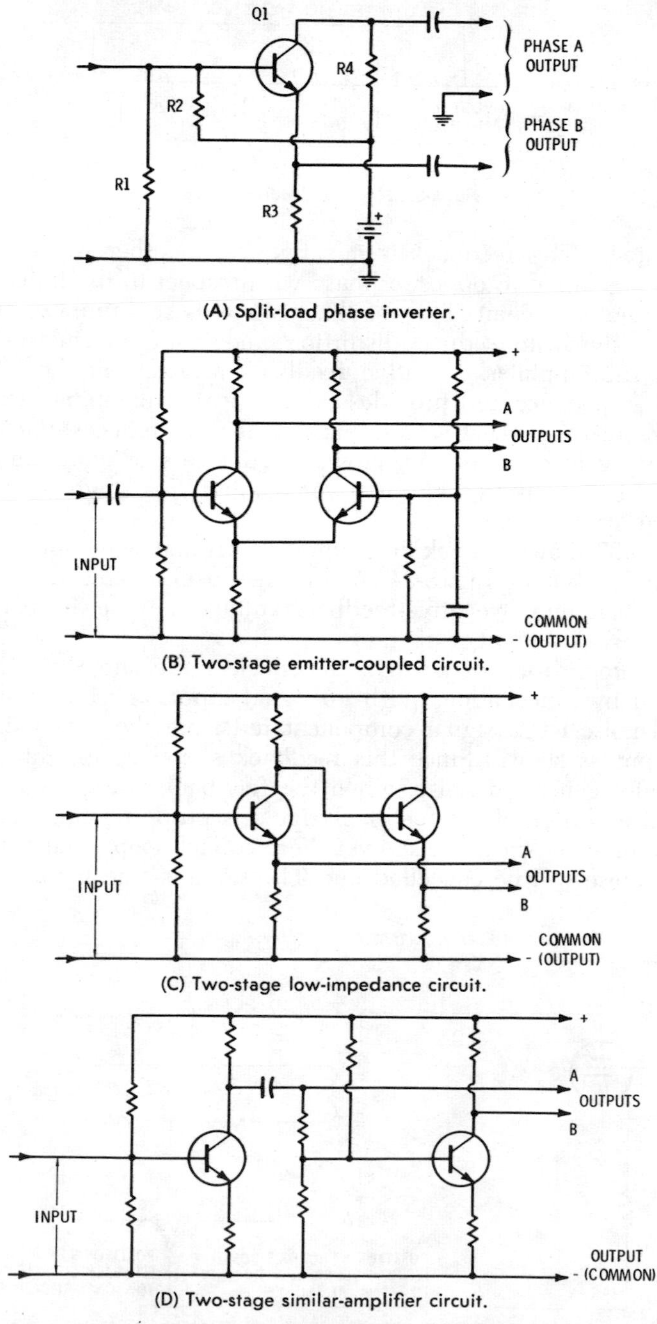

(A) Split-load phase inverter.

(B) Two-stage emitter-coupled circuit.

(C) Two-stage low-impedance circuit.

(D) Two-stage similar-amplifier circuit.

Fig. 5-32. Transistor phase-inverter circuits.

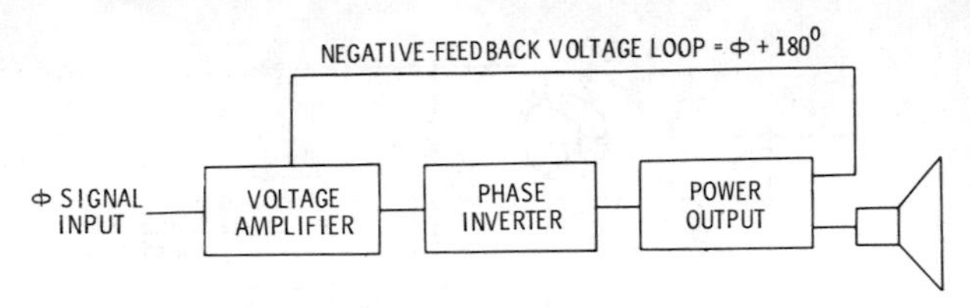

Fig. 5-33. Negative-feedback loop.

input stage. This feedback is negative, or, in other words, a cancelling signal fed in opposite phase with respect to the input signal at the feedback point. This feedback signal is superimposed on the original signal to reduce distortion and noise originating from within the amplifier. Negative feedback also lowers the effective output impedance and provides better speaker damping. The only disadvantage of negative feedback is that it reduces the overall amplifier gain considerably. However, this loss of gain can be recovered by use of an additional voltage-amplifier stage as found in modern arrangements.

Fig. 5-33 shows a block diagram of an arrangement using a negative-feedback loop. Fig. 5-34 shows a feedback circuit arrangement in one stage only. Negative-feedback voltage may be fed back over one or any odd number of stages.

With application of negative feedback, noise and distortion are reduced by cancellation. With no signal input, any internally generated noise has a signal component fed from the output back to the input as shown. Since this feedback signal is opposite to the internally generated noise, when the feedback is adjusted so that the noise portion of the feedback signal is proportionately equal to the average generated noise over the feedback loop, a large portion of the noise will be cancelled out. The same is true for hum inter-

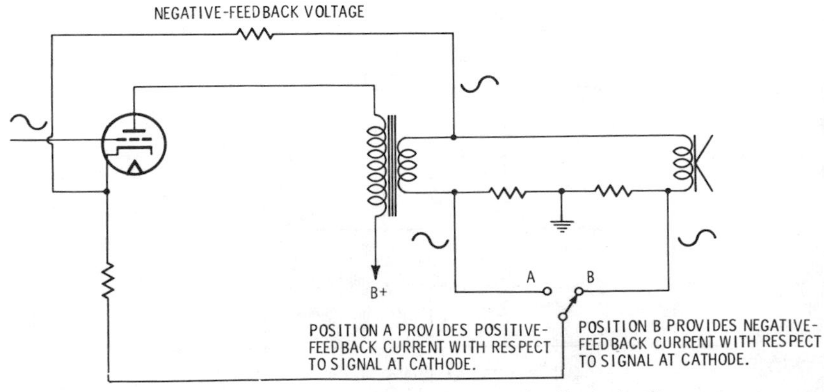

Fig. 5-34. One-stage voltage and current feedback circuit.

nally picked up, providing the phase of the hum is suitable for suppression.

The same principle of cancellation applies to distortion of the original signal originating within the amplifier at any point within the feedback loop. An internally distorted signal waveform will be pulled back toward the original signal shape. The original signal will be reduced, but additional components added internally to the original signal will tend to cancel out.

The effective output impedance of the amplifier is lowered by means of the variation in the amount of negative feedback applied in relation to the output load. The negative feedback is taken off the output of the amplifier in the same manner as the speaker takes power. When the speaker is not taking power, the relative output signal voltage rises, and consequently there is proportionately more negative-feedback voltage. More negative feedback reduces the gain and lowers the output signal, tending to compensate for the unloaded condition of the output. In this manner, negative feedback has the same effect as lowering the output impedance of the system.

Negative feedback of an amplifier is rated in dB. The feedback dB rating is equal to the reduction in gain of the amplifier caused by the negative feedback.

Positive Feedback

Just as negative voltage feedback lowers the output impedance of an amplifier system, positive-feedback voltage will increase the effective operating output impedance. Positive-feedback voltage, if applied in the same manner but opposite in phase to negative feedback, will cause system oscillation at random audio frequencies.

Current feedback applied from the output of the amplifier to the cathode of an intermediate amplifier stage will cause opposite effects on output impedance. Positive-feedback current will lower the effective output impedance, and negative-feedback current will raise the effective output impedance of an amplifier. Positive-feedback current will also increase distortion in a manner opposite to the cancellation principle of negative feedback.

Variable Damping

This feature of amplifier design does not improve amplifier performance. It is incorporated in some amplifiers to provide better speaker operation. Variation of the effective output source impedance of an amplifier, as can be provided by voltage and current feedback controls, will provide an adjustment for best loading and damping for any particular speaker arrangement. Fig. 5-34 shows how simple combinations of voltage and current feedback circuits with an amplifier can be arranged to obtain any value of output impedance and damping factor desired. In Fig. 5-34, when the switch is at position A, positive-feedback current and negative-feedback voltage are fed back to the input to lower the output impedance and increase the damping factor of the system. When the switch is at position B, negative-feedback current and negative-feedback voltage are used to raise the output impedance and lower the damping factor. Fig. 5-35 shows a circuit arrangement used in a popular amplifier kit to provide continuously variable damping designed to provide complete dynamic matching of the amplifier to any speaker according to individual taste.

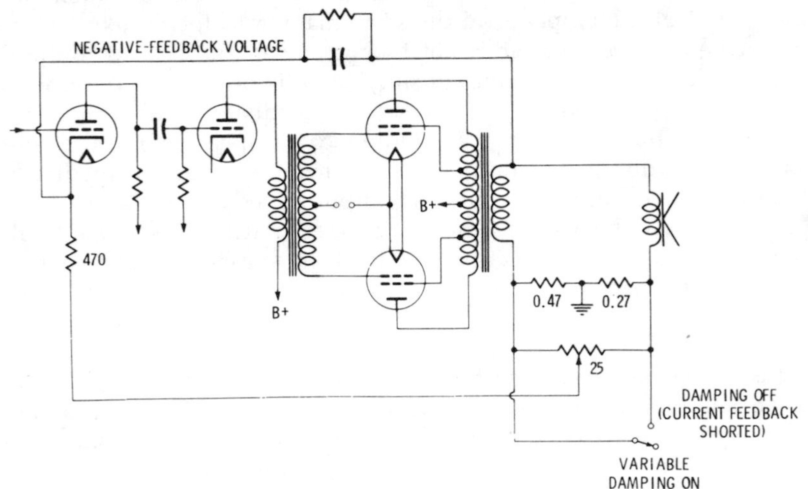

Fig. 5-35. A variable damping circuit.

PREAMPLIFIERS

Basic amplifiers usually require about 1 volt of signal into a load resistance of 100,000 to 300,000 ohms to drive them to full output. Tuners usually provide this amount of drive, so a preamplifier is not required to increase gain when only tuners are used.

High-output phono cartridges and microphones provide sufficient output (up to 1 volt) to drive some basic amplifiers, but high-fidelity phono pickups and microphones of the magnetic or dynamic type give considerably lower outputs. The lower the output, the easier it is to design quality characteristics into these pickup units.

To develop sufficient signal to drive a basic amplifier to full output, additional amplification or preamplification is required before feeding the basic amplifier. This gain is usually provided by a preamplifier, preamplifier-equalizer, or control center. Preamplifier-equalizers and control centers are also designed to provide all kinds of control adjustment, compensation, and filtering to cover about every condition one would desire, as has been described.

A low-output phono pickup feeding a simple preamplifier with one or two stages provides sufficient gain to drive a basic amplifier which needs up to 3 volts for full output, but when various arrangements of compensation filtering networks and feedback circuits are employed to develop special effects these features have circuit requirements that reduce the overall gain by an amount greater than the compensation provided. For instance, if 20 dB of treble boost and attenuation are provided, a total of 40 dB range of control is applied. If the amplifier is to provide a gain of 50 dB under flat operating conditions, then it must provide for an output level of 50 dB at the reference frequency, which is usually around 500 Hz, plus 20 dB at boost frequencies. The full attenuation of the highs or lows will also cause reduction of the overall gain at the reference frequency so that more than 50 dB of gain at the reference frequency at flat setting is required—an average design might require 10 dB of additional reference-frequency gain to cover all conditions of treble and bass attenuation-curve variations and overlap. Therefore, in order to provide for bass and treble boost of 20 dB, that much more gain (10 dB) must be added. If 20 dB of negative feedback is collectively applied, then the gain of the amplifier should be at least 50, plus 10 for overlap, plus 20 for boost, plus 20 for negative feedback as described, or a total of 100 dB of amplification to provide 50 dB of effective gain because of these provisions. There are other provisions in the more complex control units that require even more gain; thus arises the reason for more stages and complexity as more features are incorporated in the unit.

Compensation

There are several forms of compensation provided in preamplifiers and control units. One form of compensation is equalization of the highs (attenuation) and lows (boost) to compensate for emphasis and de-emphasis, respectively, of the highs and lows in the processing of high-fidelity records, as described previously. Current hi-fi records are cut with a curve like that shown in Fig. 5-3, as has been described. Equalizing circuits in preamplifiers and control centers must have provision to compensate for this by amplification compensation curves of closely equal and opposite form, as shown in Fig. 5-4. These curves show equalization for most all records cut today as per Record Industry Association of America (RIAA) standards. However, there are several other curves to which many popular records are still being made from old master recordings. These curves are shown in Fig. 5-36. Equalizers will have to provide for these as long as the older recordings are played. This requires more switching circuits. Another form of compensation is to equalize for emphasis of highs used in fm broadcasting techniques and to compensate for the loudness effect of the human hearing system, as described previously.

All these compensating effects are usually provided by various arrangements of high-pass and low-pass filters. These filters are placed in series with some channel of the preamplifier according to their use, or they are inserted in negative-feedback loops to attenuate by discrimination of feedback at certain frequencies and at the same time to reduce overall distortion.

Record equalization is usually selected by a single rotary switch control. More elaborate systems provide additional finer adjustment of rolloff and turnover frequencies to suit personal taste.

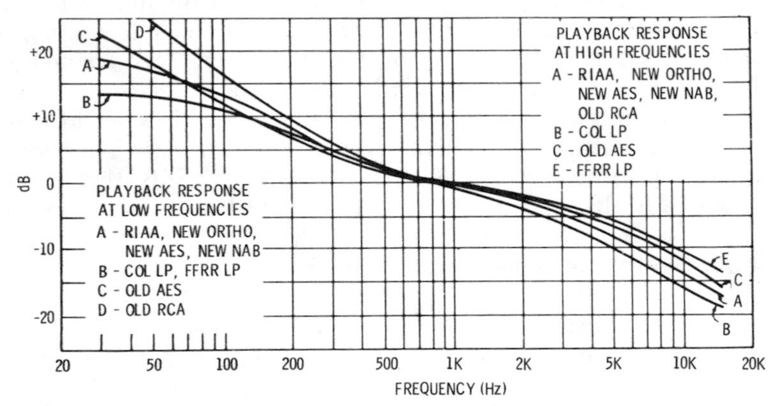

Fig. 5-36. Comparison of standard RIAA playback curve with some curves used before the present standard was adopted.

Equalization rolloff points and slopes are controlled by switching in varying amounts of capacitance, resistance, or inductance in the high- and low-pass filtering circuits as described. Almost all modern records are being cut to one standard, so this provision is mainly useful for compensating for odd records and to give considerable flexibility for experiment or to compensate for other conditions according to the critical listener's taste.

A modern high-fidelity system may have many sources of program material: phono pickup, a-m or fm tuners, television, tape, microphone, and others. Control centers are available with any degree of complexity one might desire to provide conveniently for the many kinds of inputs, to switch from one to the other, apply power to several program sources, control or alter the signal, mix programs, and in general perform the many operations used in advanced high-fidelity techniques.

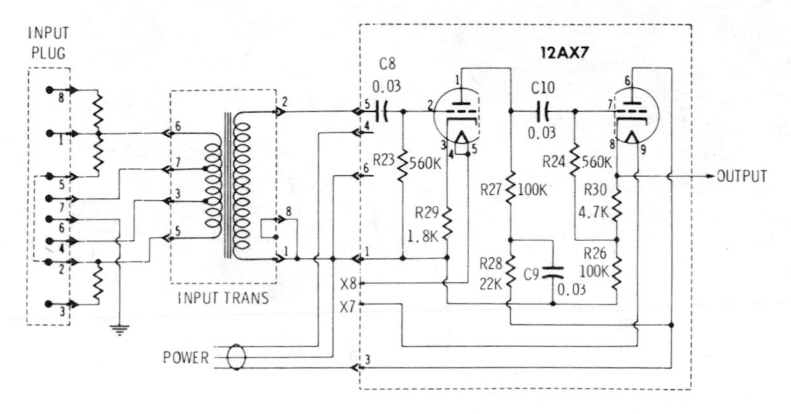

Fig. 5-37. A simple preamplifier circuit.

PREAMPLIFIER CIRCUITS

A simple tube-type preamplifier is shown in Fig. 5-37. The plug-in unit shown is the simplest form of preamplifier including only an input matching circuit, a straight amplifier working on the linear portion of its curve, and an output to match the power-amplifier input.

The purpose of this unit is to match the input device, such as a phono pickup or microphone, to the input circuits of the amplifier and to increase the input signal from such low-output devices as a dynamic microphone sufficiently to drive the amplifier to full output. The prime advantage of such a unit is its obvious simplicity.

The transformer provides a multimatch for input devices having impedances from 50 to 20,000 ohms. The output matches a 100,000-ohm or high-impedance input of a basic amplifier.

Where no control, compensation, or filtering is desired, this unit will provide all that is needed. This is especially suitable to microphone application.

Fig. 5-38 shows a circuit of a simple preamplifier specifically designed for tape. This preamplifier includes (NAB) variable equalization for tape speeds from 1⅞ ips to 15 ips. See Fig. 5-39. The maximum output into a high-impedance load is 6 volts rms. This unit must be fed to an amplifier with a 10-foot or shorter shielded lead. There is a similar model with one stage added and an output transformer to match a 600-ohm balanced line. Taps are provided for 1500-, 150-, and 6-ohm loads. A 600-ohm line up to several hundred feet long can be used.

Fig. 5-40 illustrates the use of an integrated-circuit preamplifier. The external resistor-capacitor network shown is used with the IC to provide equalization. Figs. 5-41, 5-42, 5-43, and 5-44 show pre-

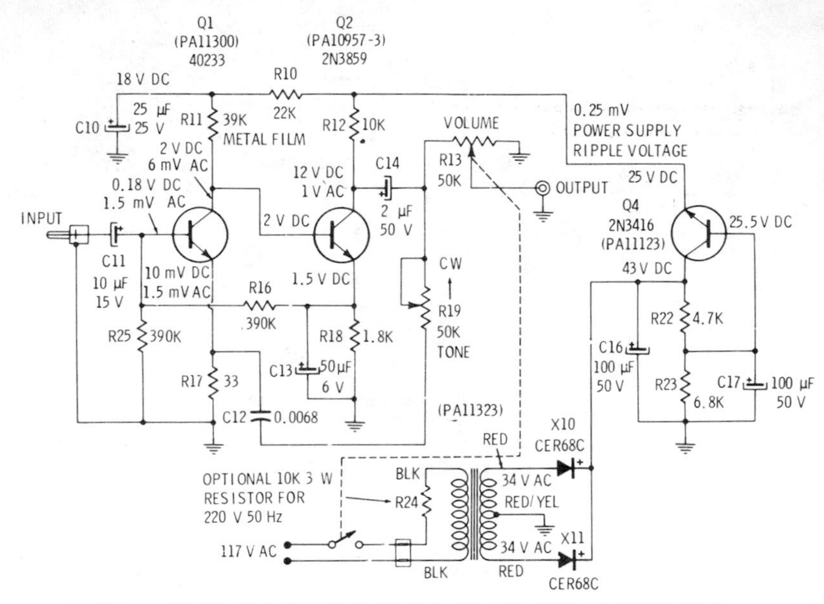

NOTE: 1 - DC VOLTAGES TAKEN WITH 20,000 OHMS-PER-VOLT METER AND SHORTED INPUT.
2 - AC VOLTAGES TAKEN AT 1 kHz WITH 2-MEGOHM AC VTVM.
3 - ALL RESISTORS ±10% CARBON UNLESS SPECIFIED.

Fig. 5-38. A tape preamplifier circuit.

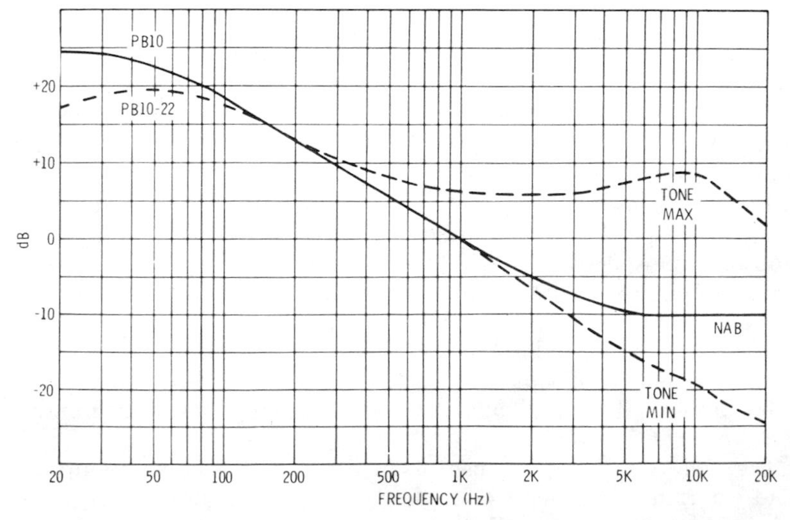

Fig. 5-39. Frequency response in series with head, for circuit shown in Fig. 5-38.

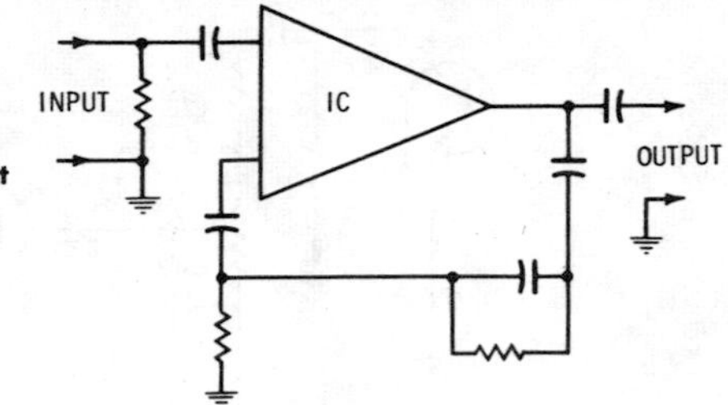

Fig. 5-40. Integrated preamplifier circuit with equalization network.

amplifier circuits of simple design, but each successive unit has progressively more complex circuitry. Fig. 5-41 is a two-stage amplifier with impedance-matching and gain characteristics similar to those of the unit in Fig. 5-37, but in addition it has fixed equalization to compensate for average pre-emphasis and de-emphasis of recordings. The purpose of this unit is to match the output of high-fidelity, low-output magnetic phono pickup cartridges, to increase the signal, compensate for record playback, and match a basic amplifier input. Equalization is achieved by use of R1 to roll off the highs, and C3 and R4 to provide low-frequency compensation. The feedback loop is negative and tends to remove any internally generated distortion or noise as well as provide attenuation for compensation.

This unit also can be used for straight public-address work or microphone input by removing the equalization circuits.

While older records had many variations in emphasis rolloff frequencies, the variations are small enough that the circuit shown in

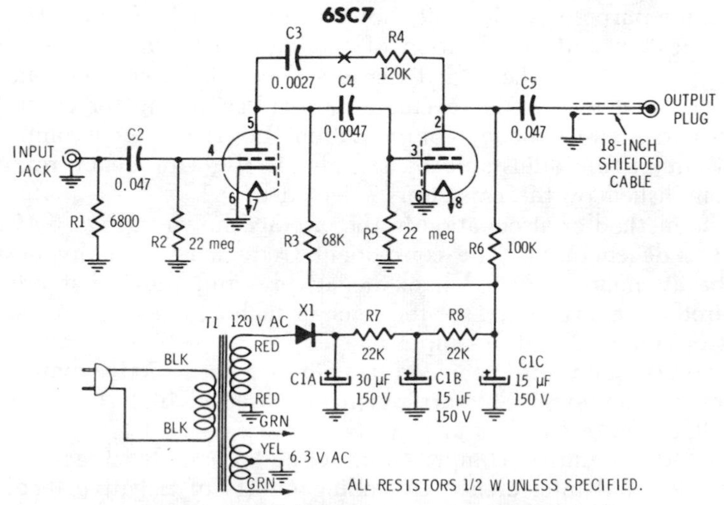

Fig. 5-41. A preamplifier with fixed equalization.

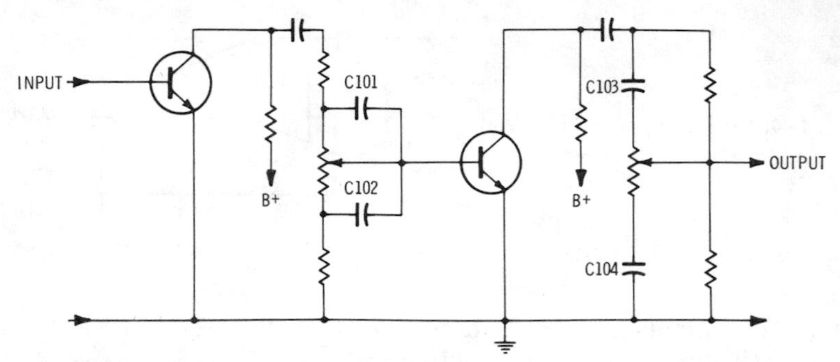

Fig. 5-42. Two-stage preamplifier with separate bass and treble controls.

Fig. 5-41 will do a reasonable job on all of them. (See Fig. 5-36.) However, if it is desired to be able to compensate closely for the original emphasis conditions, more elaborate units with several equalization filter circuits and components that can be switched in are available.

Fig. 5-42 shows a circuit with the same functions but with controls added in a transistor circuit. The tone-control network has two stages, with separate bass and treble controls. The frequencies for boost and cut are controlled by the values of C101, C102, C103, and C104. Loading of output and input must be compensated for when added according to value.

Phono Preamp and Control

The unit shown in Fig. 5-43 is similar to and used primarily for the same purposes as the unit shown in Fig. 5-41. This unit provides amplification and equalization of the signal from a hi-fi magnetic pickup, but it has the added feature of equalization selection. Examination of the circuit reveals additional switching for control of low- and high-frequency compensation. Low-frequency compensation circuits are adjustable by switch S1. High-frequency rolloff is accomplished by the capacitors selected by S2.

The method of attenuation in the circuits shown in Figs. 5-41 and 5-43 is degenerative. The component frequencies that are desired to be attenuated are fed back negatively, and those that are not desired to be attenuated are fed back in such a manner as to achieve proper emphasis and de-emphasis.

Equalization is also achieved in a simpler but effective manner by high- and low-pass filter arrangements inserted in series with the amplification circuits, as in Figs. 5-42 and 5-44.

The degenerative compensation circuit has advantages in the distortion and interference cancelling effects of negative feedback which is used over the entire audio spectrum; that is, if the signal

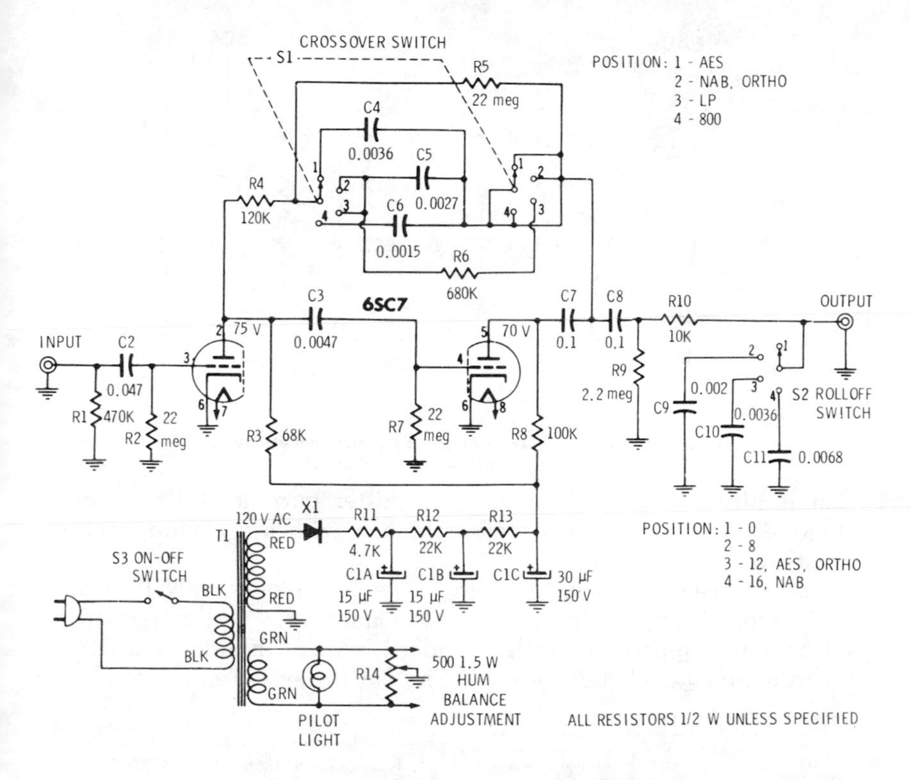

Fig. 5-43. A preamplifier with variable equalization.

is fed back so as to provide 20-dB treble attenuation, the treble is actually attenuated 30 dB and mid-range 10 dB, so that the negative feedback affects the whole spectrum.

The circuit of Fig. 5-45 shows a transistor arrangement which meets the requirements of increasing the signal level from a pickup device to about 1 volt rms. It also meets the requirement of providing compensation, if required, to equalize the input signal for a constant output with frequency when the pickup device is a tuner, magnetic microphone, phono cartridge (mono or stereo), or tape head.

The preamplifier will match the impedances of most magnetic pickups. Input impedance to the preamplifier increases with frequency in switch positions 1, 2, and 3 because of the frequency-selective negative feedback to the emitter of the first stage. The impedances of magnetic pickups also increase with frequency, but they are below that of the preamplifier.

The first two stages of this circuit have a feedback bias arrangement, with R101 feeding to the base of the first stage bias current

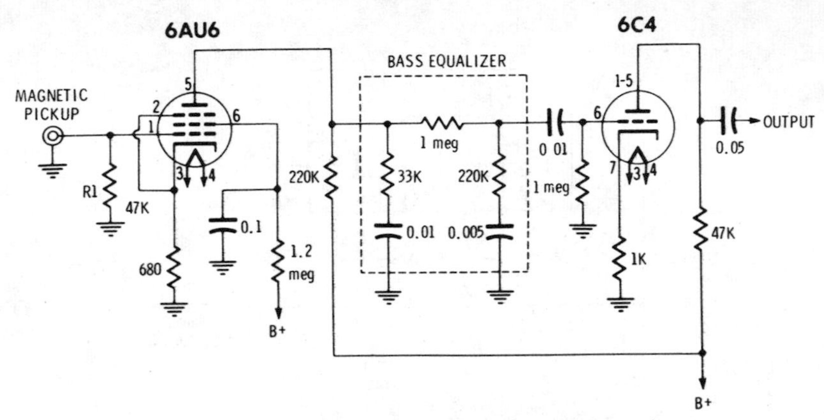

Fig. 5-44. A preamplifier with series equalization-attenuation.

that is directly proportional to the emitter current of the second stage. The output stage is well stabilized with a 5-kilohm emitter resistance.

The ac negative feedback from the collector of the 2N508 (Q2) in the second stage to the emitter of the first stage (Q1) is frequency selective to compensate for the standard NAB or the standard RIAA recording characteristics for tape. The flat response from a standard

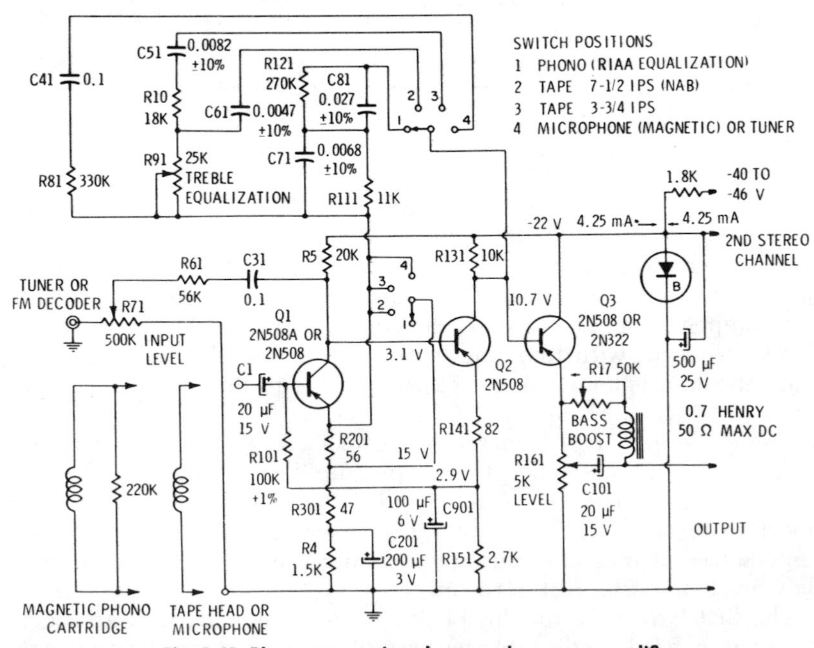

Fig. 5-45. Phono, tape, microphone, and tuner preamplifier.

NAB recorded tape occurs with treble control R91 near midposition. There is about 6 dB of treble boost with the control at 30K, and approximately 10 dB of treble cut with it at zero. Midposition of the treble control also gives flat response from a 7½-ips tape. This treble equalization permits adjustment for variations in input and output components.

A 3¾-ips recording tape head gives an equalized response with ± 1 dB variation from 60 Hz to 12 kHz. Noise level is 76 dB below reference-level output.

A good signal-to-noise ratio (S/N) can be realized with a tape-head inductance of 0.4 henry. The S/N and dynamic range are improved by R141 in the emitter circuit of the second stage, which reflects a higher input impedance from this stage and thus increases the gain of the first stage.

The voltage feedback from the collector of the second stage decreases at lower frequencies because of the increasing reactance of the feedback capacitor in series with the treble control. In the last switch position, capacitor C41 is large enough with R81 to make the voltage feedback (and thus the gain) constant across the audio spectrum. This flat preamplifier response can be used with a tuner, fm decoder, or microphone. The input impedance to the preamplifier in switch position 4 is about 5.5 kilohms, and a 300-microvolt input level gives a 1-volt output (70-dB gain). This sensitivity and input impedance give excellent performance with low- and medium-impedance magnetic microphones. The noise is 65 dB below the 1-volt output level. A magnetic pickup should be left connected at the preamplifier input while using the tuner or fm decoder. This tuner input has a sensitivity down to 250 millivolts.

The RIAA feedback network (first switch position) has capacitor C71 for decreasing the amplifier gain at the higher frequencies. This eliminates the need to load a magnetic cartridge with the proper resistance high-frequency compensation. An input level of 7 millivolts gives a 1.5-volt output.

The emitter-follower output stage of the preamplifier gives a low-impedance output for a cable run to a power amplifier (transistor or tube) and acts as a buffer so that any loading on the preamplifier will not affect the equalization characteristic. The preamplifier output should not be loaded with less than 5 kilohms, preferably about 15 kilohms.

Control-Center Preamplifier Circuit

It has been shown that to obtain more control features and flexibility, more gain is required and therefore more stages, circuits, and components. Quality design becomes increasingly difficult, especially at the input stage of a high-gain preamplifier. This stage is probably the most critical because of noise conditions. Any noise additional to the original signal produced or picked up in the first stage will be amplified in direct proportion to the total amplification of the input signal. If the input signal is low, as it is from magnetic and dynamic phono pickups, the first-stage noise and hum pickup must be proportionately very low not to cause noticeable interference. Hum pickup from ac-operated filaments is such a problem when low-output magnetic pickups are used that several manufacturers of high-grade preamplifiers have a special rectifier and filter to provide

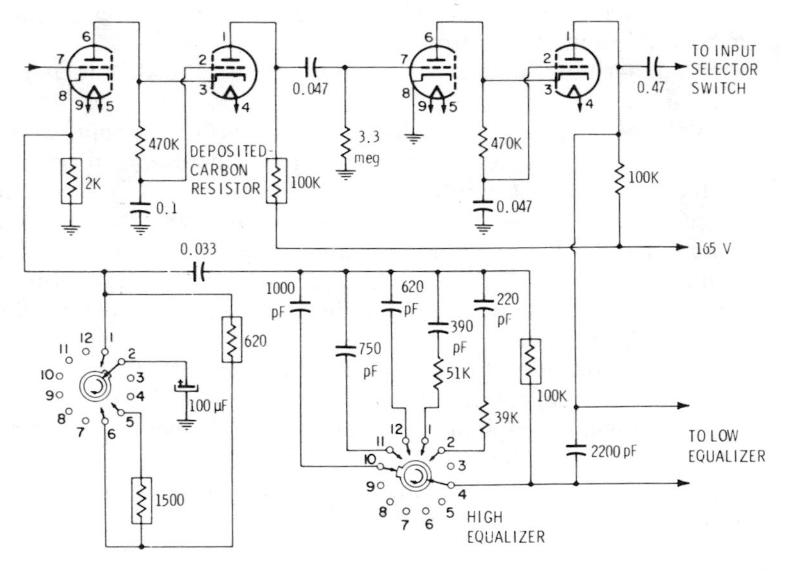

Fig. 5-46. Application of low-noise deposited-carbon resistors to reduce input-stage noise.

current for the first or more stages in the preamplifier. Still further precautions are taken in designs by having a separate chassis for the power supply and a provision for dc filament supply. The real advantage of this is again experienced in the input stage. Further noise reduction is attempted by use of special low-noise resistors in loading-circuit controls and networks where carbon resistors do develop and add noise to the signal as it passes through. Circuits where these special resistors are applied—plate, cathode, and compensating networks—are shown in the schematic diagram of Fig. 5-46 from a Bogen preamplifier design. The rectangular resistor symbols represent low-noise deposited-carbon resistors.

The input section of preamplifiers and controls also contains the switching circuits for changing various input sources as desired. One interesting arrangement having considerable flexibility is found in the master control shown in Fig. 5-47. This unit provides for seven inputs. The input stages provide the proper impedance match to the input unit. How this is accomplished is also shown in Fig. 5-47.

This unit has been conservatively designed on the following principles: (1) that an equipment having higher than necessary output capabilities will give better fidelity when working on a smaller portion of its overall response curve, and, (2) that process quality control by use of close-tolerance components in all critical circuits will ensure that each unit produced will be very similar in quality of operation to a laboratory model.

Naturally this type of unit will cost more than an ordinary model, but it has features and specifications as follows:

Hum: Inaudible at full gain (20 volts output, whereas usually less than 5 volts is required to drive most basic amplifiers).

Noise: 70 dB or more below average low-level signal input from magnetic phono pickup (10 mV).

Intermodulation Distortion: Less than 0.1 percent at normal output (2 volts).

Frequency Response: ±1 dB from 20 Hz to 60,000 Hz.

Controls:

A. Switching
   (1) Low-level microphone
   (2) Two low-level magnetic cartridges
   (3) Fm or a-m tuner
   (4) Television
   (5) Phono pickup cartridge
   (6) Tape playback or monitoring (monitor as you record)

B. Equalization compensation. Six accurate equalizer turnover positions. Six accurate treble rolloff positions.

C. Loudness compensation. A continuously variable loudness control which is so designed as to compensate for the Fletcher-Munson hearing characteristics at various levels. (See Figs. 1-1 and 5-48.)

To obtain consistency and quality, the circuit contains over a dozen accurate low-noise resistors, a dozen low-tolerance silvered mica capacitors, and a high-Q toroidal coil for cutoff filter circuits. Filaments operate on direct current which has three stages of hum filtering. A separate plug-in miniature power supply contributes to reduction of hum to below the thermal noise level so as to be inaudible and not measurable at full gain and with full bass boost. A variable high-frequency cutoff filter is provided for use with tuners to eliminate a-m interstation beat whistle (10 kHz). This circuit also provides positions to quiet undesirable record-surface noise, and tape and fm-tuner hiss.

There are several circuits common in principle to almost all the foregoing control centers. Among these are the bass and treble controls designed to provide from 15 to 20 dB boost or attenuation of bass or treble frequency components separately. This kind of circuit is shown in Fig. 5-47. Treble control R26 (Fig. 5-47) and associated components provide adjustment of the high frequencies by variable high- and low-pass filtering. Bass control R27 and associated components provide a similar effect in the bass regions. These controls are arranged in high- and low-pass filter networks and made variable about a design center which constitutes the flat position when the controls are on dead center. Turning to the left droops and turning to the right boosts high- and low-frequency notes for each control, respectively.

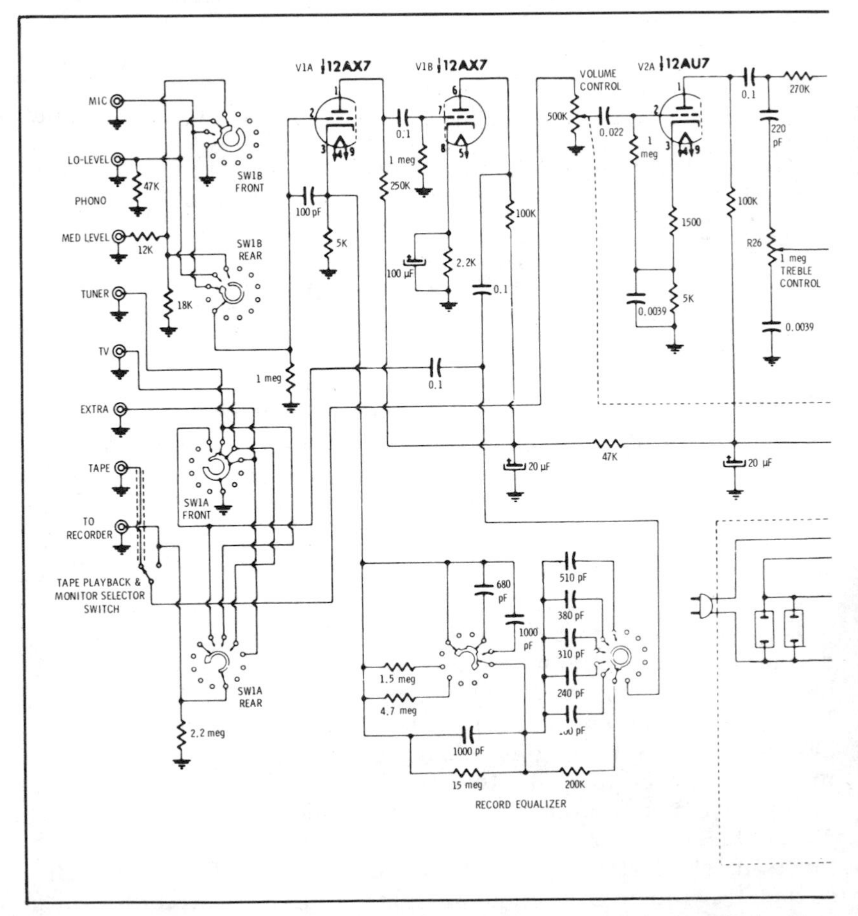

Fig. 5-47. A single-channel laboratory-

Another feature common to most advanced control centers is the convenience of two outputs: one to feed a tape recorder for recording one's own program material, and one to feed the amplifier system. These outputs are usually fed through a cathode-follower circuit arrangement in order to reduce the output impedance. The advantage of the high-impedance input and low-impedance output characteristics of a cathode-follower output circuit is that the low-impedance output characteristic permits the use of long connecting lines between the preamplifier and the basic amplifier without danger of hum pickup or loss of high-frequency components of the signal.

In preamplifiers having direct-coupled output from the plate of the last stage, a short shielded lead, preferably shorter than one or

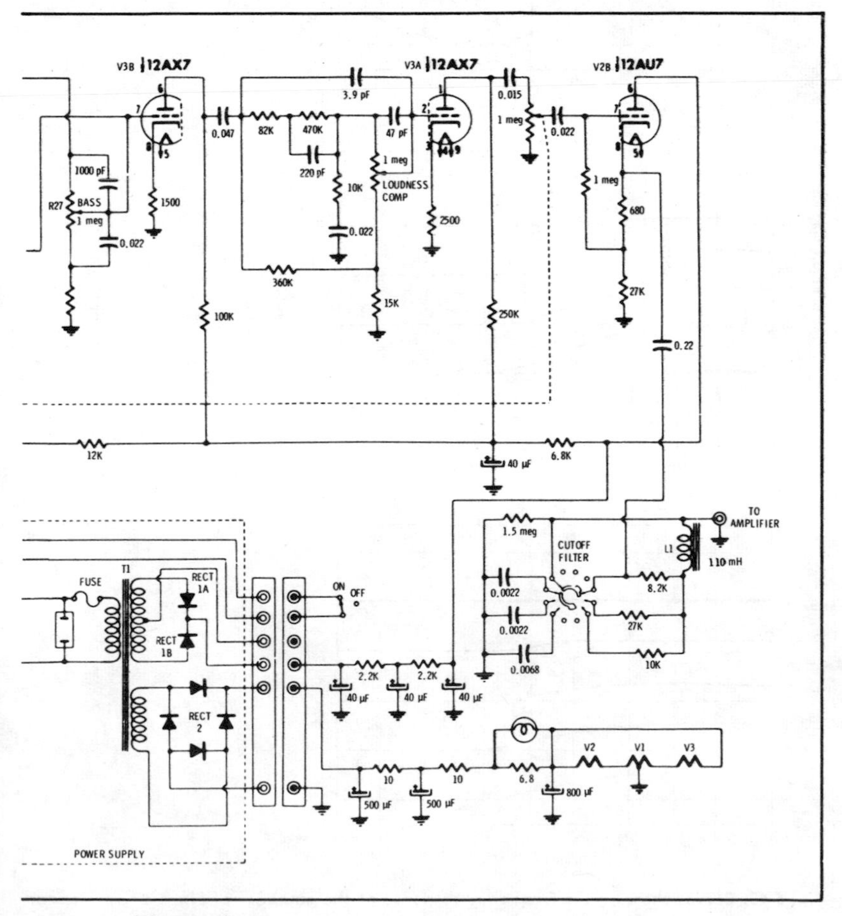

type preamplifier equalizer.

two feet, is desirable because this lead will be sensitive to surrounding magnetic fields which may induce undesired noise components. Also, the capacitance of the lead may attenuate the highs.

A loudness control is usually furnished as a switch for a common setting or as a variable control to compensate for different degrees

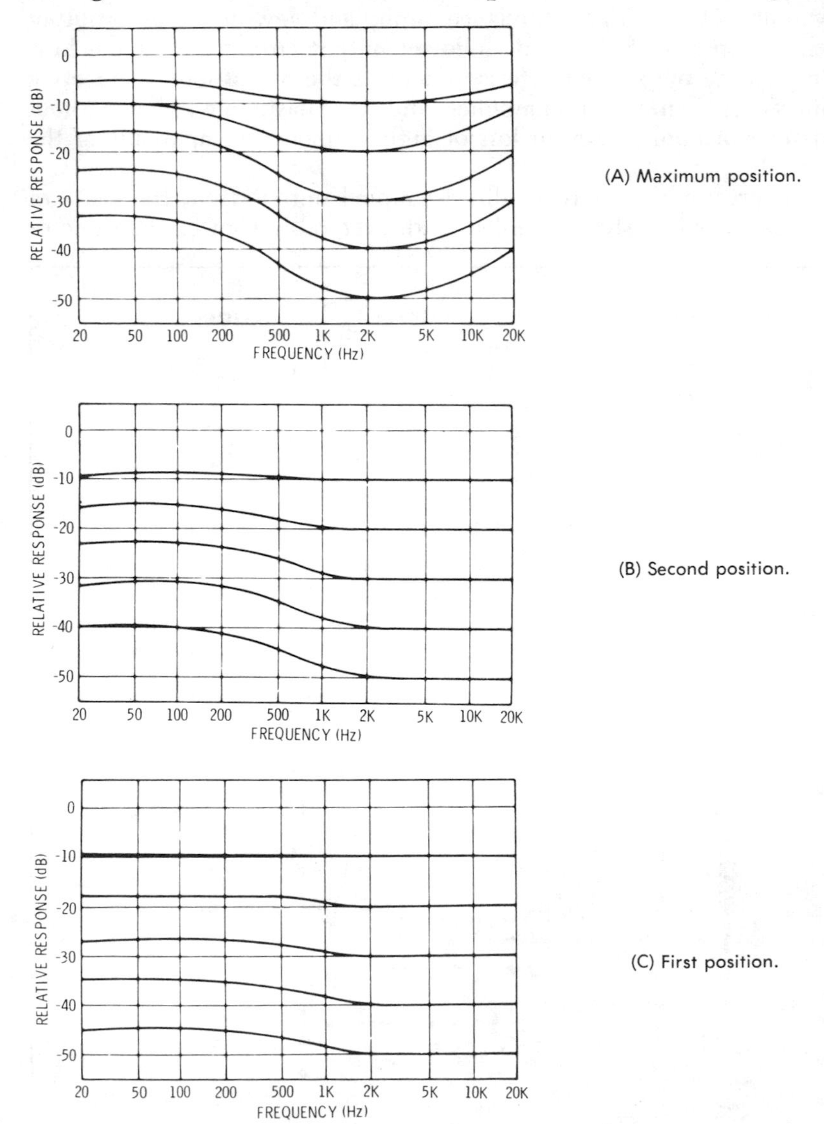

Fig. 5-48. Curves showing the frequency response of the amplifier of Fig. 5-47 corresponding to various positions of the loudness control.

of human hearing response. Refer to the Fletcher-Munson curves at different audio volume levels, as shown in Fig. 1-1.

Common to all units is the volume control, either coupled to a loudness control or separate. This control is usually inserted either in one or two grid circuits or on the output side of a cathode-follower output stage. The latter has the advantage of simple uncritical design in that it does not affect frequency response in any way.

The grid-connected ganged volume control as used in the circuit of Fig. 5-47 prevents overloading by too strong a signal before it reaches the last stages of the system.

Integrated Circuits for Preamplifier Service

A schematic diagram of an integrated-circuit single-channel preamplifier is shown in Fig. 5-49. The circuit of the integrated amplifier array is based on use of one half of an RCA 3048 or RCA 3052 16-lead JEDEC package. Tone controls and provision for balance with a second amplifier have been added. Amplifier A1 provides

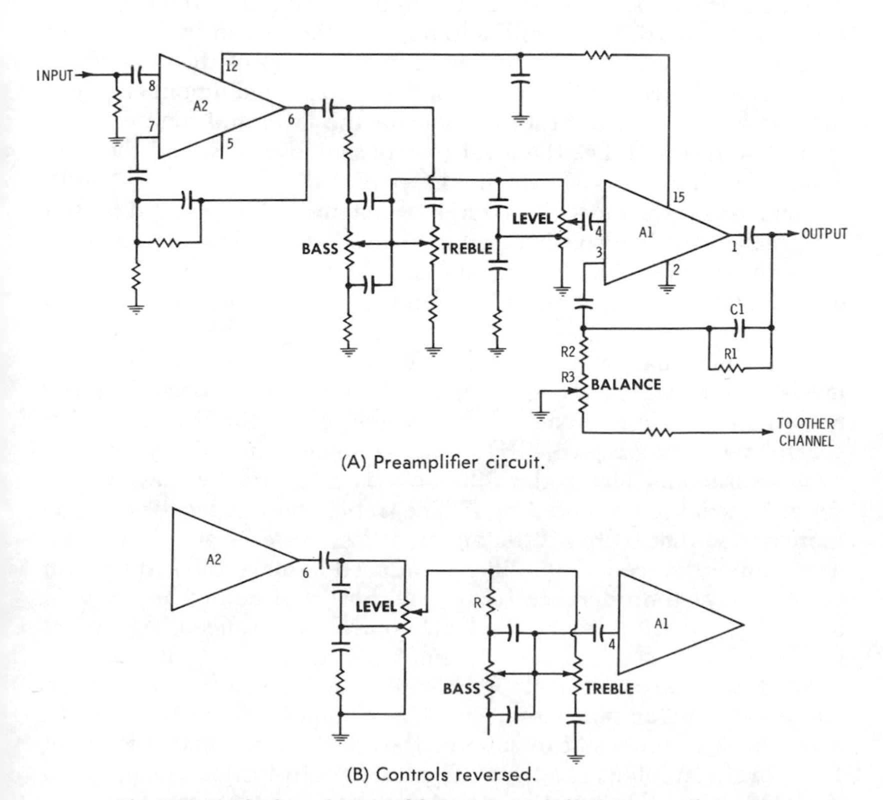

(A) Preamplifier circuit.

(B) Controls reversed.

Fig. 5-49. Single-channel preamplifier using one half of integrated circuit.

a flat response in the audio range, but is rolled off at about 20 kHz by capacitor C1.

Because each amplifier of these integrated circuits has an independent feedback point, it is possible to vary the gain for balancing when a second similar channel is added (for stereo). With this arrangement, the gain of one channel is increased while that of the other is decreased as the balance control is varied. The result is negligible change in level throughout the range of the control.

Resistor R1 in Fig. 5-49A acts in parallel with the feedback resistance already on the integrated-circuit chip to reduce the gain. Resistors R2 and R3 reduce the negative feedback supplied by R1. This same amount of gain could have been programmed by omission of R1 and a corersponding increase in the values of R2 and R3. Analysis of differential-amplifier stages has shown that the equivalent noise source resistance comes from both inputs. Therefore, the low resistor values achieved by the use of R1 result in a decrease in the noise output by about 4 dB. This noise reduction is important when the level control is at the minimum setting and the signal-to-noise ratio is 0 dB. It is possible to improve this circuit under certain operating conditions by alteration of the position of the controls in the circuit. To achieve this, the equalizer and second-amplifier stages may be kept intact, but the positions of the level and tone controls should be reversed; i.e., the level control should be first and the tone controls following, as shown in the partial diagram of Fig. 5-49B. In some cases, the design of complete systems is complicated by the fact that many loudspeakers tend to overload at low frequencies. In addition, there may be acoustic coupling between the input and output that causes an unstable microphonic condition at full gain.

With the arrangement of Fig. 5-49B, it is possible to realize a system which has a great deal of bass boost at normal listening levels, but in which the gain at the bass end is restricted as the maximum level is approached. The value of resistor R in the tone-control network (Fig. 5-49B) is made smaller in this system. At average listening levels, the difference is made up by the reduced series impedance presented by R. The treble-boost capacitor must be increased so that there is little apparent loss in treble at low settings. The gain of the second amplifier is reduced to make the system gain equivalent at the reference frequency, but in so doing the net gain at the bass end is less. As the level control is advanced, the sensation is one of a small change in emphasis from lows to highs.

Feedback Level Control—Fig. 5-50 shows the use of integrated circuits in a preamplifier that employs feedback volume control. In a feedback-volume-control circuit, the gain, rather than the input level, of the amplifier is varied. The level control is located between the output and the inverting input between the two integrated cir-

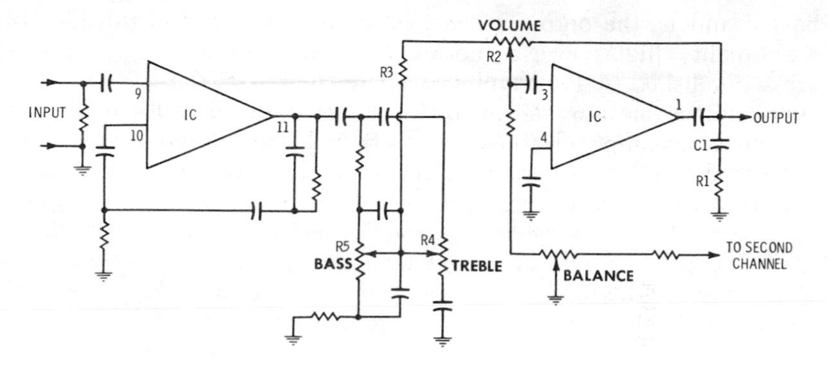

Fig. 5-50. Integrated-circuit preamplifier with feedback volume control.

cuits. At minimum volume, the entire output is fed back to the input. With this amount of feedback, some external stabilization is required; C1 and R1 are used to provide this stabilization.

The maximum gain level of the second amplifier stage is determined by the ratio of R2 and R3. Adjustment of R2 also varies the ratio of feedback resistance to source resistance. The input impedance to the second stage varies from R3 at maximum volume to R3 + R2 at minimum volume. Adjustment of R2, therefore, varies the loading on the preceding tone-control circuit. The circuit shown in Fig. 5-50 exhibits less bass boost at maximum volume than at lower levels, as does the circuit in Fig. 5-49B. To maintain bass boost at higher levels, it is necessary to scale the impedances of the tone-control circuit to lower values.

At minimum volume, the feedback-volume-control circuit effectively places the noise source for the second stage at the output of the preamplifier. Under these conditions, the source resistance seen by the power amplifier is reduced.

The feedback-volume-control circuit requires a special taper of the volume-control potentiometer. A linear taper acts rather like a switch in that it provides very little volume as the control is rotated up to about 90 percent of its rotation. The level then rises very quickly to maximum. The correct taper is a counter-clockwise logarithmic type, i.e., one in which the rate of change of resistance is very fast at first, and then slows down as maximum rotation is approached.

Stereo Integrated Circuits—Silicon monolithic integrated circuits have been designed specifically for stereo preamplifier service (Fig. 5-51). Each package consists of four identical, independent amplifiers that can be connected to provide all the amplification necessary in a dual-channel preamplifier for a high-quality phonograph system. When a signal source is connected to the inputs of amplifiers B and D, the output of each channel may be used to drive a high-quality, high-power audio amplifier; all intermediate functions are acomplished by interconnection with the integrated preamplifier circuits. The top amplifier array is schematically identical with the bottom. Each amplifier of this unit is tightly specified for equivalent output noise under a variety of test methods. The IC's shown will perform to RIAA test specifications for equivalent input noise using one test method for amplifiers A and C, and an appropriately different method for amplifiers B and D. These circuits are supplied in a 16-terminal dual-in-line plastic package and may be operated over a temperature range of −25°C to +85°C.

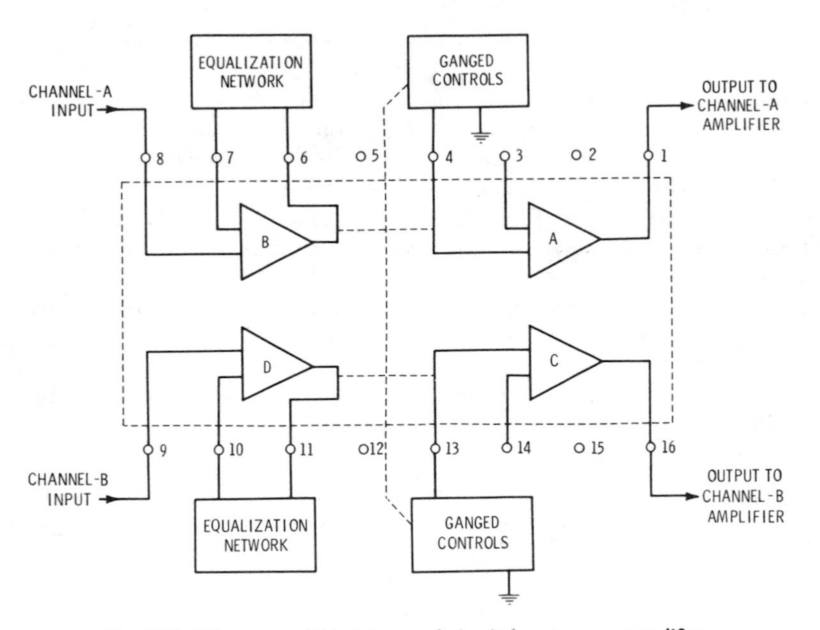

Fig. 5-51. Silicon monolithic integrated circuit for stereo preamplifier.

Fig. 5-51 shows a block diagram of the stereo (2-channel) preamplifier integrated-circuit amplifier array. Each of the amplifiers provides two stages of voltage gain. The input stage is basically a differential amplifier with a Darlington transistor added on one side. The output stage uses a combination of three transistors connected in an inverting configuration.

Input signals to the amplifiers in the array are normally applied to the noninverting input terminal to the base of a Darlington input transistor. A 0.1-meghom resistor supplies bias current for this transistor. The voltage drop across the resistor is small because the base current of the Darlington transistor is very small.

Each amplifier in these arrays may be viewed as an ac operational amplifier in which a fixed resistance is permanently connected between the output and the inverting input. In amplifier D, this resistance is provided by a series combination of resistors.

The amplifiers in the integrated-circuit arrays are normally operated in the noninverting configuration; it is important, therefore, to minimize the capacitance from output to input.

The gain of the first amplifier section of the preamplifier provides an increase of the signal by more than 40 dB. The gain of the first amplifier, however, is not so high that the amplifier overloads at maximum signal levels. The first integrated circuit is connected to the equalizing and compensating networks. The second integrated circuit adds more gain and provides for connections to the ganged controls desirable in stereo operations.

STEREO AMPLIFICATION

The overall amplifying functions of a stereophonic high-fidelity system can also be divided into two or more channels, each having three main parts, as illustrated in Fig. 5-52. The first part, frequently physically separate, is called the *preamplifier* and is similar to single-channel systems. Although it amplifies the signal, its primary jobs are not amplification but: (1) to compensate for frequency characteristics of source material (mainly discs and pickups), (2) to provide adjustment of balance among the signal components of high, low, and medium frequencies (and for speaker compensation and listener taste), and (3) to provide control of loudness and other effects.

The second and third amplifier parts in Fig. 5-52 are usually combined physically into what is known as just "the amplifier." The

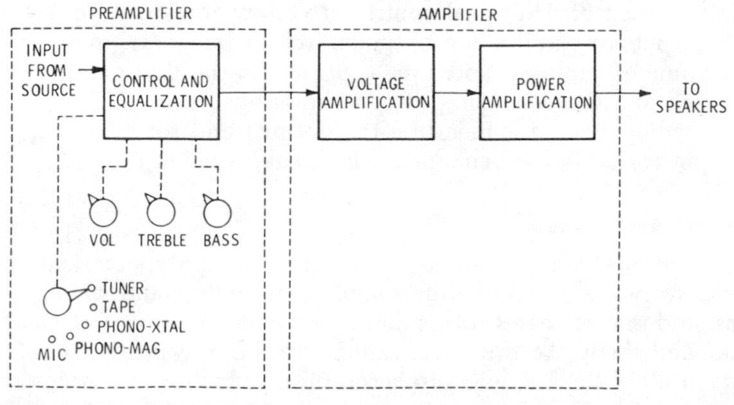

Fig. 5-52. Major divisions and functions of one channel in stereo amplifying system.

voltage amplifier brings the signal voltage up to a level sufficient to drive the power amplifier. In addition, it provides phase splitting so that two equal signal components, 180 degrees out of phase with each other, are available to drive grids of the push-pull power amplifier. The power amplifier is very similar to the ones found in single-channel systems.

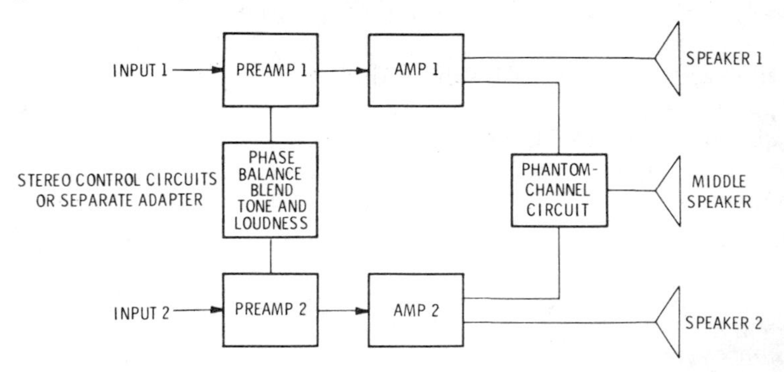

Fig. 5-53. How two amplifier controls are combined for stereo.

Basically, stereo amplifiers are different from ordinary amplifiers because they must amplify two or more separate chanels rather than just one. The first stereo systems used two complete single-channel amplifiers, separate physically as well as electrically. The block diagrams in Figs. 5-53 through 5-58 illustrate how two or four amplifiers, or amplifier channels, are combined for stereo. Special features differing from monophonic arrangements may be summed up as follows:

1. Use of two or four amplifying channels, and combination of these channels physically into relatively compact arrangements.
2. Provision of additional controls, to allow adjustment of relative output amplitudes for optimum stereo effect. In many cases, volume, loudness, and tone controls for the two or four channels are ganged to simplify operation.
3. Arrangements for blending the outputs and for providing output for additional channels and speakers when desired.

Physical Arrangements

The simplest way to provide two or four separate electrical channels is to provide two or four complete amplifiers and speaker systems, and in some cases this is done. However, this makes the system bulky and costly. Accordingly, various ways of physically combining parts or all of the amplifiers to accomplish reduction in size and cost have been devised.

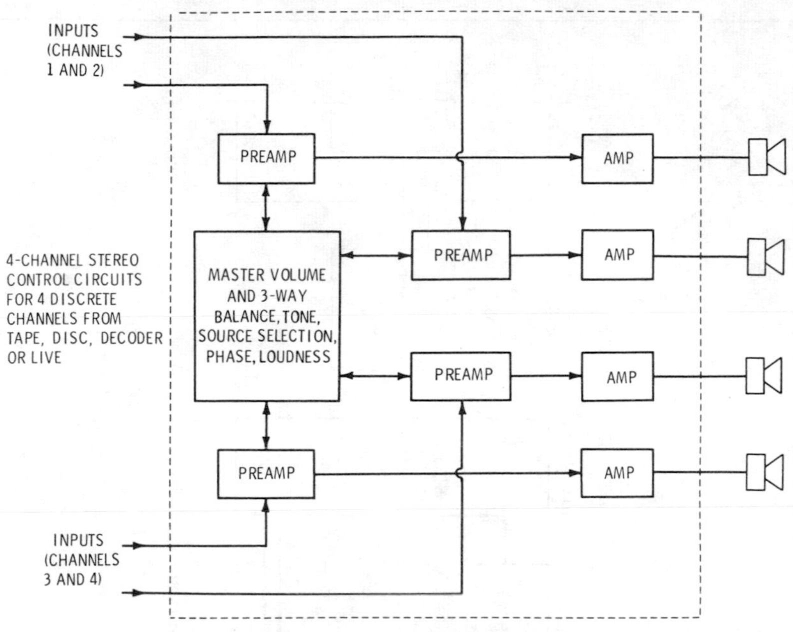

Fig. 5-54. Discrete four-channel arrangement (4-4-4).

Just mounting two or four complete amplifiers on one chassis is sufficient to reduce weight and size considerably. A power supply of twice the current capacity is considerably less than twice as large or heavy, and the use of one, instead of two chassis (even though it is a little larger than either of the two it replaces), makes the weight and size much less.

One interesting way that has been used to combine the two or four channels physically is the use of dual components for all stages

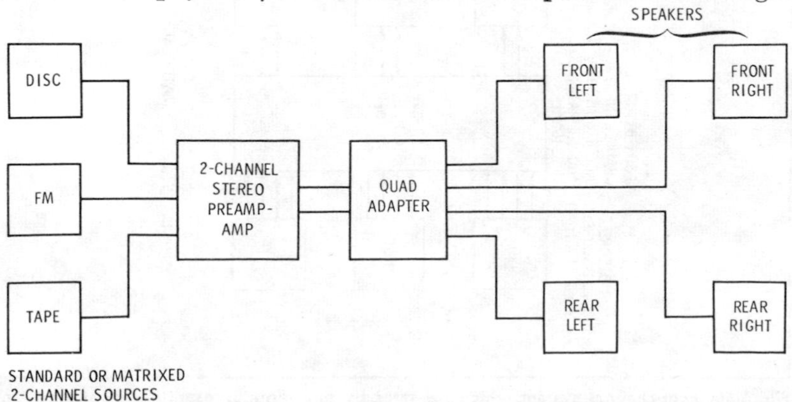

Fig. 5-55. Conversion of two-channel system to four-channel derived stereo (2-2-4).

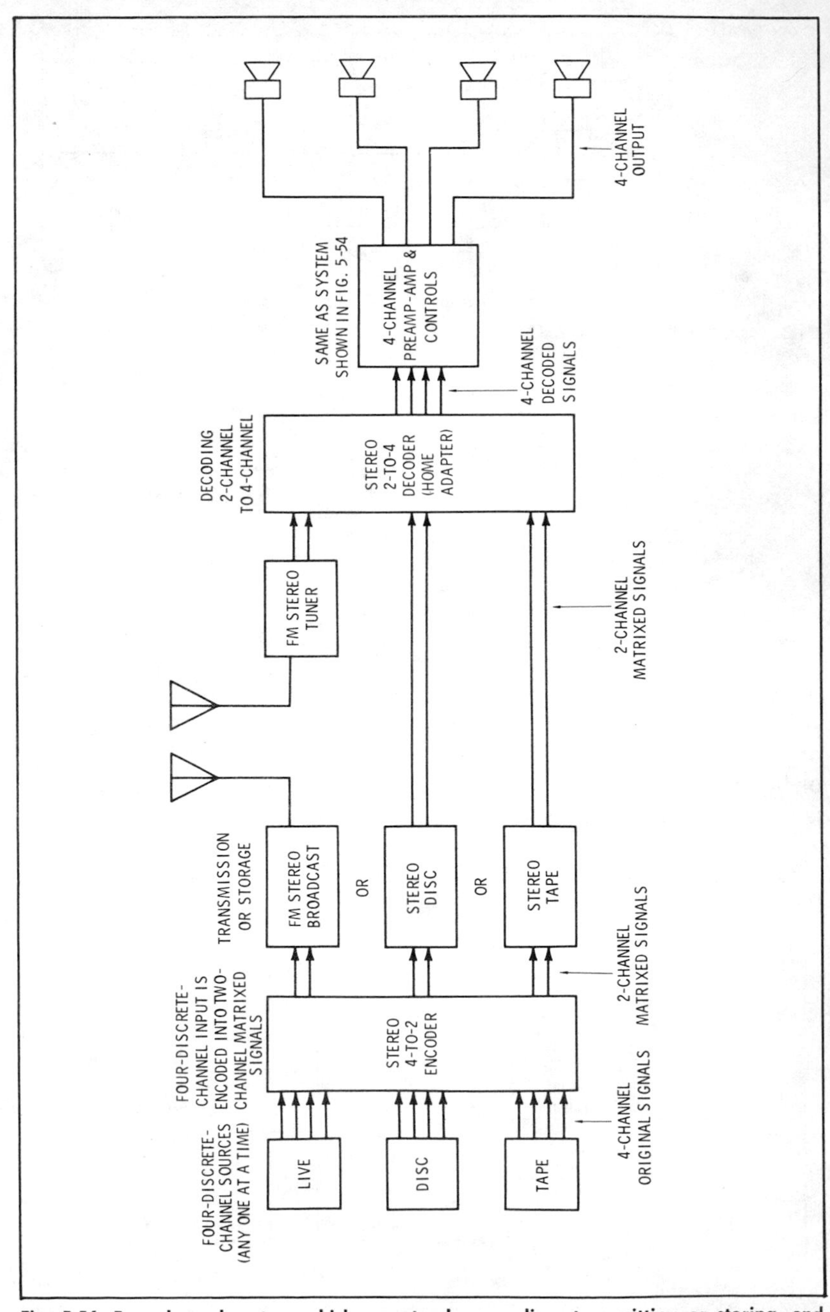

Fig. 5-56. Four-channel system which operates by encoding, transmitting or storing, and decoding program material (4-2-4 system).

except the power amplifier. Just as many functional sections are used, but because each pair of voltage-amplifier stages is in the same integrated-circuit package, space is conserved.

Fig. 5-54 shows the arrangement commonly used to process four-discrete-channel sound from any source. The sources may be microphones, tapes, discs, or a decoder. (The decoder system arrangement is shown in Fig. 5-56.)

Four-channel open-reel tape can provide a fine-quality, impressive performance in the home, but for average operation in home or car, tape in cartridges is most popular. Latest techniques provide for very good reproduction from cartridge tapes when Dolby circuits and chromium-dioxide and similar tapes are used.

Fig. 5-55 shows the system arrangement for a derived four-channel sound system (2-2-4 system). In this system, the sound is decoded, or separated, after amplification.

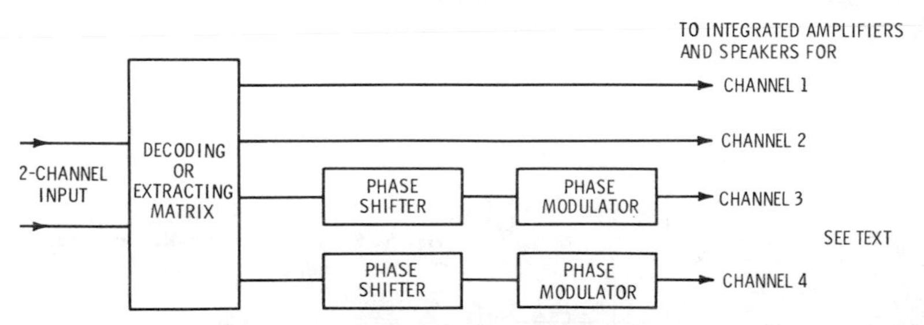

Fig. 5-57. Four-channel synthesized-sound system.

Fig. 5-56 shows the common 4-2-4 system for encoding four-discrete-channel signals into two encoded or matrixed signals and then transmitting or recording the two matrixed signals for further use. To reproduce a likeness of the original sound, it is necessary to decode the two matrixed signals into four discrete signals and then to control and amplify the four signals in the manner shown in Fig. 5-54.

The stereo effect of two- and four-channel sound enhances the effect to the listener in varying degrees depending on the technique used and the program material. Generally speaking, four-discrete-channel sound can provide excellent results under all conditions. The 4-2-4 arrangements are considered next best, followed by the 2-2-4 systems. However, two-channel stereo with top-level components properly selected, installed, and operated will provide almost everything anyone could want for the home.

Fig. 5-57 shows a system arrangement for producing synthesized sound. To produce four channels from two, the synthesizer extracts selected portions of the two-channel signals and then phases and/or delays the extracted components to develop simulated space, distance, and reflection effects somewhat similar in quality to the effects of direct and reflected sounds in a large hall. This system may also provide switching to eliminate the phasing and delay operations and may provide straight decoding of encoded signals in a manner similar to the system shown in 5-56. When operated in the synthesized-sound mode, this system is a "sophisticated 2-2-4 system"; however, when this system is operated as a straight decoder, it is operating as a 4-2-4 system.

Fig. 5-58 shows a common method of converting two-channel stereo into a four-channel system using adapter-decoder-amplifier combinations available. These units vary, but generally they provide for decoding of two-channel matrixed signals, controls for balance and phasing of the four-channel output, plus outputs for four separate channel amplifiers or outputs for two channel amplifiers and provision for internal amplification of the other two channels, as shown.

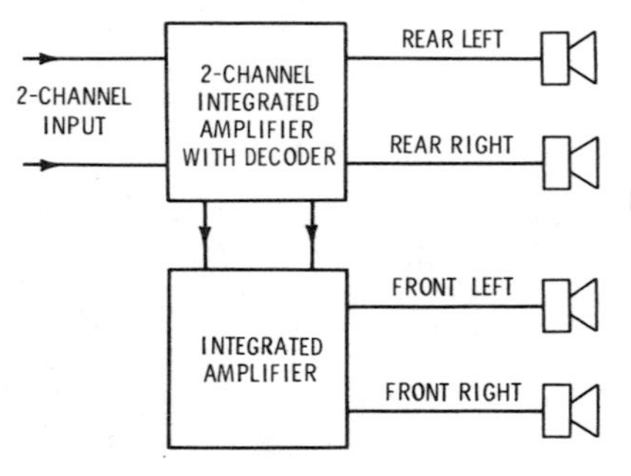

Fig. 5-58. Use of combination decoder and rear-channel amplifiers.

Control Functions

When two or more amplifying systems are used together for stereo, they cannot provide the desired stereo effect unless their operation is coordinated. Although the pioneers of stereo employed completely separate amplifiers, this required operation of multiple sets of controls, including loudness, treble, bass, and equalization selection. Having all these controls separate makes a flexible setup for controlled laboratory experiments, but is not practical for the home stereo system. Therefore, in most commercial stereo equipment, at least some of these controls are combined for greater ease of operation. Let us review the controls having special significance in stereo.

1. *Balance Control*—This is a combination of the volume controls of the two or four channels. It has additional significance compared with monophonic volume controls because in stereo not only the

volume of the sound but also the relative volumes of sound from the two or four channels are important. A balance control for two-channel stereo is usually a ganged potentiometer with one section for each amplifier. When the shaft is rotated clockwise, the output of one channel increases while that of the other channel decreases; at about midposition the outputs are the same. Adjustment of this balance control takes care of differences in source-equipment outputs, gain of the amplifiers, difference in efficiencies and directive properties of the speaker systems, speaker locations, room acoustics, etc. If a stereo program is being reproduced, adjustment of the balance control should make the overall source location appear to move toward the left, right, or center, depending on which speaker output is being increased.

In four-channel stereo control centers, the volume control is usually a master volume control with a four-gang potentiometer to adjust the overall volume of the four channels. In addition, there are three additional sets of ganged balance controls, one for adjustment of the volume balance of the left and right front channels, one for adjustment of the balance between the left and right rear channels, and one to adjust the relative volume balance of the front and rear sets of channels to each other. The master volume control is used to set the overall volume of the four channels, and the balance controls are set to improve the overall effect by balancing the power levels applied to the various speakers so that the desired stereo effect is achieved.

In some cases, both balance and individual volume controls are included. The individual controls are "one-time" adjustments to set the balance control. A monophonic signal is fed through the amplifiers, and the listener stands at an appropriate location for listening; then, with the balance control set at midposition, the individual loudness controls are adjusted until the sound seems to come from a central location with respect to all speakers, and the volume is at the desired level. The volume controls can then be left set at this level and the balance control used to compensate for different types of program material.

2. *Master Gain Control*—Even though the volume controls are adjusted for comfortable volume during the balance adjustment, the listener may still want to adjust overall volume without disturbing balance. This is the purpose of a master gain control. One way of providing such control is to gang the loudness controls through a mechanism that allows either individual or ganged operation. Such a mechanism is now used on some amplifiers. The shafts for the controls are concentric, like those on some television controls. By pulling out on the inner shaft, the operator unlocks the controls for independent adjustment. After the balance adjustment is completed, the

shaft is pushed in again, and the combination used as an overall loudness control.

3. *Blend Control*—This is used to "dilute" the stereo effect. It allows some of the L signal to be fed to the R or center channel or some of the R signal to be fed to the L or center channel. This control is useful mainly when pickup microphones are spaced far apart and there is too strong a left, right, or "hole-in-the-middle" effect in playback.

4. *Quasi-Stereo Switch*—This allows switching either the L or the R signal alone into all channels. The purpose of this is to use all channels, and thus all speaker systems, for a monophonic source; those who have tried this say it makes for more pleasing listening than with a single channel. However, with matrixed (sum and difference) connections, such as those used in tuner multiplex, this is automatically done with monophonic material. Therefore, the main use of the quasi-stereo switch has been with a conventional tuner or monophonic pickup or tape head wherein a single channel is connected to only one of the amplifier input circuits.

5. *Cartridge Paralleling*—This is very important for playing monophonic discs with a stereo cartridge. As explained earlier, rumble in a turntable has a strong vertical component which can be picked up by a stereo cartridge. However, the rumble signal of one channel is out of phase with that of the other channel. Connecting the two channels in parallel for monophonic playback causes the rumble signal to cancel out. Thus, although in monophonic playback both channels reproduce the same monophonic signal, they should be paralleled for rumble reduction. In addition, the signals from the two channels combine to produce a higher-energy output when the parallel connection is used.

6. *Phase-Reversing Switch*—This reverses the phase of one of the stereo signals with respect to the other by simply reversing connections of the two leads carrying the signal of one channel at the input. Although this should seldom be necessary after initial setup of the stereo system, some audiophiles like to be able to change phase. They feel that in some studio programs and in some home recording experiments errors in phasing will crop up.

7. *Input Equalization*—Both stereo channels require the same input equalization for record and pickup characteristics as do single-channel systems. For this reason, input selectors are usually ganged, with one control for both channels.

Two-Channel Stereo Adapters

The controls described in the foregoing discussion for the most part require connection into the circuits of both channels of the stereo system. This means that for the use of two separate amplifying systems not designed for stereo the owner must break into the circuit somewhere and insert these controls.

To simplify the use of separate monophonic systems for stereo, several types of adapters have been developed and are available commercially. A block diagram of such an arrangement is illustrated in Fig. 5-53. The adapter connects to both preamplifiers, either at the outputs or at a jack called "Tape Output." This jack ordinarily allows the signal from the disc or tuner into the preamplifier to be tapped off and fed to a tape machine for recording. In this case, the output can be applied to the stereo adapter.

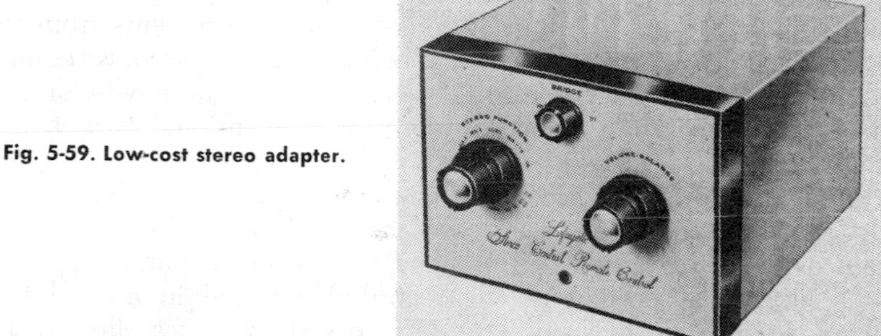

Fig. 5-59. Low-cost stereo adapter.

Courtesy Lafayette Radio Electronics Corp.

An adapter contains the essential controls that are not used in a monophonic (single-channel) system and also provides for coordination of the two signals. The two stereo signals leave the adapter unit and are fed to two standard single-channel amplifiers arranged for stereo. A typical adapter is pictured in Fig. 5-59.

Four-Channel Adapters

Four-discrete-channel systems using four-channel tape, four-channel disc (modulated multiplex disc in the 4-4-4 system mode—see Fig. 5-54), or four live inputs provide optimum four-channel service, but these systems are not compatible for use with current fm broadcast equipment. Therefore, adapters for derived or matrixing systems by which four-channel information can be transmitted or stored in two-channel mediums have been developed. These are compatible with current stereo techniques in all modes (disc, fm, tape, etc.). Four-channel stereo systems may be built by adding components such as adapters, additional amplifier units, and speakers to two-channel systems.

Such adapters may be separate units that produce derived or decoded matrix sound. Or, adapters can be obtained that are combined in several ways with two additional integrated amplifiers (some may be had complete with four built-in channels), controls, and one or more decoders.

Derived-Four-Channel Sound Equipment (2-2-4)—A simple and effective adapter arrangement to convert conventional two-channel stereo sound to derived-four-channel operation (2-2-4) with most conventional two-channel amplifiers uses the Quadaptor system and components shown in Figs. 5-55 and 5-60.

The Quadaptor system serves as a junction circuit arrangement between the outputs of a conventional two-channel stereo amplifier or receiver, and a four-loudspeaker system. It is a passive device, requiring no additional power.

Utilizing the derived system, this unit provides simple but essential circuitry to provide four different signal components from two stereo channels. Accurate electrical balance in the system is required to attain maximum separation. The unit also provides switching for listening to the front speakers alone as conventional two-channel stereo.

The Quadaptor system does not add anything to the original program, nor should it in any way materially alter the overall content or distortion levels of the signals. It is not a synthesizer.

A substantial portion of the recorded material in normal two-channel sources is masked. The derived system makes effective use of all of the signal information, including the masked signal, by extraction of portions of signals from the total signals that have been recorded. This simple technique of ambience recovery in derived sound may reveal an added "presence" or realism to present stereo two-channel output from disc, tape, or stereo fm broadcasts, according to the program material.

The theory of operation of derived sound is described in Chapter 2. Installation is described in Chapter 8.

Matrixed-Four-Channel Sound Equipment (4-2-4)—Electro-Voice, CBS, Sansui, and other makers have systems of four-channel sound that contain true matrixing systems. The home receivers of these systems are designed or adapted to decode the material that has been encoded before broadcast or before recording by the process shown in Fig. 5-56.

As discussed in Chapter 2, two-channel fm transmission, disc, or tape may be used to carry or store the matrixed information in two-channel matrixed format. A decoder at the beginning of the reproducing end of the system converts the two channels back into four with half the separation (more or less) possible within the two-channel system used to transmit or store the matrixed information signals. However, the quality of reproduction of properly arranged and operated systems should be equal to the quality of equivalent

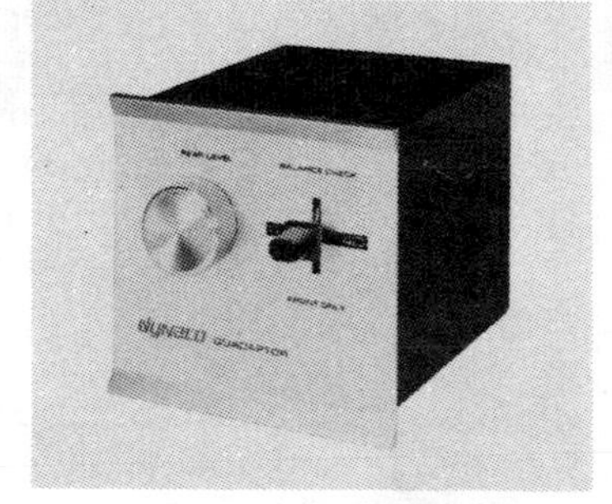

(A) Quadaptor unit.

Courtesy Dynaco Inc.

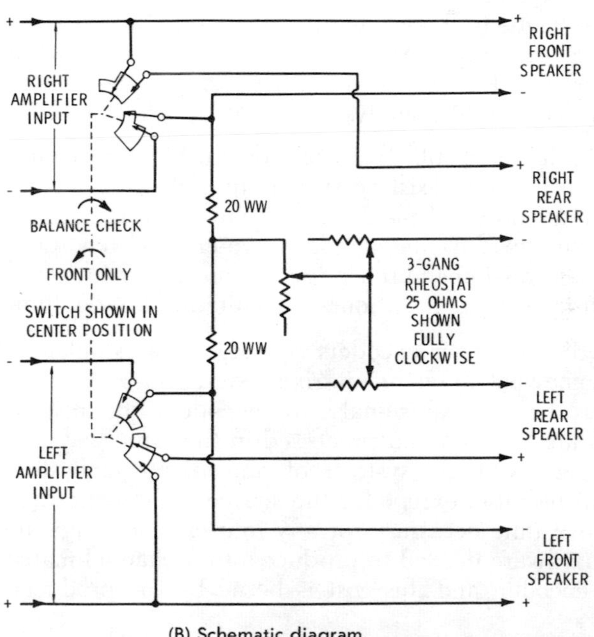

(B) Schematic diagram.

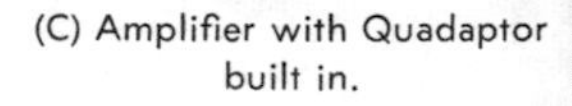

(C) Amplifier with Quadaptor built in.

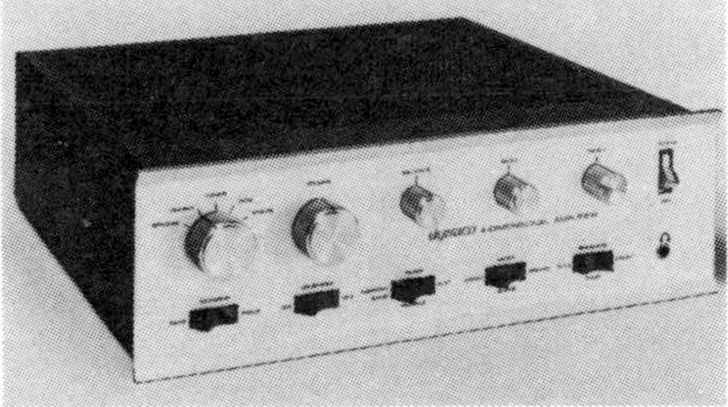

Courtesy Dynaco Inc.

Fig. 5-60. Dynaco Quadaptor.

two-channel equipment, except for the enhancement that four-channel operation may provide. In the matrixed systems, the total separation between any combination of the four output channels cannot exceed the total separation between the two matrixed signals. However, to improve performance, it is customary to increase the separation between the two front speakers at the expense of the degree of separation between the two rear speakers and the separation between front and back. The psychoacoustic relationship is similar to that of live programs in that the ear is considerably more sensitive to and able to differentiate between forward sounds than rear sounds (front source dominance). In addition, image shift of sound to either side of the listener is vague much as secondary vision is vague to the viewer of side images.

The Electro-Voice professional encoder shown in Fig. 5-61 is the heart of the E-V Stereo-4 matrix compatible four-channel 4-2-4 system. This system allows:

1. Cutting and playback of compatible four-channel matrixed records using existing stereo equipment.
2. Production and playback of compatible four-channel matrixed tape recordings on standard two-channel recorders.
3. Transmission and reception of four-channel matrixed fm broadcasts using conventional stereo-multiplex equipment.

The CBS and other encoders operate under similar principles but with somewhat different matrixing designs.

After the original signals are encoded, the resulting matrixed signals are handled and processed in the same way as conventional stereo signals. These systems of transmission and storage are economical because, except for the source equipment and some added mix-down time necessary for any four-channel origination, the only extra hardware needed to produce a four-channel matrixed program is the encoder, and this cost is borne by the producer rather than

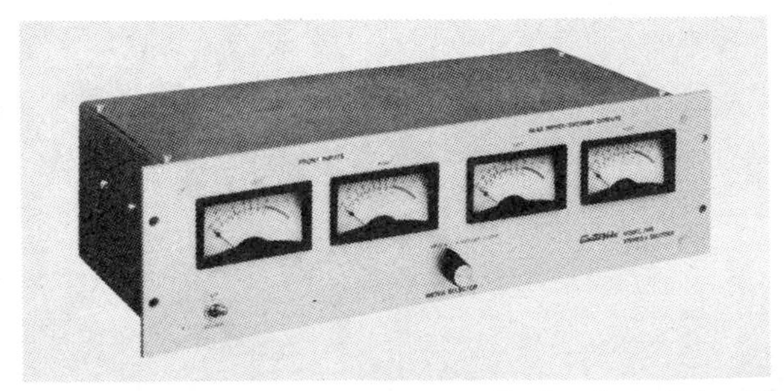

Courtesy Electro-Voice, Inc.

Fig. 5-61. Electro-Voice Model 7445 Stereo-4 encoder.

the consumer. The Stereo-4, SQ, QS, and other systems give the listener complete freedom of choice in building toward four-channel reproduction. Since these matrixed signals are compatible, they may be reproduced in conventional mono or stereo on presently owned equipment. This same equipment can be used and simply added to, to achieve four-channel reproduction.

The matrix system (4-2-4) provides an effective system by using two complementary pieces of equipment, an encoder at the recording studio or broadcast station and a decoder in the home. This system also permits four channels of information to be stored and retrieved with the standard two channels available on tape, disc, and fm broadcast. This is accomplished with negligible change to those specifications which affect audio quality, i.e., frequency response, signal-to-noise ratio, and distortion.

The encoder incorporates a matrix circuit which acts upon the four original stereo channels in a predetermined way to create two composite signals (channels) which may be cut on a record or broadcast by a conventional stereo fm station. The underlying principle of the matrix circuit is algebraic summation; i.e. the four original channel signals are added and subtracted in the matrix network to achieve the two desired composite (matrix) signals. These two composite signals may be reproduced without decoding on conventional two-channel stereo equipment or on monophonic equipment, the same as with present two-channel recordings, or they may be decoded into four channel signals and played back through any four-channel system.

Input and output impedances of the encoder are usually 600 ohms, transformer coupled. On the Electro-Voice unit shown in Fig. 5-61, four front-panel meters allow monitoring of either the four input signals or the two encoded output signals. Internal gains are adjusted so that a 0-VU signal level into all four inputs produces a 0-VU level on the two encoded outputs. Indicated 0 VU is +8 dBm.

For minimum equipment investment and to maintain FCC requirements for proof of performance in broadcasting, the encoder may be installed between a four-channel tape deck and a two-channel console input. A standard four-channel hi-fi tape deck or recorder is sufficient for this purpose.

Encoded matrixed material may be processed additionally without degradation of the recovered four channels. High-quality complementary processing methods, such as signal-to-noise stretchers, may be employed on the encoded signal with equally good results after deprocessing. Noncomplementary processing, such as compressors or limiters or competing equipment lines using different matrix design or arrangements, should be so connected as to act equally on all stereo channels simultaneously, or the gain and other relationships between the two encoded channels will be upset.

The Stereo-4 decoder shown in Fig. 5-62A has a matrix complementary to that of the Stereo-4 encoder shown in Fig. 5-61 to convert the two encoded signals back into four. The two signals entering the decoder are added and subtracted in a predetermined complementary way to derive signals similar to the four original stereo channels which entered the encoder. These in turn are fed to four independent amplifier channels and speakers. The SQ decoders shown in Figs. 5-62B and 5-62C operate in a similar manner and are designed to decode the CBS SQ matrixed signals. The decoder shown in Fig. 5-62D is designed to decode all current

Courtesy Electro-Voice, Inc.

(A) E-V Stereo-4 decoder.

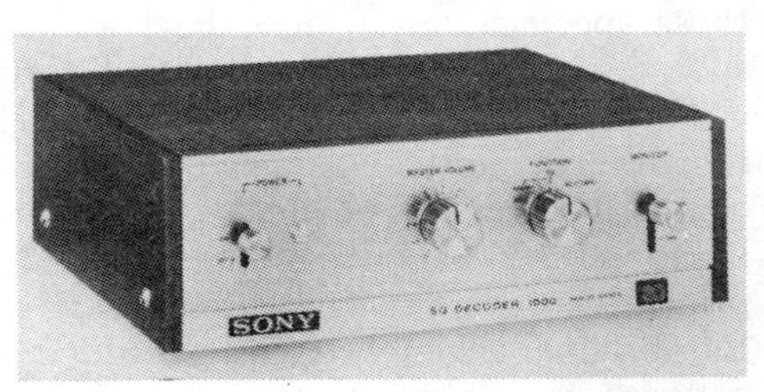

Courtesy Sony Corporation of America

(B) Sony SQ stereo decoder.

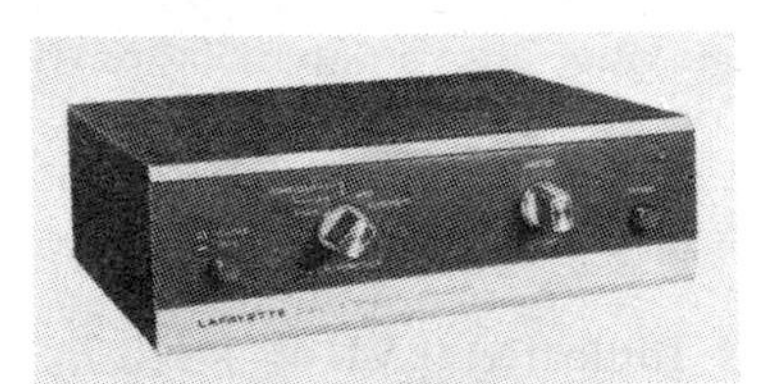

Courtesy Lafayette Radio Electronics Corp.

(C) Lafayette SQ decoder.

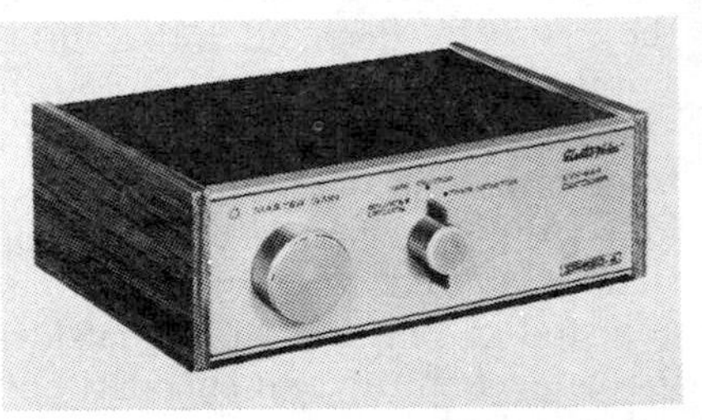

Courtesy Electro-Voice, Inc.

(D) Electro-Voice universal decoder.

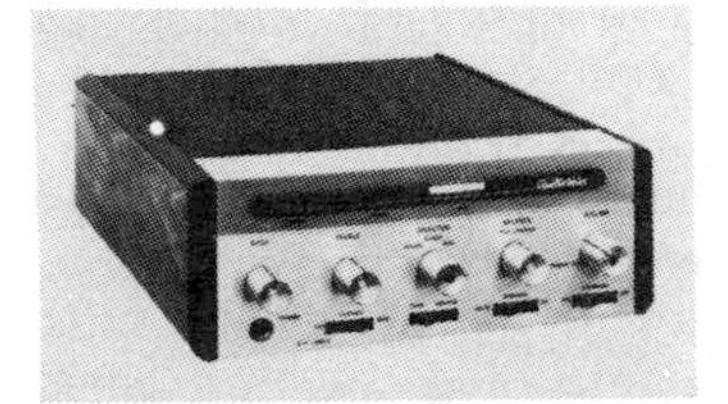

Courtesy Electro-Voice, Inc.

(E) E-V decoder with amplifiers.

Courtesy Kenwood

(F) Kenwood "Surround Sonic" Quadrixer.

Fig. 5-62. Examples of equipment

matrix formulas, including Stereo-4, SQ, and QS, and is offered as a universal decoder.

The second outstanding advantage of the matrix system is that the decoder matrix circuitry may also provide significant enhancement to ordinary stereo recordings and broadcasts, even though they have not been specially processed with an encoder.

A four-channel matrix decoder circuit is so adjusted that it acts to separate into four channels any two-channel stereo signal fed to it. Fortunately, due to the way modern two-channel (unmatrixed) recordings are made with multiple microphones and multitrack recorders, when all these signals are "mixed down" to two channels, a considerable amount of positional information remains which is not discernible when reproduced in only two channels. (Much of this information is there but masked.) This extra positional information often resembles what the matrix encoder would have produced for the recording. This extra information occurs unpredictably and varies from one recording to the next, but it can be said that virtu-

Courtesy Sansui Electronics Corp.

(G) Sansui Quadphonic synthesizer.

Courtesy Sansui Electronics Corp.

(H) Sansui four-channel rear amplifier.

Courtesy Fisher Radio

(I) Fisher decoder–amplifier–cartridge-player.

for four-channel conversion.

ally all modern stereo recordings do contain some degree of these additional signals. The decoder is selective and separates some of this masked information from the front channels and distributes these signal components to the rear channels, thereby allowing signals that would have been masked at the front to be heard from the rear. Thus the decoder acts upon this information in such a way that it delivers a different signal to each of the four speakers. The net effect is that most recordings exhibit a considerable amount of "four-channel information" which becomes audible or discernible only when they are played through the decoder and the signals are distributed to the four speakers.

Fig. 5-57 shows a block diagram of the internal arrangement of a type of decoder-adapter with synthesizers built in to provide simulated spatial and reverberation effects for program material that is, or is not, encoded. There are several variations of equipment for this system available, some of which are shown in Fig. 5-62.

Fig. 5-62E shows an Electro-Voice adapter-amplifier-decoder which provides easy expansion of an existing two-channel stereo system into four-channel operation. This unit contains two complete integrated amplifiers for the rear channels, a universal decoder, and full switching and control facilities for four-channel operation. The unit may be connected as shown in Fig. 5-58.

The Kenwood Surround Sonic Quadrixer is shown in Fig. 5-62F. This unit was designed to convert an existing two-channel stereo system to a four-channel stereo system by functioning in conjunction with the existing two-channel system and with an additional two-channel amplifier and two additional speakers (four channels of amplification and four speakers altogether). In addition, this system can provide synthesized four-channel sound from two-channel sources, whether the material has been encoded or not, by reverberation or time delay adjusted to suit room acoustics.

Fig. 5-62G shows the Sansui QS-1 four-channel adapter with Quadphonic synthesizer. The QS-1 makes use of a reproducing matrix (decoder) and selective phase modulation technique to develop four-channel signals from two-channel input. The QS-1 provides high and low output levels to the amplifiers, two- and four-channel monitor circuits, input level control, four VU meters, four-channel ganged master volume control, and three balance controls for (1) front left to front right, (2) rear left to rear right, and (3) front to back. The synthesizer control has seven positions to provide (1) 2 Channel only, (2) Solo, (3) Concert Hall 1, which is not emphasized or treated and is essentially flat in response, (4) Concert Hall 2, which provides exaggerated concert-hall effects (selection and phasing of selected frequency components), (5) Surround Normal, to obtain a normal surround stereo effect, (6)

Surround Quarter Turn, to reposition the sound effect by 90°, and (7) Surround Half Turn, to reposition the sound by 180°.

Fig. 5-62H shows the Sansui QS-500 adapter–rear-channel-amplifier combination unit, designed to convert existing two-channel stereo systems to four-channel operation with the addition of the QS-500 and two rear speakers. The unit includes the two-channel amplifiers and controls and metering for front, rear, and front-to-rear balance. This unit also provides a decoder and a synthesizer. Fig. 5-58 shows a diagram of how to arrange this unit and two speakers with a complete two-channel system to provide a complete four-channel system with decoder-synthesizer. The other provisions of this unit are essentially the same as those of the QS-1 shown in Fig. 5-62G.

Fig. 5-62I shows a Fisher unit with two decoders (2-2-4 and 4-2-4), cartridge player, and integrated rear amplifier (50 watts). Installation of a decoder is covered in Chapter 8.

Although a few compact low-cost stereo preamplifiers are available (such as the one shown in Fig. 5-63A), most stereo preamplifiers are of the more sophisticated type illustrated by the examples shown

(A) Compact preamplifier.

Courtesy Shure Brothers Inc.

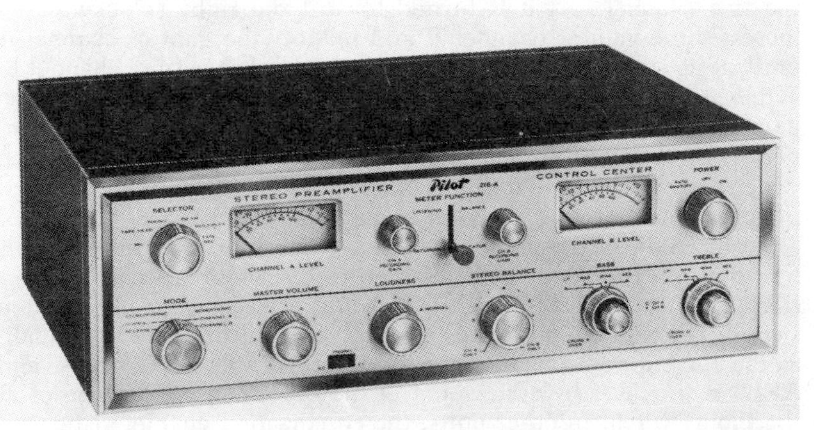

Courtesy Pilot Radio Corp.

(B) Preamplifier and control unit.

Fig. 5-63. Two types of stereo preamplifier.

in Figs. 5-63B and 5-64. To illustrate the features of a typical preamplifier control center and clarify the meanings of the various controls, let us take the example of the unit illustrated in Fig. 5-63B.

The SELECTOR control in the upper left corner selects the input and impedance required for proper load for the source device. It has positions for microphone, tape head, phono pickup cartridge, fm and a-m tuner, tuner with multiplex, and tape-recorder inputs. The distinction is made between a direct connection to the tape head for tape playback, in which proper loading and equalization for the head itself are necessary, and the input to the preamplifier from the output of the playback amplifier of the complete tape recorder.

The MODE switch is in the lower left corner of the unit in Fig. 5-63B. This switch allows manipulation of the two channels of the preamplifier so they can be used for normal stereo operation (Normal), with the two inputs swapped (Reverse), with just channnel A, or with just channel B.

Just to the right of the MODE switch is the MASTER VOLUME control. This is a dual control, which simultaneously adjusts the gain of both channels. Because the MASTER VOLUME control adjusts volume without allowing for the loudness characteristic of the human ear, a LOUDNESS control (just to the right) is also included. This is really a switch, providing stepped increases or decreases of volume in both channels simultaneously. At each step, an equalizing network compensates for the response characteristics of the listener's ear (see Fig. 5-48).

To the right of the LOUDNESS control is the STEREO BALANCE control. In its center (up) position, it allows approximately equal gains in the two channels. As it is turned toward the right (clockwise) it increases the gain of channel B and reduces the gain of channel A until, at its extreme position, output is obtained only from channel B. When it is turned to the left (counterclockwise) it increases the gain of channel A and decreases that of channel B in the same manner. This control is used for adjustment for optimum stereo effect, which may be realized at different adjustments for different types of program material and different stereo systems.

To the right of the STEREO BALANCE control are the tone controls for adjusting the degrees of bass and treble boost. This unit differs from many others in that the BASS and TREBLE controls are step switches instead of continuous controls, and in that here the equalization for different playback characteristics (LP, NAB, RIAA, and AEB) is provided by adjustment of the boost controls instead of at the input switch. In other units, the equalization characteristics are selected by the input selector, which will have such labels as ceramic, crystal, RIAA, etc., and continuous adjustment of the bass and treble boost is provided.

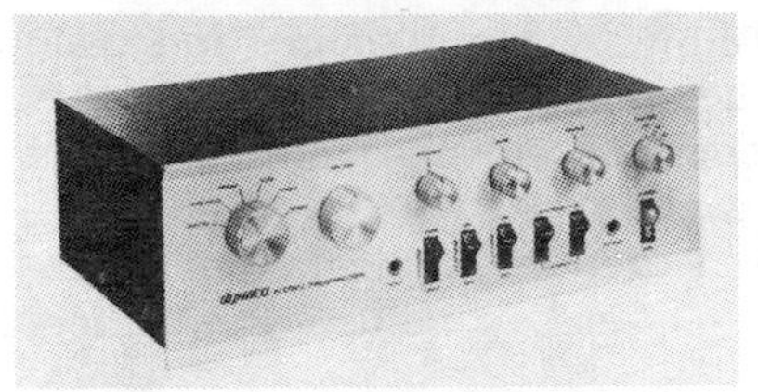

Courtesy Dynaco Inc.

(A) Dynaco preamplifier-control center (assembled or kit).

Courtesy JVC America, Inc.

(B) JVC all-silicon "SEA" stereo system preamplifier.

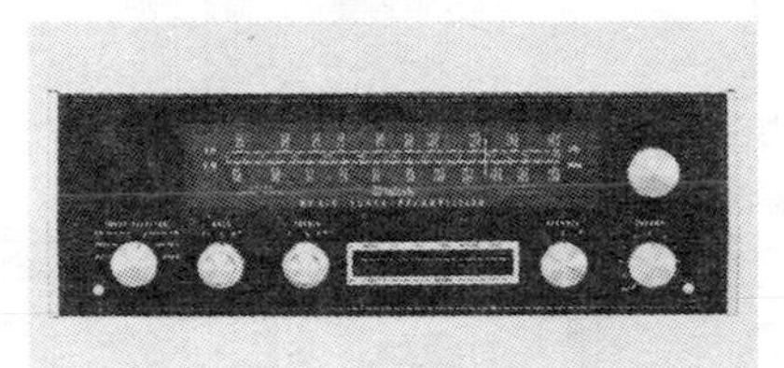

Courtesy McIntosh Laboratory Inc.

(C) McIntosh combination tuner-preamplifier-control center.

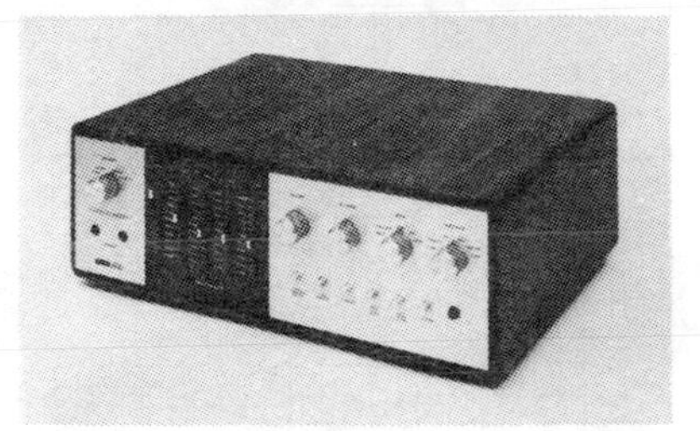

Courtesy Harman-Kardon, Inc.

(D) Harman-Kardon preamplifier/equalizer control center.

Fig. 5-64. Examples of stereo preamplifier-control units.

Two output-level meters are shown in the upper portion of the panel in Fig. 5-63B. These are used to check the balance between the two signals during either playback or recording. The switch located between the meters transfers the meters to register for recording or playback, and the controls adjust recording gain and playback gain independently. The control in the upper right-hand corner is used to turn the unit on and off or provide connection for automatic shut-off. The later is used when the phono turntable mechanism is interconnected with the preamplifier to shut off all power when the records of a stack have all been played.

Fig. 5-64A shows a high-quality stereo preamplifier which may be purchased as a kit. Fig. 5-64B shows a stereo preamplifier which provides for finite control of frequency response and balance with a wide range of input capability. Fig. 5-64C shows a unit containing a stereo preamplifier, control centers, and built-in tuners. A block diagram of this unit is shown in Fig. 5-65.

Amplifiers

Power amplifiers for stereo reproduction have the same requirements (electrically) as amplifiers for monophonic reproduction. They must have sufficient output to operate the speaker system at a desirable level, low harmonic and intermodulation distortion, frequency response from 20 or 30 Hz to 20 kHz or more, and good response to transients. If any of the requirements of high fidelity for monophonic amplifiers are compromised, the full benefit of stereo reproduction cannot be realized.

The primary differences between stereo amplifiers and high-fidelity monophonic amplifiers are physical in nature. First, there

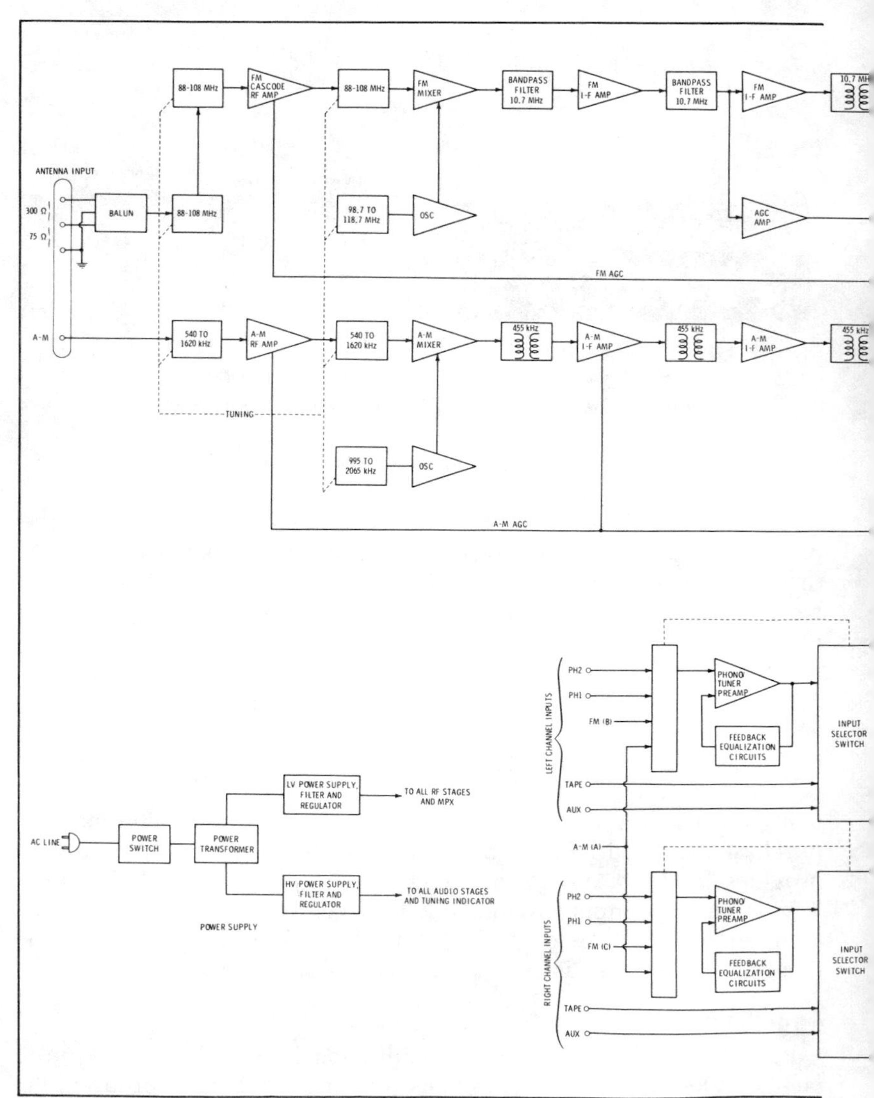

Fig. 5-65. Block diagram of McIntosh

must be two or more separate amplifier circuits or amplifier channels. These are usually combined on one chassis. Of course, there is nothing to stop one from buying two or three separate amplifiers, but the compactness of combination units is an advantage.

Because all of the manual controls can be provided in the preamplifier which feeds the amplifier, most amplifier units are designed

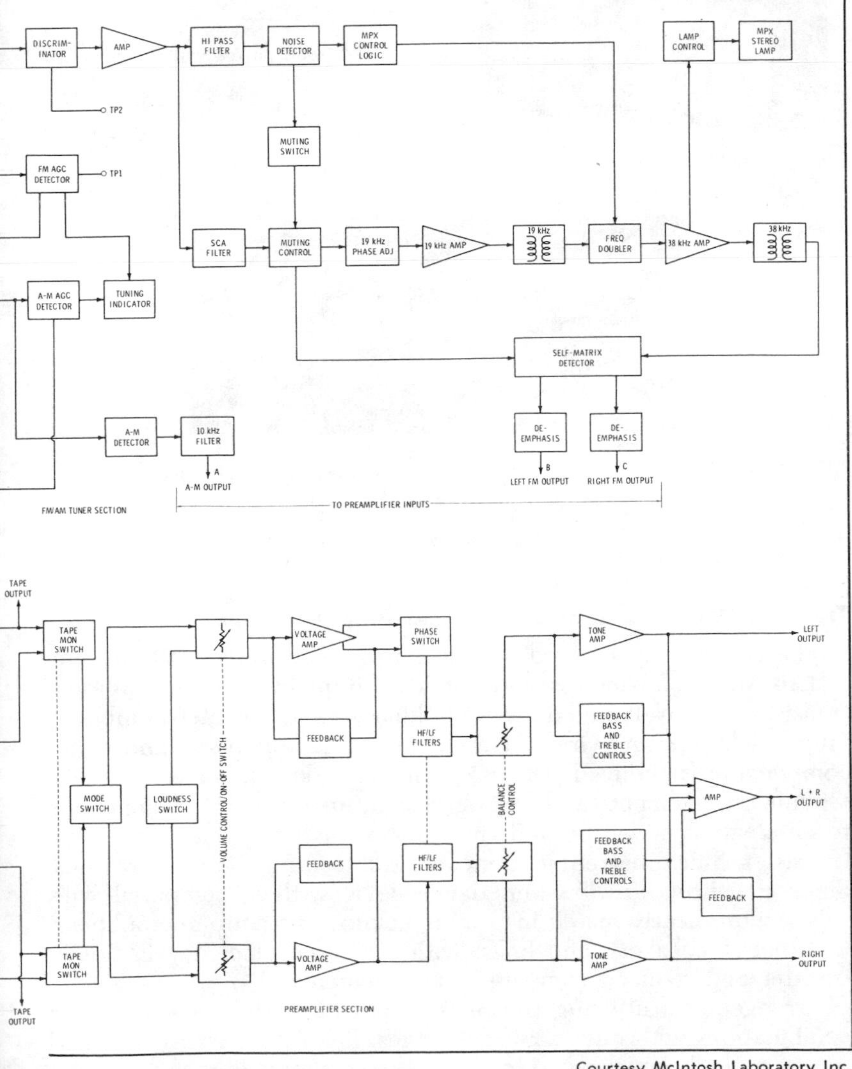

Courtesy McIntosh Laboratory Inc.

stereo preamplifier-tuner.

with only a limited number of controls, and these controls need adjustment only once, or at the most, just occasionally. Examples of stereo power amplifiers are illustrated in Fig. 5-66. Each can be concealed completely in a cabinet, with all the controls located on the preamplifier panel.

(A) Marantz power amplifier only.

Courtesy Marantz Co., Inc.

Courtesy McIntosh Laboratory Inc.

(B) McIntosh power amplifier only.

Fig. 5-66. Typical stereo power amplifiers.

Transistor Stereo Control Centers and Amplifier Units

The trend in hi-fi stereo design is toward multiple units on one chassis, and compactness without loss of quality. For this reason, transistors are well suited for high-fidelity control-center amplifiers. Hum pickup from wiring, transformers, microphones, and other components is reduced because transistors do not have filaments. Circuit arrangements and layouts can be made more efficient, due to elimination of wiring and due to the small physical size of the transistor. Shielding requirements are reduced for the same reasons. Since transistors are low-impedance devices, they, compared with tubes, more nearly match magnetic pickups, dynamic microphones, tape decks, speakers, and lines, with increased efficiency in power transfer and reduced component requirements.

Arrangements utilizing transistor application and transistor/tube combinations with nuvistor input stages, field-effect transistors, and silicon and silicon-planar transistor types have made many new designs possible, and the state of the art continues to advance. Transistors allow reduction in size. In some cases, transformers can be eliminated.

Properly designed circuits for transistor use or transistor/tube combinations with negative feedback can provide for high-fidelity equipment the required wide frequency response and low distortion, equaling that of circuit arrangements utilizing only vacuum tubes.

It is difficult to attain faithful reproduction of all signals with transformer-coupled circuits. To have good response at low frequencies, transformers are physically large and expensive.

Transistors have made possible practical development of transformerless circuits for audio frequencies, since the transistors for audio are basically low-voltage, high-current devices. The emitter-follower stage, in particular, offers interesting arrangements, since it has low inherent distortion and low output impedance.

Hi-fi stereo preamplifier-amplifiers utilizing transistors and integrated circuits throughout are shown in Fig. 5-67.

Combination Units

The trend toward combining all units except speakers into one compact cabinet is reducing cost and increasing the adaptability of hi-fi equipment to all sizes of homes. For the most exacting requirements in hi-fi, separate units still have some advantage, but the latest designs of stereo "compacts" are hard to beat. For average use, they are more than adequate, and even lower-cost models provide exceptional quality and require a minimum of space.

Courtesy McIntosh Laboratory Inc.

(A) McIntosh MA 5100.

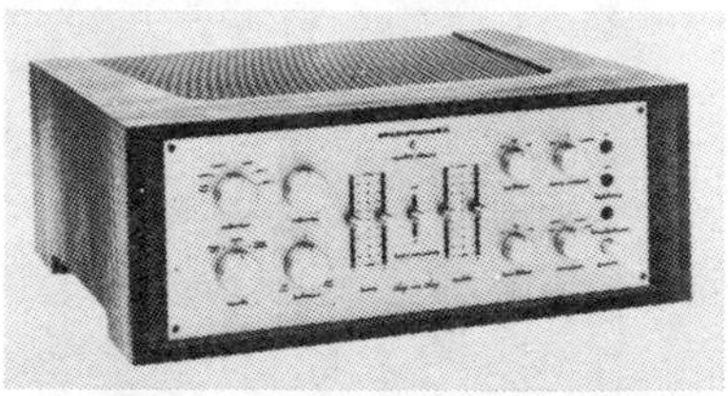

Courtesy Marantz Co., Inc.

(B) Marantz Model 30.

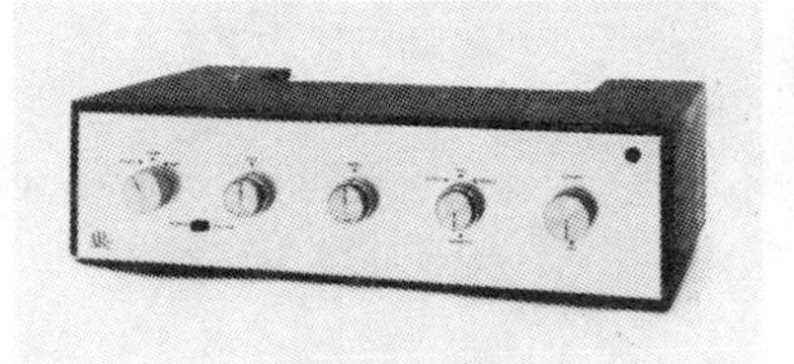

Courtesy Acoustic Research, Inc.

(C) AR stereophonic amplifier.

Courtesy Heath Co.

(D) Heathkit AA-2004 (four channels).

Fig. 5-67. Transistor stereo preamplifier-amplifier units.

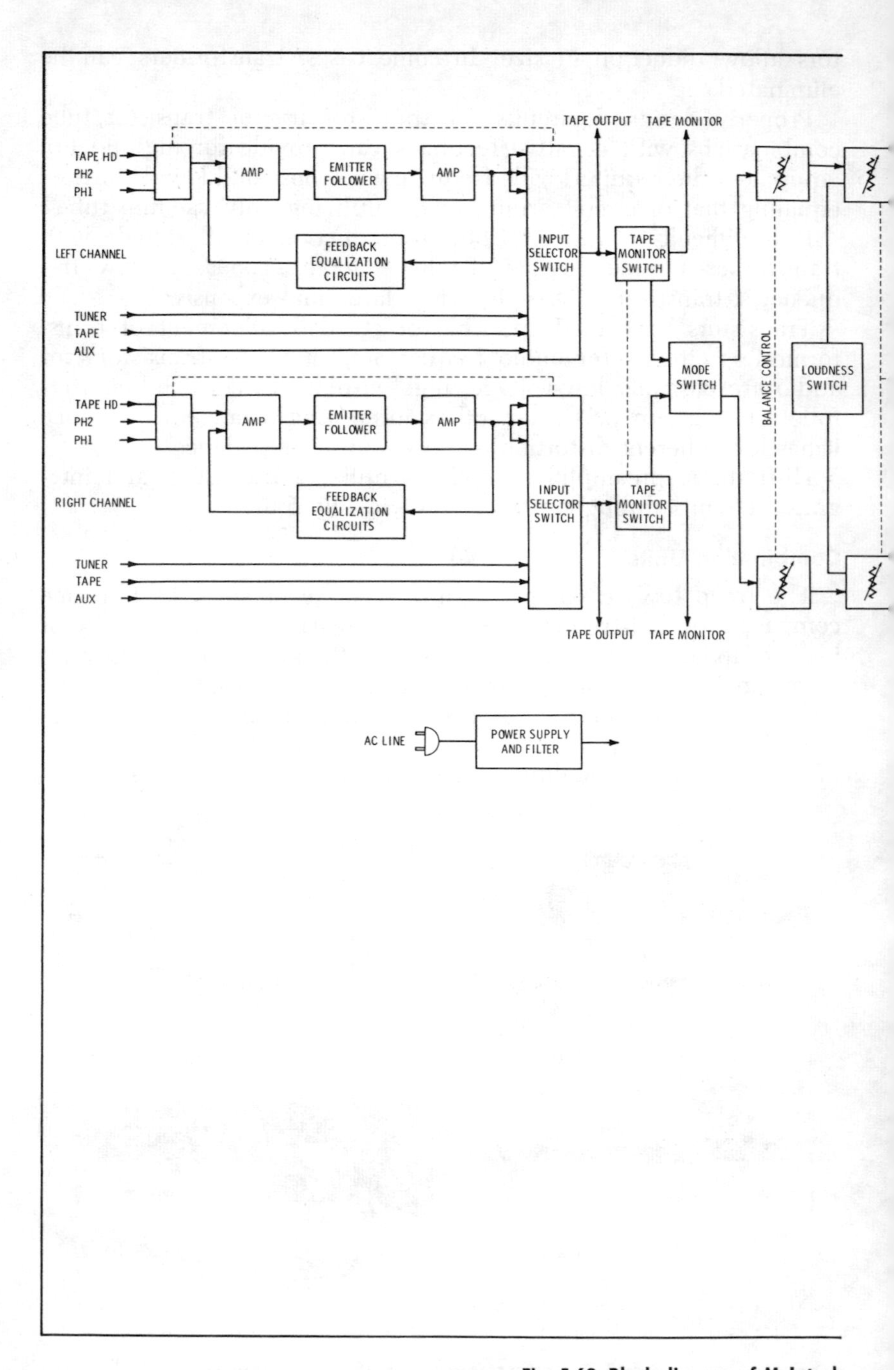

Fig. 5-68. Block diagram of McIntosh

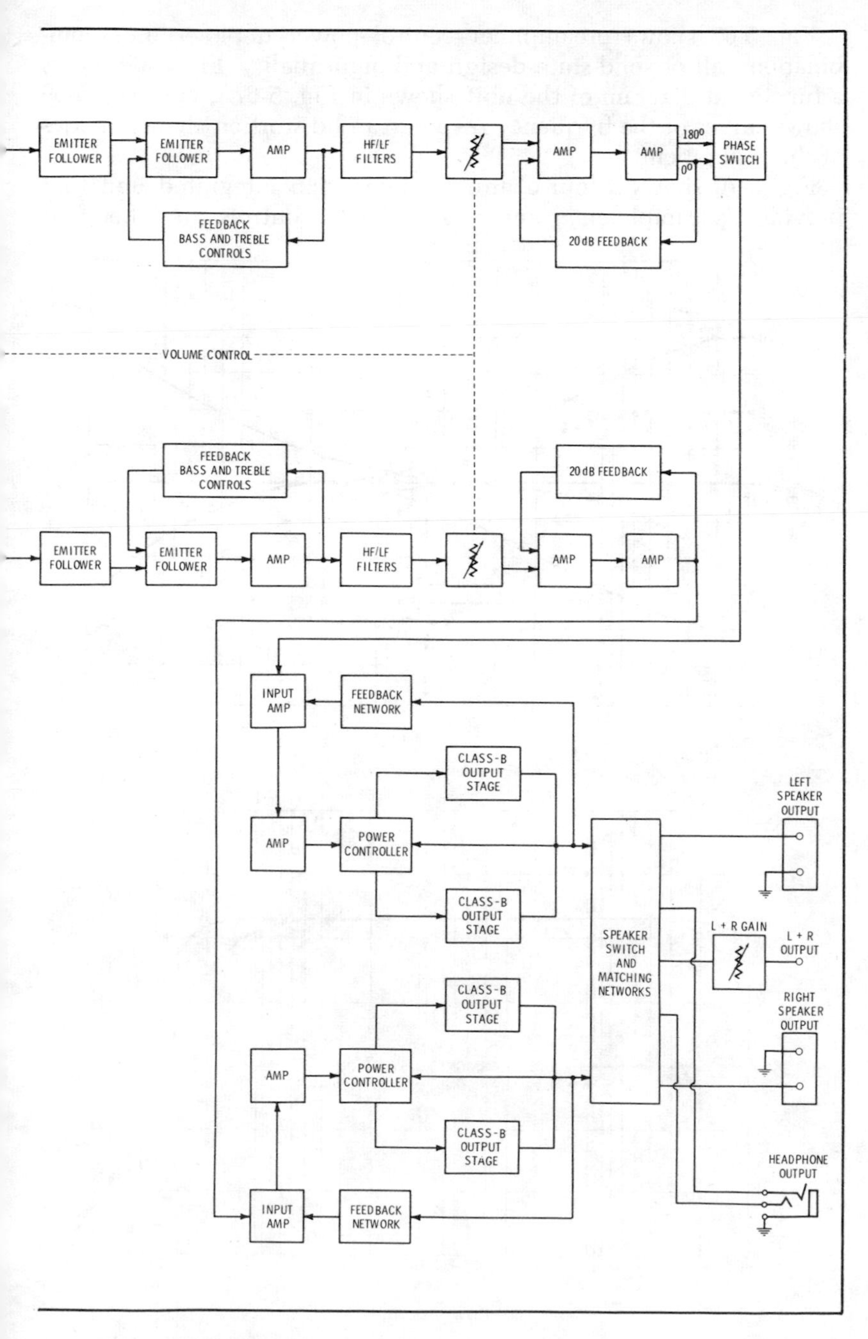

MA 5100 preamp-amplifier.

Fig. 5-67 shows preamplifier–control–power-amplifier unit combinations, all of solid-state design and high quality. Fig. 5-68 shows a functional diagram of the unit shown in Fig. 5-67A, and Fig. 5-69 shows curves of the frequency response and distortion characteristics of this equipment.

Fig. 5-70 shows a four-channel combination integrated unit that provides preamplifiers, power amplifiers, controls for the four

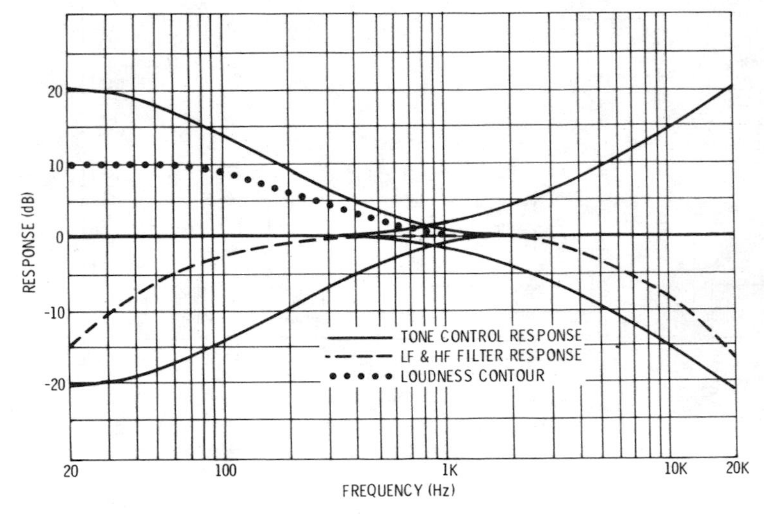

(A) Frequency response.

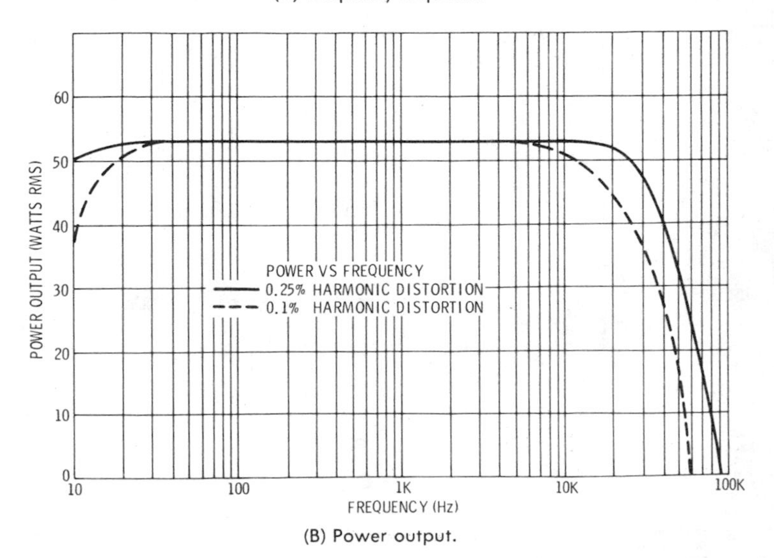

(B) Power output.

Fig. 5-69. Performance curves for

channels, and interrelated controls including four-way balancing among all four channels. This unit can be adapted to 2-2-4, 4-2-4, and 4-4-4 systems, and it provides for enhancing with a hall effect and quadrixer.

Fig. 5-71 shows two-channel integrated receivers which provide everything needed for hi-fi broadcast reception except the speakers. In addition, there is a complete set of controls and preamp-amplifier

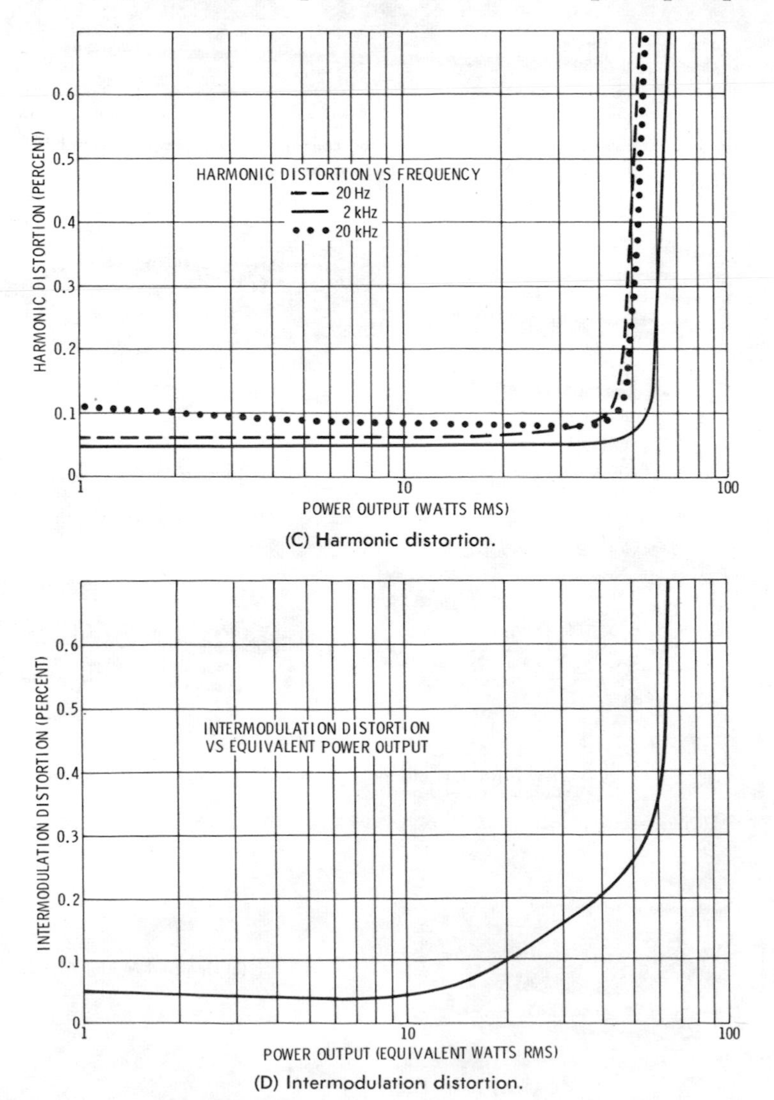

(C) Harmonic distortion.

(D) Intermodulation distortion.

McIntosh MA 5100 preamp-amplifier.

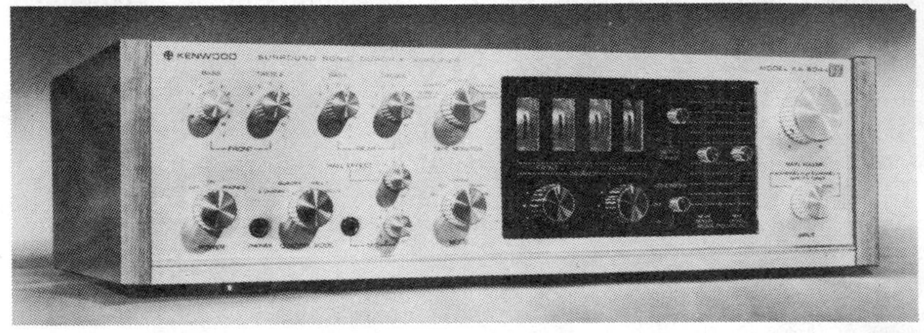

Courtesy Kenwood

Fig. 5-70. Kenwood four-channel combination control center-preamplifier-amplifier with decoder and synthesizer.

Courtesy Marantz Co., Inc.

(A) Marantz Model 22.

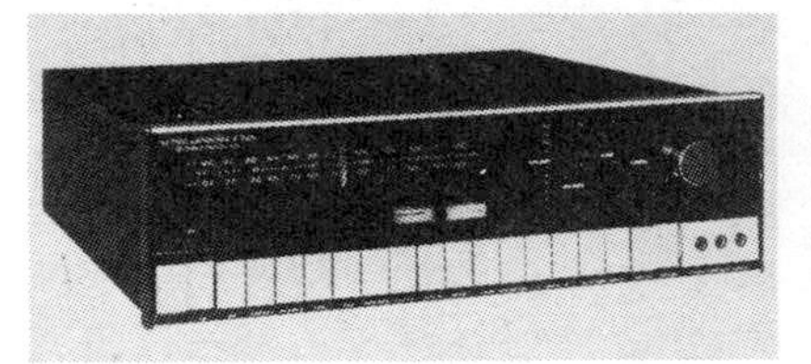

Courtesy Altec Lansing

(B) Altec Lansing Model 725A.

Courtesy Matsushita Electric Corporation of America

(C) Panasonic Model SA-6500.

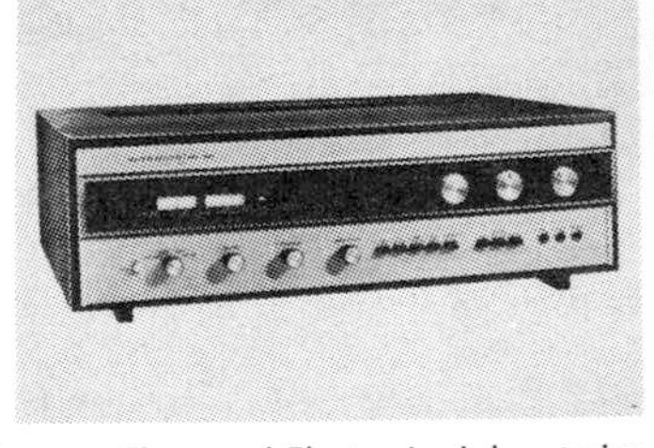

Courtesy Sherwood Electronics Laboratories

(D) Sherwood Model SEL-200.

(E) Heathkit Model AR-15.

Courtesy Heath Co.

Fig. 5-71. Two-channel integrated receivers.

facilities for phono, tape, tv, record, and playback inputs and outputs.

Fig. 5-72 shows four-channel receivers with similar and advanced features. These units have provision for all kinds of inputs, outputs, monitoring, and control of the four channels. The unit in Fig. 5-72C comes with a built-in universal decoder that converts 4-2-4 programs directly through its own four channels. The receiver shown in Fig. 5-72D provides all of the foregoing features together with decoding and forms of program enhancement such as synthesizing.

Fig. 5-73 shows a system arrangement with many of the features that now can be included in a four-channel system package. The sample receiver shown will accept two phonos, one for standard

Courtesy Fisher Radio

(A) Fisher Model 701.

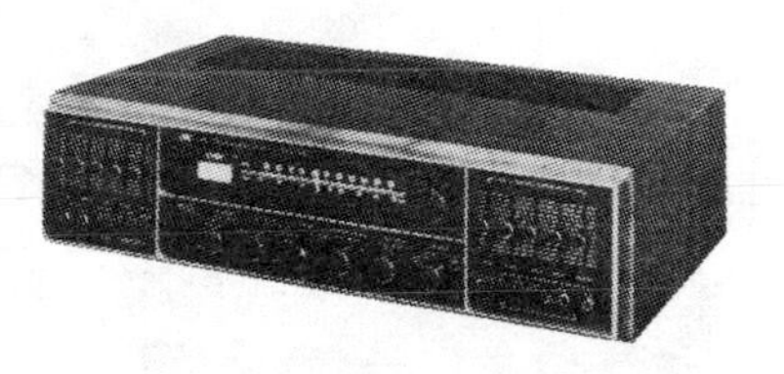

Courtesy JVC America, Inc.

(B) JVC receiver.

Courtesy Electro-Voice, Inc.

(C) Electro-Voice receiver.

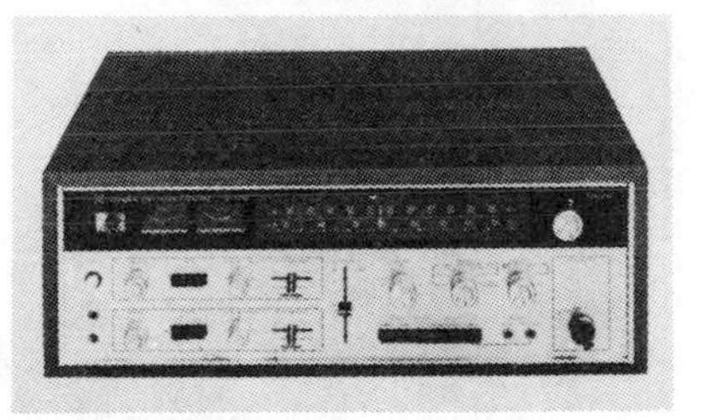

Courtesy Sansui Electronics Corp.

(D) Sansui Model QR-6500.

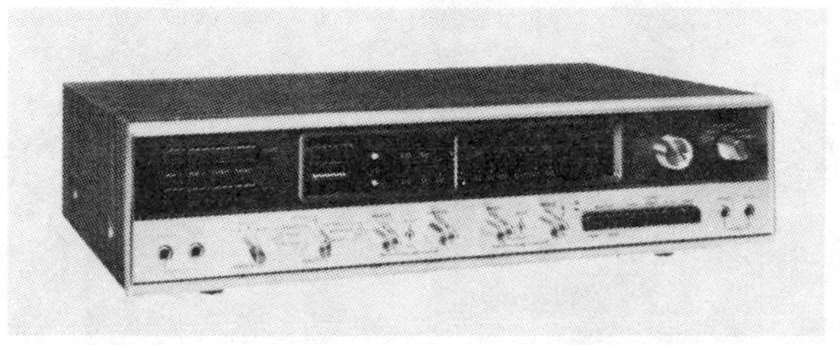

Courtesy Lafayette Radio Electronics Corp.

(E) Lafayette receiver.

Fig. 5-72. Four-channel receivers.

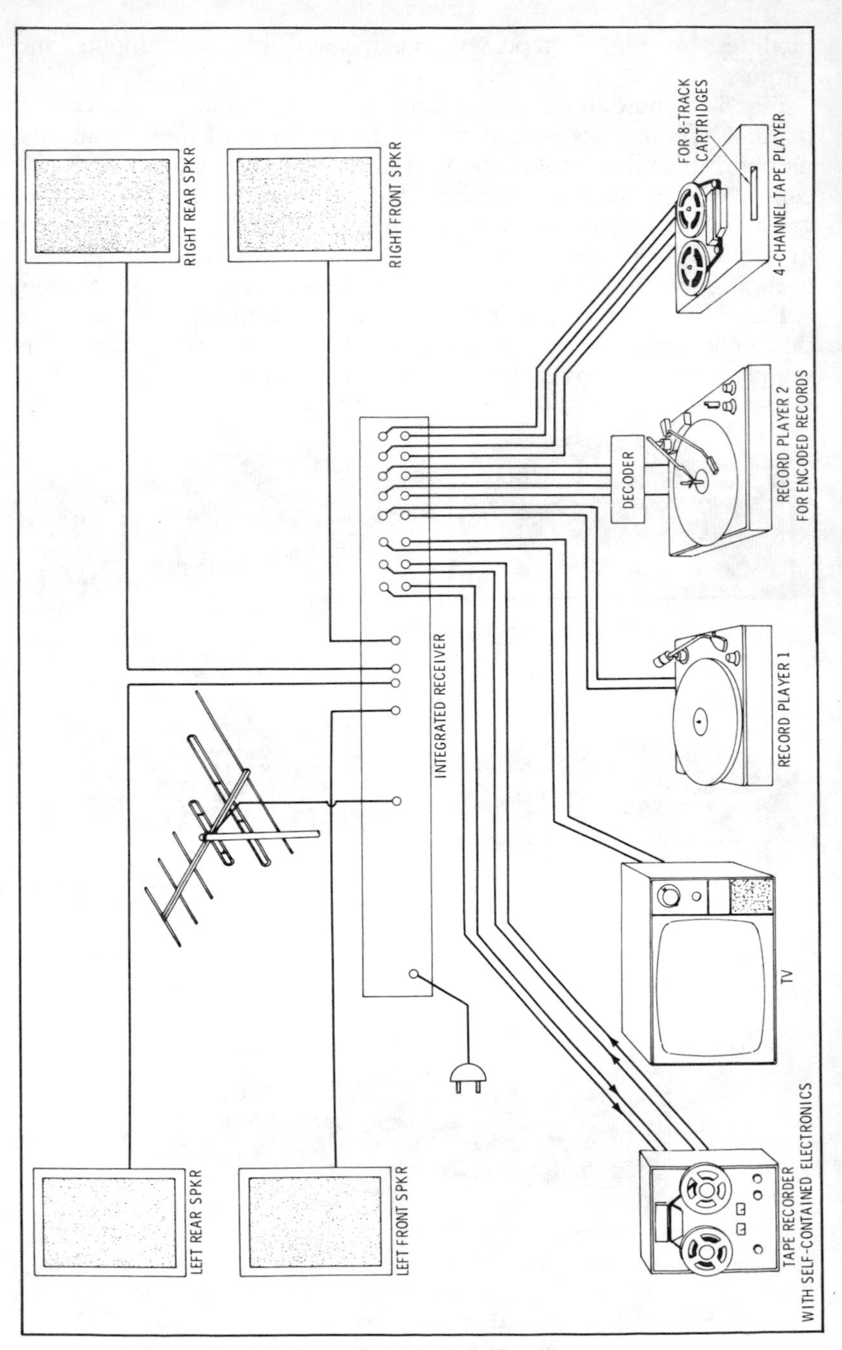

Fig. 5-73. Example of four-channel stereo system.

Courtesy Harman-Kardon, Inc.

(A) Harman-Kardon Festival 235.

Courtesy JVC America, Inc.

(B) JVC Model 4450.

Courtesy Matsushita Electric Corporation of America

(C) Panasonic Model SC-8700.

Courtesy Sansui Electronics Corp.

(D) Sansui Model MQ-2000.

Courtesy AKAI America, Ltd.

(E) AKAI "true surround sound" package.

Courtesy Fisher Radio

(F) Fisher Model 40.

Fig. 5-74. Stereo package compacts.

records and one for encoded records such as the Columbia SQ records which contain four channels of information matrixed into two channels. The decoder shown between record player 2 and the receiver has two inputs and four outputs. If the Electro-Voice or Sansui units shown in Figs. 5-72C and 5-72D, respectively, are used

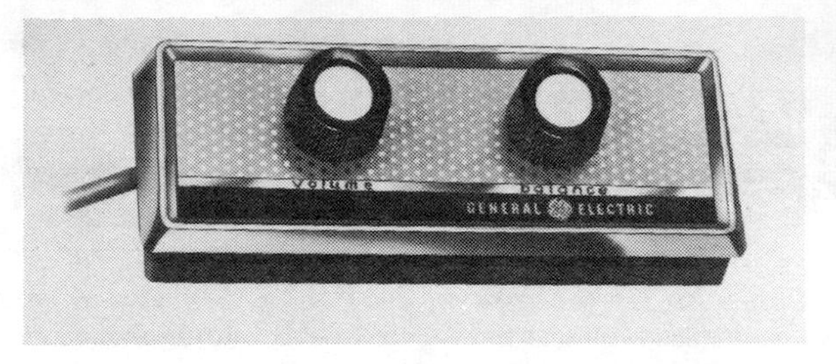

Courtesy General Electric Co.

Fig. 5-75. Remote-control unit used with stereo preamplifier.

and properly adjusted, the decoder is not necessary as it is built in. Connections may have to be made to adapt to the phono, however.

The sample receiver shown in Fig. 5-73 provides for recording two-channel programs, storage, and replay. Television sound may be played over the system in mono to gain the benefit of the audio quality of a hi-fi reproduction system, which is generally superior to the average tv-set audio system. In addition, there is provision for two- and four-channel tape playback from both reels and cartridges.

The outputs to the speakers may be adjusted and balanced to suit the listener's taste and the room acoustics. All of the above features can be obtained at relatively low cost if sufficient study and time are given to the selection and installation of the system.

Fig. 5-74 shows several complete package compacts which will provide complete hi-fi two- or four-channel systems according to the desire of the user. Sometimes it is preferable to buy the compact separate from the speakers, as is possible with the units shown in Figs. 5-74D and 5-74F. In this manner it is possible to obtain equipment sized to fit the room requirements, quality to suit the taste, and prices to fit the pocketbook.

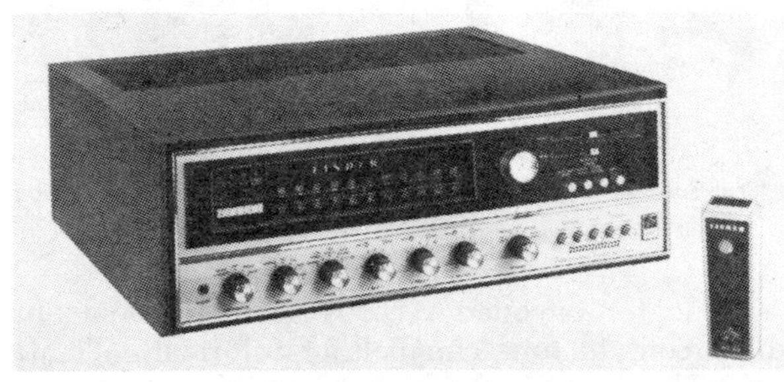

Courtesy Fisher Radio

Fig. 5-76. Fisher four-channel receiver with wireless remote control of station selection.

Courtesy Kinematix, Inc.

(A) Kinematix stereo balance.

Courtesy Lafayette Radio Electronics Corp.

(B) Lafayette stereo audio analyst.

Fig. 5-77. Typical stereo balance meters.

Many listeners like to be able to control volume and balance from the listening position, since this is the only place adjustment can be made for the proper effect. To provide for this need, manufacturers are making remote-control units available. A typical remote-control unit is illustrated in Fig. 5-75. This unit has a 30-foot cable, the other end of which plugs into the tape input and output jacks of the preamplifier. Thus when the controls are set for the tape-record position, the outputs of the amplifier must pass through the cable to the remote-control unit and back to the preamplifier before passing on to the amplifier. In the process, the volume and balance may be controlled by the remote unit.

Other units available include radio-controlled station-selector and volume-control devices (Fig. 5-76) that can be operated from anywhere in the room and which have no connecting wires. These devices do not enhance the quality, but they may come in handy for adjusting a four-speaker system, which has different effects for different listener positions and different positions of speaker placement. Such devices are also desirable for listeners who have difficulty in getting about.

Balance and Tuning Instruments

One of the functions important for the maintenance of good stereo effect is the preservation of balance in the system. If you have a preamplifier with volume-level meters, these meters can be used to check balance when a test tone is applied to both input channels at once. However, most systems do not have built-in meters. Therefore, an external metering device which will indicate when balance is achieved is useful. Two examples of commercially available balance meters are shown in Fig. 5-77. These meters can be attached to the outputs of the amplifiers and adjustments made until the meter needle is centered—indicating the two signals have exactly the same amplitude. It should be noted that this balancing is not normally done with the balance control; rather, fixed-setting adjustments (screwdriver adjustments) are set for balanced output with the balance control in its center position. Then, the balance control can be adjusted later, during actual operation of the system, to compensate for such things as speaker types and placement, room acoustics, etc.

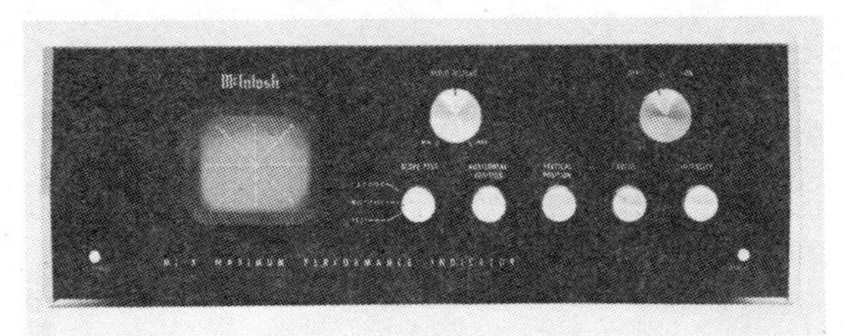

Courtesy McIntosh Laboratory Inc.

Fig. 5-78. McIntosh maximum performance indicator.

A common cause of distortion is multipath reception of the fm signal, which is caused by reflected signals from the same transmitter arriving at the receiving antenna slightly later than the direct signal. This can usually be corrected by antenna rotation or replacement. To check this condition, an indicator using a 3-inch cathode-ray tube is available (Fig. 5-78). This instrument also may be used to check signal strength and indicate correct tuning position more accurately than a meter; it will indicate the nature of the signal (mono or stereo), the relative phase of the two channels, which channel is being used if only one is used, channel balance, and separation. For those who want the ultimate in fm stereo reception, this instrument is a valuable tool. There are receivers available with similar features built into one integral unit.

6

Speakers

The speaker system constitutes several links in the overall high-fidelity equipment chain. These links are illustrated in the block diagram of Fig. 6-1. Each link must be considered not as a separate entity but in its relationship to the links which precede and follow it.

The amplifier output stage may be considered the energy source which supplies the driving power to the speaker system. Because most speaker electrical input circuits have a low impedance (2 to 20 ohms), most systems must employ an output transformer to convert this low impedance to the value into which the output amplifier should work, ordinarily about 1500 to 6000 ohms. As will be explained, the transformer works both ways. The load on the speaker system is reflected back to the amplifier stage and influences its operation, and the amplifier impedance is reflected forward to contribute to the load on the speaker driver and radiator.

The speaker system itself can be divided into three functional parts:

1. The electromagnetic part, consisting of the voice coil and field magnet. Audio-frequency electric current in the coil causes mechanical motion of the cone or diaphragm on which it is mounted. This part is often referred to as the *driver* or *motor* of the system.
2. The mechanical part, on which the driving coil is usually mounted and which is set into mechanical motion by the audio-frequency electric current in the driving coil.
3. The acoustic part, which transmits the sound energy developed by the mechanical part to the room or other area served by the system, in the most efficient and faithful manner possible.

This takes the form of a baffle or enclosure, with a horn being a form of enclosure.

A complete understanding of the operation of speaker systems requires a sufficient view of the whole flow of sound energy from the output amplifier stage to the listener, as depicted in Fig. 6-1.

SPEAKER DRIVERS

This section will cover the definitions and types of drivers. The speaker driver is that portion which converts electrical energy from the output transformer to mechanical energy in the diaphragm or cone radiator. The driver is also sometimes called the *motor* because, like electric motors, its input is electrical and its output mechanical.

A number of different types of speaker drivers have been tried during the history of sound-system development. Those sufficiently successful to be commercially available include the following:

1. Moving-coil dynamic driver
2. Crystal drivers
3. Capacitor drivers

Of these, by far the most popular and useful in high-fidelity applications is the dynamic moving-coil type, so our discussion will be primarily about that type. Capacitor and crystal drivers are sometimes employed in high-frequency (tweeter) portions of dual systems, and these are also given brief mention.

It should be mentioned here that the performance of direct-radiator speakers is very importantly influenced by the type of baffle or enclosure used. Since baffles and enclosures are a subject in themselves and must be discussed after speakers, the explanation of these factors must await a later chapter. All comparative data in this section will assume that direct-radiator speakers are employed with an infinite baffle. A definition of an infinite baffle is given in the chapter on baffles and enclosures.

Moving-Coil Dynamic Drivers

The principle of the dynamic speaker driver is based on the interaction of two magnetic fields. One field is relatively strong and steady; the other, developed by the passage of an audio-frequency signal current through the voice coil, varies with the instantaneous amplitude of the sound to be reproduced.

Electrodynamic Type—The basic construction features of an electrodynamic moving-coil type of driver are shown in Fig. 6-2. The strong, steady magnetic field is generated by a large field coil wrapped around the core. This core is mounted in a frame of mag-

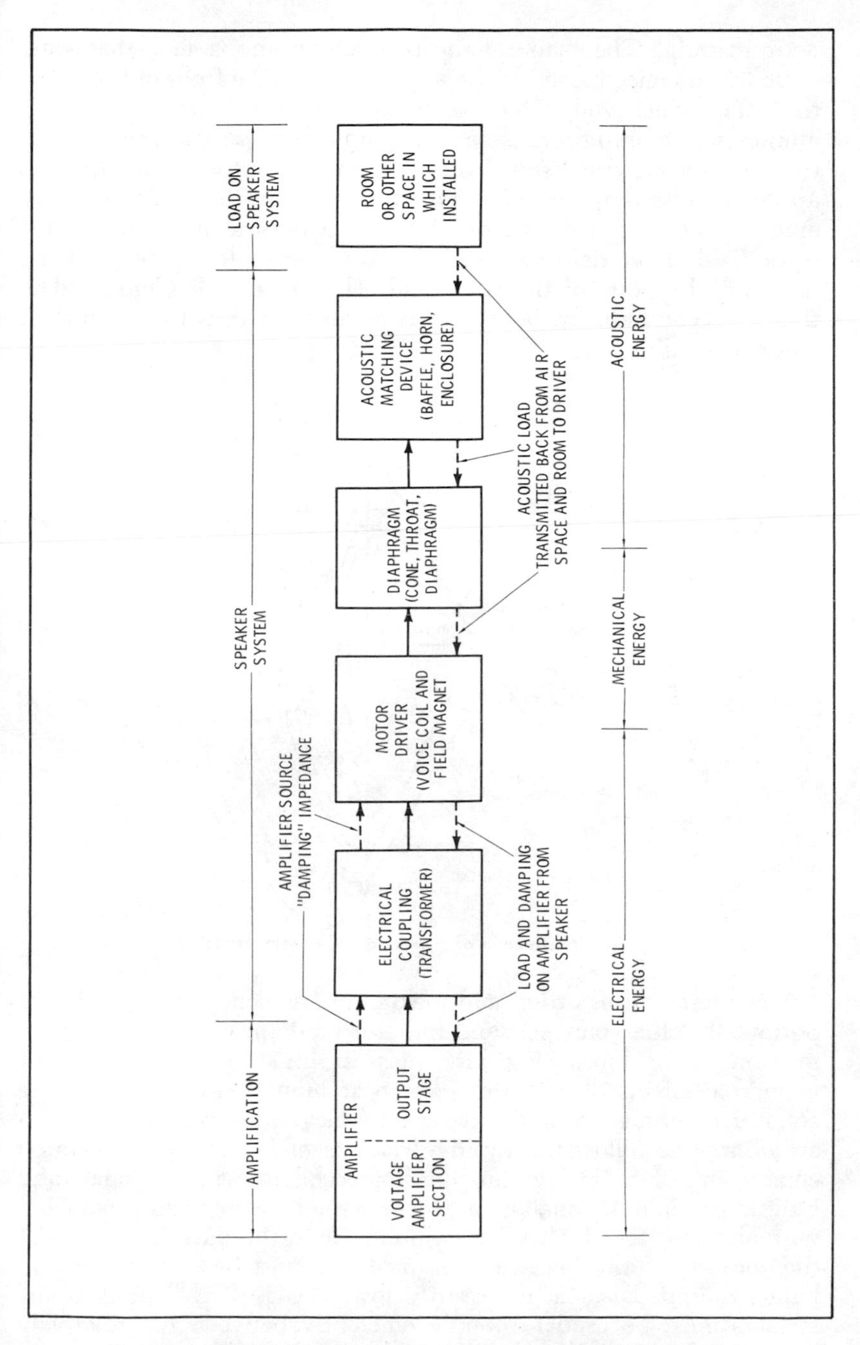

Fig. 6-1. Sections and functions of the output of a high-fidelity system.

netic material. The shape of the core and frame is such that magnetic flux is concentrated in the annular gap in the front of the structure. The voice coil, which is wound on a cylinder of fiber or aluminum, fits into the annular air gap in the core and frame structure. The af electrical signal output from the output transformer is applied across the voice coil. The af current in the voice coil generates a varying magnetic field which works against the strong static field of the field coil, and the resultant motor force produces mechanical motion of the voice coil. The voice coil is mounted to the cone-type radiator; hence the cone also moves with it and radiates the sound.

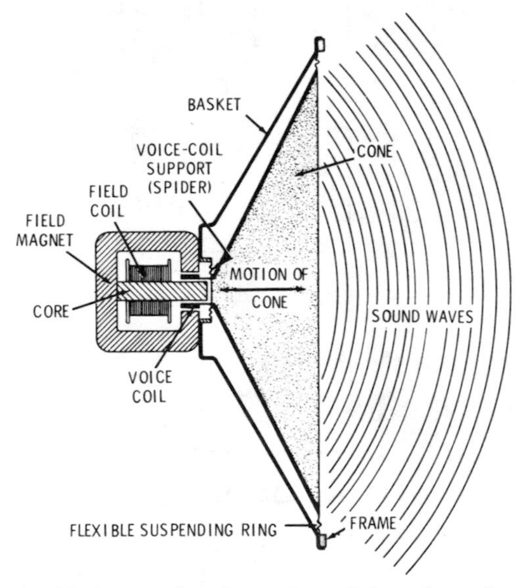

Fig. 6-2. Construction of a moving-coil dynamic speaker.

For minimum distortion and maximum frequency range, it is important that the voice coil and radiator or diaphragm have a minimum mechanical mass. Excessive mass in this structure would result in inertial effects that become worse at high frequencies, and the cone or diaphragm would have a tendency to distort physically in an attempt to follow the rapid variations of high-frequency sound components. For this reason, the voice coil is made as small and light as possible. It usually consists of a single layer of fine enameled wire about ½ to 2 inches long wound along the outside surface of the voice-coil form. Because the voice coil must be so compact and light, its impedance is necessarily low; this is why output transformers must be used to couple efficiently between the relatively high-impedance output-amplifier tubes and the dynamic speaker.

Leads from the voice coil are usually cemented to the middle portion of the cone surface, then brought out to terminals mounted on the basket of the speaker structure.

The impedance rating of the voice coil, in commercially available units, may be any value from 2 to 16 ohms. This impedance is not that of the voice coil alone but includes the effect of acoustic loading on the cone or diaphragm and mechanical effects in the structure. These factors tend to resist motion of the voice coil and thus raise the impedance "looking into" the voice-coil terminals. These latter effects are the largest part of the rated impedance, and the effects of the self-inductance and resistance of the voice coil are relatively small. The impedance rating is also specified for some standard frequency, usually either 400 or 1000 Hz. For the foregoing reasons, a simple resistance test of the voice coil will show a relatively low resistance, compared with the impedance rating of the speaker as a whole. Further details about the acoustic and other loading effects which help make up the impedance looking into the voice coil are discussed later, under "Enclosures and Baffles."

Permanent-Magnet Type—The foregoing discussion was devoted to the dynamic driver which employs a field coil, as shown in Fig. 6-2, to derive the strong steady flux required in the gap containing the voice coil. This type of driver is referred to as the electro-dynamic type. The magnetic-field flux is provided by a permanent magnet instead of a coil in what is known as the permanent-magnet dynamic speaker. The constructional features of this type of speaker are illustrated in Fig. 6-3.

Practical permanent-magnet dynamic speakers were made possible by the development of high-grade magnetic materials, particularly alnico. Very powerful magnets, which hold their magnetic properties indefinitely with little loss, can be made from this material. A round piece of this magnetized material is mounted between the core and the frame of the magnet structure in the dynamic-driver unit, as shown in Fig. 6-3. It is thus effectively in series with the

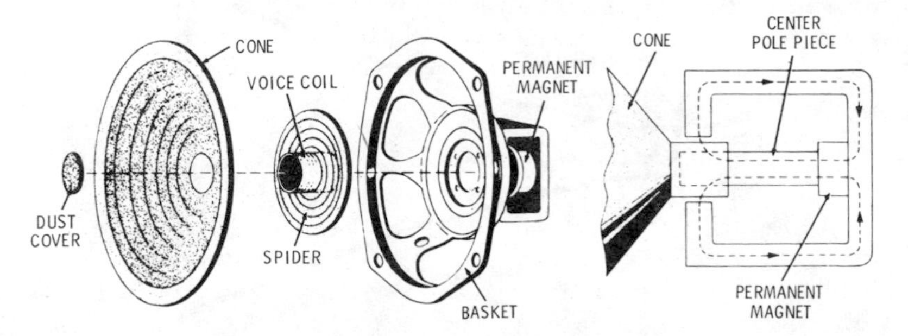

Fig. 6-3. Construction of a permanent-magnet speaker.

other iron in the magnetic circuit, producing flux in the same way as the many turns in the field coil of the electrodynamic type. The obvious advantage of such an arrangement is that no field-coil supply is required. Also, because the permanent magnet is lighter than a field coil providing the same amount of field, the overall weight of the speaker is reduced below that of an equivalent electrodynamic unit.

Since the permanent magnet supplies the fixed field for operation of the driver, the output power of the speaker is limited by the available flux and the size of the permanent magnet. In addition to power-handling capability, the size of the magnet also affects frequency response because the lower-frequency components contain most of the power in the audio signal, and they are attenuated in driver units with small magnets. Permanent-magnet speakers in general contain magnets weighing from 2 to more than 20 ounces. Types desirable from a high-fidelity standpoint are those with 6-ounce or heavier magnets, depending on the power requirements for the speaker.

The use of the dynamic driver is by no means limited to cone radiators. It is also frequently employed with horn-type radiators, as illustrated in Fig. 6-4.

Magnetic-Armature Type

At one time, magnetic-armature speakers were the most popular type in use, but they have now been almost completely superseded by the more efficient and better-quality dynamic type. In the magnetic type, the coil is fixed and consists of many turns of wire around a soft-iron core. The armature, which is either the radiator or a diaphragm fastened to the radiator, is also of soft iron and is mounted in a gap in the iron-core magnetic circuit around which the coil is wound. One of the prime advantages of this arrangement is the fact that the coil, being fixed, can be large enough to match

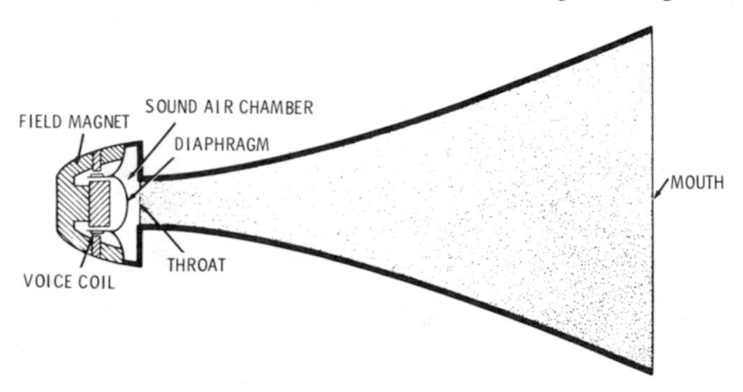

Fig. 6-4. Cross-sectional view of a horn-type reproducer.

the impedance of the output tube or tubes without an output transformer. Inertial and mechanical-motion limitations and nonlinearity of flux distribution have all but made this type obsolete, and it is mentioned here only for the purpose of completeness and perspective.

Crystal Drivers

Rochelle-salt crystals have the property of becoming physically distorted when a voltage is applied across two of their surfaces. This property is the basis of the crystal type of speaker driver. The crystal-driver type of speaker is illustrated schematically in Fig. 6-5. The crystal is clamped between two electrodes across which the audio-frequency output voltage is applied. The crystal is also mechanically connected to a diaphragm. The deformations of the crystal caused by the audio-frequency signal across the electrodes cause the diaphragm to vibrate and thus to produce sound output.

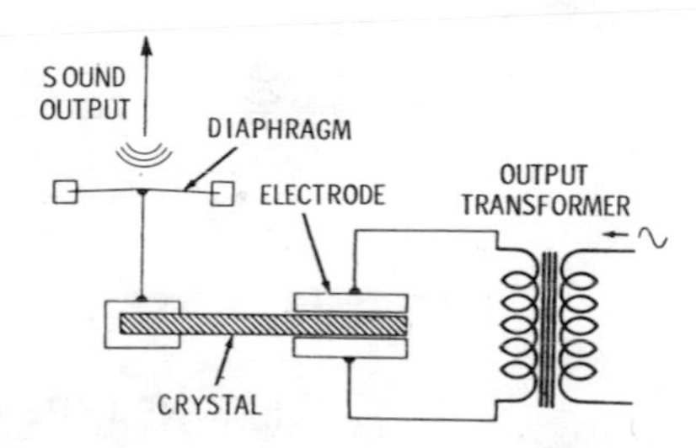

Fig. 6-5. Schematic representation of a crystal-type speaker.

In general, crystal speakers have been impractical for reproduction of the full audio-frequency range because the input impedance is almost completely capacitive. Thus it is difficult to couple power into them. At high audio frequencies, the reactance becomes lower and the relative amount of power smaller. Consequently, crystal units have found some use in tweeters (the high-frequency portion of dual-speaker units to be discussed later). Even as such, they offer no obvious or outstanding advantages over moving-coil dynamic types, and thus they are rarely encountered.

Capacitor Drivers

Another principle which has been applied to speaker drivers is that of electrostatic force. When a potential is applied between two metal plates, the resulting electrostatic field produces a force that tends to pull the plates together (because opposite charges attract). If like charges are applied to the plates, a force which tends to push the plates apart is created.

The principle is usually applied in a push-pull arrangement, as illustrated in Fig. 6-6. A pair of plates is connected to a balanced audio-frequency output source as shown. Another plate, mounted between these two plates, is free to move and drive a diaphragm. The movable plate is polarized with a positive charge, as shown in the diagram.

When the polarity of the audio-frequency signal is such as to make the top capacitor plate negative, the positive armature plate is attracted to it and is repelled by the positive capacitor plate. The use of a push-pull arrangement reduces harmonic distortion, which is inherent in most capacitor-type speakers.

As is the case for the crystal speaker, the input impedance to the capacitor speaker is almost a pure capacitance; accordingly, similar problems of coupling power into it are encountered. Although a few are in use, they are very much in the minority.

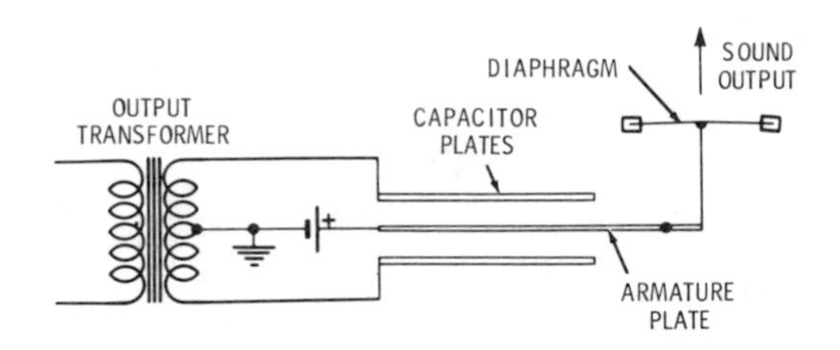

Fig. 6-6. Schematic representation of a capacitor-type speaker.

CONE-TYPE RADIATORS

Of the functional blocks of the speaker system illustrated in Fig. 6-1, we have discussed the driver only. The next step is the diaphragm which converts energy of mechanical motion into energy of air motion, called *acoustic energy*. There are two commonly used forms of diaphragms: (1) the cone type and (2) the horn type. The cone type acts as both diaphragm and radiator because it not only converts the mechanical energy of the driver into acoustic energy but also at the same time couples this energy into the room or area where the listeners are located. On the other hand, the horn-type diaphragm provides only mechanical-to-acoustic conversion; its acoustic energy output must be fed to the throat of a horn which couples it to the listening area. First, we will consider the cone-type diaphragm and radiator, which is more common than the horn-type diaphragm.

It is the purpose of the cone or any diaphragm-radiator combination to convert mechanical energy from the driver into acoustic energy in the listening area. The conversion must be such as to provide the greatest amount of acoustic power output for a given electrical power input, with a minimum of distortion of the output sound waveform. Although the speaker cone is a power transducer, output quality is more important than output power. Although efficiency (ratio of output acoustic power to mechanical power input) should be as high as possible, the modifications necessary to keep distortion to desirable low levels make the majority of speakers inherently low-efficiency devices. Most speakers have overall efficiencies ranging from 5 to 15 percent; some very elaborate systems approach 40 percent.

For higher efficiency in transfer of the mechanical energy to the air, it is desirable that the greatest possible area of contact be made between the radiator and the air. Since it is desired that the air mass be alternately moved forward and backward (and not up and down or sideways), it is natural to envision a large flat sheet driven by the speaker driver, as shown in Fig. 6-7A. The greater the area of contact, the better the air mass loads the driver unit. Unfortunately, such a flat structure is not mechanically practical, because when it is constructed light enough for good high-frequency response it does not retain rigidity over its entire surface. To retain better overall mechanical rigidity with comparable large-area air contact, the cone type of radiator has been used; this type is shown in Fig. 6-7B. It has been found that, with such a shape, a relatively large area of air may be activated with a relatively high ratio of strength to weight.

Treated paper is universally used in cone construction. The more rigid the paper, the greater is the sound output obtained, but the poorer is the frequency response. Soft, blotterlike cone materials improve uniformity of response in the low- and medium-frequency ranges but give poor response at high frequencies. Soft cones are also better for transient-response rejection. High-fidelity speakers often use a two-piece cone of different materials, as will be explained later.

The size of the cone is important because it influences both the low-frequency response and the power-handling capacity. The larger the cone diameter is, the greater is the power capacity for all

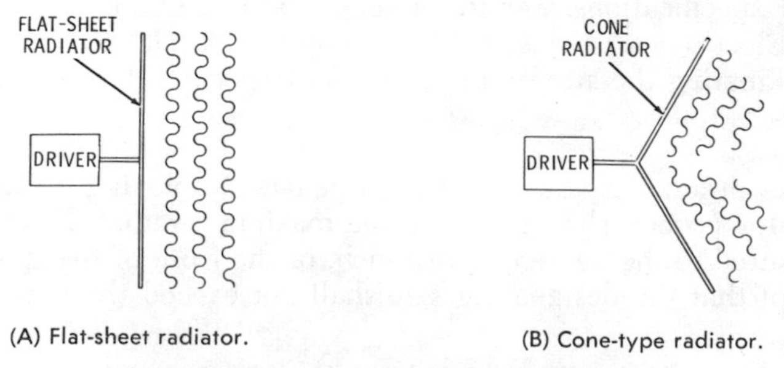

(A) Flat-sheet radiator. (B) Cone-type radiator.

Fig. 6-7. Representation of two types of radiators.

frequency conponents combined, and the better is the low-frequency response. However, such improvements are not necessarily derived from larger cones unless the voice coil is appropriate. The acoustic impedance offered to the cone rises as the cone is made larger; the voice-coil impedance must then also be made larger for proper energy transfer and efficiency. The larger the cone is, the lower is the lowest useful frequency of operation. But frequency range is not the only factor improved by increase of cone size. Because the major portion of the ordinary af signal power is in the low-frequency components, the overall power-handling capacity is also improved, as mentioned previously. The increase in the frequency range at the low end of the spectrum by an increase of cone size is illustrated in Fig. 6-8. Note that these curves show a response peak just before the response falls off at the low-frequency end of the range. This peak occurs at the resonant frequency of the speaker, which will be explained later.

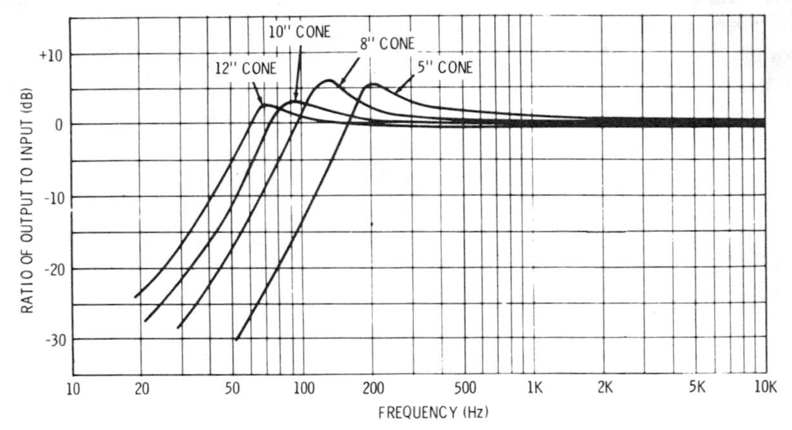

Fig. 6-8. Effect of cone size and resonance on low-frequency response.

The designation of the cone diameter is one of the most important speaker specifications. For this reason, the definition set up by the Radio-Electronics-Television Manufacturers Association* (RETMA) for designating the size of a speaker is of special interest. This definition is as follows:

> The designating size of a loudspeaker employing a circular radiator (cone) shall be twice the maximum radial dimension, measured to the nearest eighth inch, of the front of the speaker, except that the designating size shall not exceed the maximum

*Now Electronic Industries Association (EIA).

diameter of the unsupported portion of the vibrating system by more than 25 percent.

The size of the cone is also important in the choice of an enclosure in which the speaker is to be mounted. Enclosures are designed to operate with speakers of specified characteristics which depend mostly on size; that is why a given enclosure is stated to be used only with a speaker (or speakers) of a given size. Of course, it is assumed that the overall design of the speaker is consistent with high-fidelity performance with the nominal cone size. A large cone with a small voice coil is not considered adequate for high-fidelity output.

The shape of the cone also influences performance. It has been found that a circular cross section as illustrated in Fig. 6-9A gives the best performance. Ellipitical cones (Fig. 6-9B) tend to have a lower acoustic impedance than those of a circular cross section, and it is thus more difficult to couple power into them with good efficiency. For this reason, elliptical cones are not used in high-fidelity equipment. Also important is the shape of the flare of the cone. Straight sides are most common, but they tend to concentrate high-frequency sound components in the small area surrounding the axis of the cone. Better distribution of the high-frequency components is obtained by use of a curved flare (illustrated in Fig. 6-9C) in the cone sides, and some speakers are manufactured with this shape. However, this is a more difficult manufacturing process than that of the nonflared speakers, and these speakers are therefore more expensive than straight-sided versions. In dual-speaker arrangements in which a separate high-frequency driver unit is employed, the operation of the large cone at high frequencies is not so important,

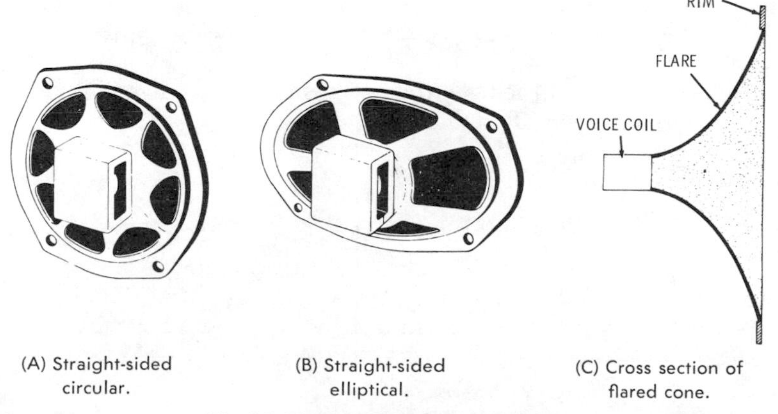

(A) Straight-sided circular.

(B) Straight-sided elliptical.

(C) Cross section of flared cone.

Fig. 6-9. Three variations in cone shape.

and the straight-sided cone may be as good as the flared one. In speakers in which the response of a single cone is extended over the full desired range, the curved, flaring shape shown in Fig. 6-9C is often employed.

The higher the frequency is, the smaller is the portion of the cone around the center which is used for radiation. In fact, for many cones, the highest frequencies in the audible range are radiated only by the voice coil itself. The portion useful at a given frequency is approximately a fixed quantity independent of the overall size of the cone; therefore, the larger the cone, the smaller the percentage of area employed at high frequencies. This accounts for the general fact that the larger the speaker is, the poorer the relative high-frequency response is and the better the relative low-frequency response is.

A generalized speaker response curve is shown in Fig. 6-10. First, there is the resonant peak at about 100 Hz (often at lower frequencies). Below resonance, response falls off rapidly. Above resonance and up to about 1000 Hz, the whole cone acts as a unit, all parts of it vibrating in phase. Response in this region is about constant. Above 1000 Hz, breakup occurs; that is, parts of the cone vibrate independently of each other, as shown in Fig. 6-11. In this portion, the response increases gradually until losses and impedance in the system increase sufficiently to cause final drop-off at the high-frequency end of the range. In some speakers, efficiency at high frequencies is improved by incorporating into the cone a more flexible material in the form of circular rings or corrugations coaxial with the cone to allow greater flexing (see Fig. 6-12). This makes it easier for the small inner portion of the cone to operate independently at high frequencies, but the structure will also transmit low-frequency vibration so that the whole cone will act as a unit at the lower frequencies.

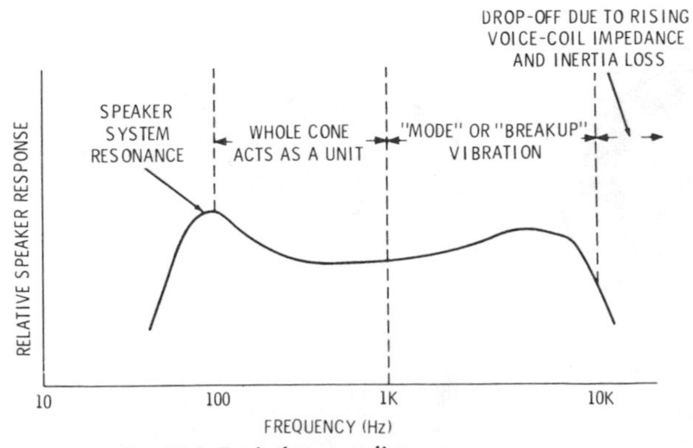

Fig. 6-10. Typical cone-radiator response curve.

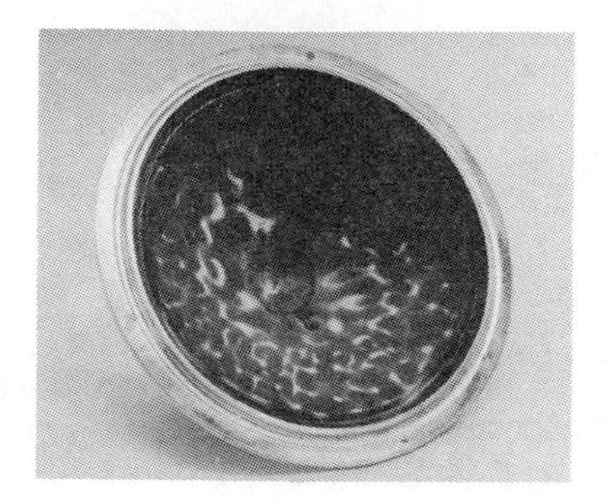

Courtesy North American Philips Corp.

Fig. 6-11. Typical cone breakup at high frequencies.

The cone is mounted to the speaker frame in two places: (1) at the outer edge or rim and (2) at the center, near the voice coil. These mounting agencies are called *suspensions* of the cone assembly. The stiffness of the suspension affects the frequency response and other performance features, as is explained later. The inner suspension near the voice coil is often referred to as the *spider* because of the physical resemblance of some versions to a spider. Spiders may be divided into two main groups. One group employs a phenolic or plastic sheet cut out so that the voice coil will be suspended by relatively narrow cross members, as illustrated in Fig. 6-13A. The other type, now more common, is a piece of solid, flexible material with circular corrugations, as illustrated in Fig. 6-13B. The outer suspension is sometimes just an extension of the cone structure itself, where it fastens to the metal rim of the basket. In other cases, the material at the outer edge of the cone is feathered (made thinner) or corrugated to provide increased flexibility. Some

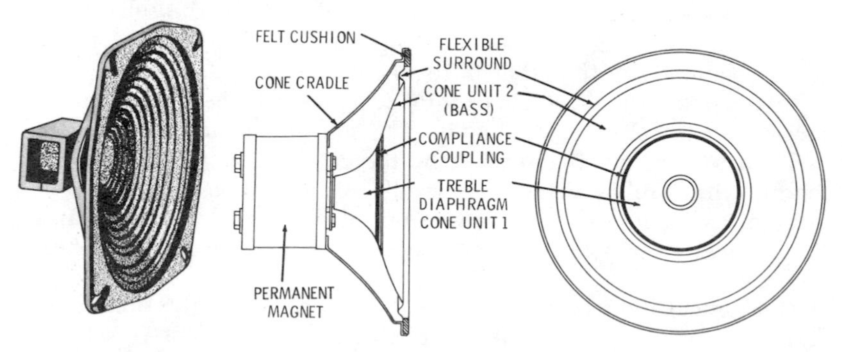

(A) Multiring unit.

(B) Hartley-Turner two-unit speaker.

Fig. 6-12. Use of corrugations in cone to improve efficiency at high frequencies.

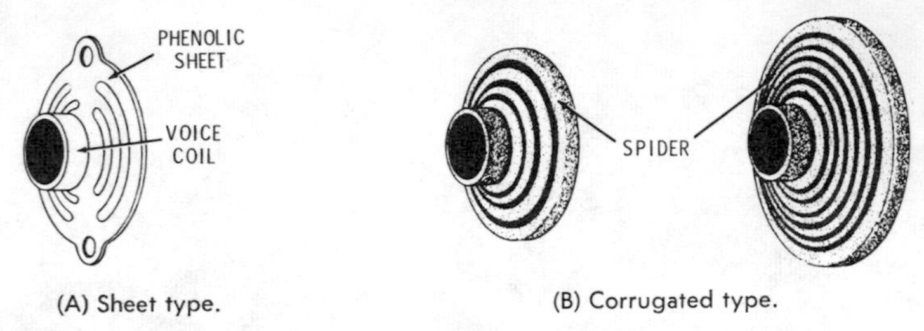

(A) Sheet type.

(B) Corrugated type.

Fig. 6-13. Two types of spiders.

speaker units have been manufactured with suspensions of leather or other damping materials. This soft, flexible material is excellent for minimizing transient distortion at low frequencies.

SPEAKER IMPEDANCE

It has been pointed out previously that the impedance looking into the voice coil is not merely the self-impedance of the coil itself but a combination of the self-impedance and the more important reflected acoustic impedance. A parallel may be drawn with a transformer or electric motor. Each of these devices draws a small current when operating unloaded, indicating a relatively high input impedance. When the transformer secondary is loaded by an electrical resistance or the motor shaft is coupled to a mechanical load, the input current rises and the input impedance of each device is lowered in proportion. In other words, the load impedance has been reflected into the input circuit in each case, whether it is an electrical load in the transformer or a mechanical load in the motor.

The voice-coil winding is similar to any other coil in that it has resistance (of the wire used in the winding), inductance, and a small amount of capacitance (distributed, between turns). The resistance and reactance of the coil combine to form the self-impedance of the winding, without any impedances coupled into it from its association with the other parts of the speaker.

In a speaker, the principle of total impedance is the same as in our motor and transformer analogies. The self-impedance of the voice coil (in a vacuum) is modified by the reflected impedance of the load on the diaphragm. It may be difficult at first to think of a diaphragm as having an impedance; this will be made more clear if the following analogies between mechanical and electrical systems are considered.

Mechanical inductance is called *inertance*. When the diaphragm starts to move some air, that air resists the force tending to set it in motion, due to its inertia. After the air is in motion, it tends to stay in motion when the diaphragm stops or reverses its motion. The degree to which the air tends to stay at rest or in motion is a measure of its inertance. In the electrical analogy, it is the inductance of a circuit which provides electrical inertia, and it is the current which tends to stay at rest or in motion in proportion to the amount of inductance present. Mechanical inductance, when applied to the air, is also referred to as *acoustic inductance*. The term "inertance" is more especially applied to the acoustic system and the air in contact with the diaphragm. The mechanical inductance of the cone and voice-coil structure and of its suspensions is also a factor in the input impedance to the voice coil, and is reflected back to it with the acoustic inertance.

Mechanical capacitance is called *compliance*. This is the "springiness" or "give" of the mechanical assembly or the air. The best example of a mechanical capacitance is a spring. Force applied to a spring stores work (force times distance) in the spring. Then, when the spring is released, the stored energy is released. This is exactly what happens electrically in a capacitor in which energy is stored by the flow of current into the capacitor by application of a voltage. The applied voltage is analogous to the applied mechanical force, and the resulting current is analogous to the motion or change of displacement of the spring. When speaking of mechanical systems, we call this effect *mechanical compliance*. Although the speaker does not contain a spring, the cone suspensions do act as springs and offer to cone motion resistance which increases as cone displacement increases. The suspension compliance is the main capacitive effect, although the "springiness" of the air load and the cone and voice-coil structures during flexing add other capacitive factors. When applied to the air, the capacitive effect is known as *acoustic compliance*.

Mechanical resistance is friction. It is the resistance force developed between two surfaces, two layers, or two or more groups of particles within a material when they rub together. In a speaker of the dynamic type, there are no material surfaces which rub together (under normal operating conditions). Purely mechanical resistance arises in the friction within the cone and suspension materials when they flex during operation. The useful resistance component is that of the acoustic load. The latter is developed by the friction of the particles and layers of the air surrounding the cone or diaphragm when they bear upon each other or along the mechanical surfaces of the speaker assembly when motion is imparted to the air in the form of acoustic vibrations.

Mechanical components of impedance (explained in the foregoing) have the same relationships among themselves as exist among their counterparts in the electrical circuit. Power is dissipated only by the resistance component. The inertance and compliance produce mechanical reactance which varies with frequency in the same way that electrical reactance varies.

ACOUSTIC IMPEDANCE AND RESONANCE

The resistances and reactances of the system (including acoustic, mechanical, and electrical effects) combine in the effective impedance looking back into the output transformer. This combination is best visualized by means of an equivalent circuit, illustrated in Fig. 6-14. The diagram of the corresponding portions of the system in the upper portion of the figure help symbolize the corresponding physical locations in which the impedance factors appear. The efficiency of power transfer can be seen to be dependent on the proportion of the impedance represented by R_A, which represents

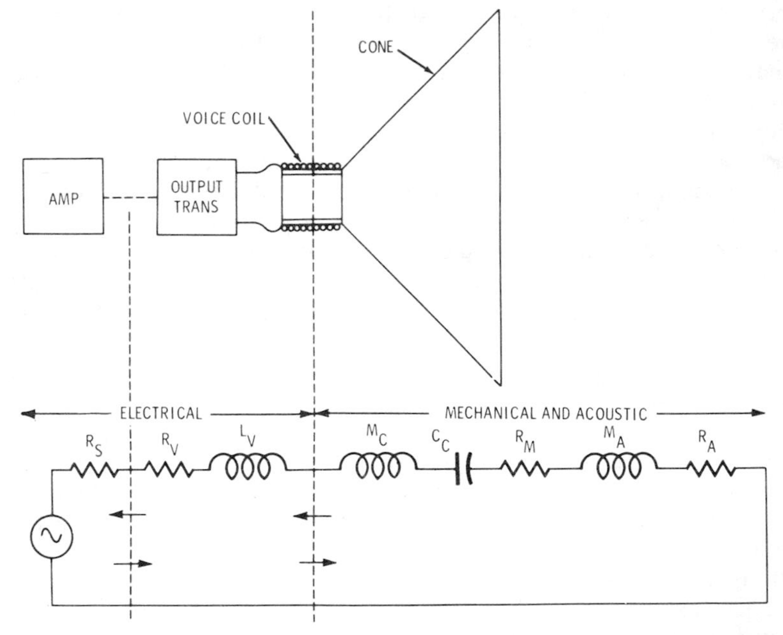

R_S - EQUIVALENT RESISTANCE OF SOURCE (INCLUDING PLATE RESISTANCE AND TRANSFORMER WINDING RESISTANCE)

R_V - EQUIVALENT RESISTANCE OF VOICE COIL

L_V - INDUCTANCE OF VOICE COIL PLUS LEAKAGE INDUCTANCE OF TRANSFORMER

M_C - MASS OF CONE AND VOICE COIL ASSEMBLY

C_C - COMPLIANCE OF SUSPENSION

R_M - MECHANICAL RESISTANCE OF CONE AND SUSPENSION

M_A - MASS OF AIR LOAD

R_A - FRICTIONAL RESISTANCE OF AIR LOAD

Fig. 6-14. Equivalent circuit of a speaker.

actual acoustic power dissipated in overcoming air friction and in radiating the acoustic power.

We have observed that there are two types of mechanical and acoustic reactance in the speaker system. They are mass or inertance, corresponding to electrical inductance, and compliance, corresponding to electrical capacitance. As in a purely electrical system, the capacitive reactance which is the compliance C_C resonates with the combined inductive effects L_V, M_C, and M_A at some frequency. This frequency is known as the *resonant frequency* of the speaker. At the resonant frequency, all reactance is cancelled out of the system, and output and efficiency increase greatly over what they are for other frequencies. Rather than being beneficial, such an increase in efficiency is actually detrimental because it occurs only in the vicinity of the resonant frequency. If the effect of speaker resonance is not reduced considerably or if the resonant frequency is not made lower than the lowest frequency to be employed, extremely annoying frequency, amplitude, and transient distortions result. The system is then highly sensitive to signals at or near the resonant frequency. Every time changes in signal amplitude occur rather suddenly, the system tends to self-oscillate at the resonant frequency even though it is excited by sound-signal components of other frequencies. The reader has probably heard this effect in sound systems which, when listened to from a distance, give the impression of producing nothing but a "booming" noise. The "boom" is a result of the reaction of the speaker system at its resonant frequency.

The curves of Fig. 6-8 show the effect of speaker resonance on response. The response rises to a peak at the resonant frequency; then it falls rapidly at lower frequencies. The shape of the response curve shows how the resonant frequency can be used as an indication of the limit of low-frequency response. Undesired sharpness of the resonant peak can be lessened by electrical, mechanical, or acoustic damping. Damping is the addition of a resistive load.

One method of providing damping is through design of the output stage of the amplifier. The latter should have as low a source impedance (R_S in Fig. 6-14) as possible. The amplifier impedance is reflected through the output transformer and is effectively connected across the voice coil. If this reflected impedance (resistance) is low enough, it reduces the Q of the resonance of the speaker and thereby reduces the severity of the response peak at the resonant frequency and the sharpness of the drop-off below it. The reduced Q also minimizes transient distortion because the tuned circuit represented by the speaker has less "flywheel effect." The impedance of the output amplifier stage is lowered by use of low-impedance tubes and by the use of negative feedback. Triodes have a much lower plate impedance than tetrodes and beam tubes, and that is one important reason for the preference of some for triodes.

Another method of providing speaker damping is through design of the speaker enclosure, as will be explained in the chapter on baffles and enclosures.

HORNS AND HORN DRIVERS

A horn is a tube so flared (tapered) that the diameter increases from a small value at one end called the *throat* to a larger value at the other end called the *mouth*. A basic horn-driver combination is illustrated in Fig. 6-4. Horns have been used for centuries for increasing the radiation of the human voice and musical instruments.

The horn does acoustically what the cone does mechanically. It couples the small voice-coil area to a large area of air. In this way, the horn acts as an acoustic transformer and converts the relatively high impedance at the throat and driver. The horn is a fixed physical boundary for its enclosed column of air and does not vibrate itself. Acoustic energy fed to its throat must therefore be obtained from a vibrating diaphragm which converts mechanical motion from the driver voice coil (or other armature) to acoustic energy. Although the cone-type radiator acts as both diaphragm and radiator and transduces from mechanical to acoustical energy, the horn acts only as a radiator, with both input and output energy being acoustic.

We have seen that the high-frequency response drop-off of a cone-type radiator is caused by the inertial effect of the mass of the cone. Because the transformation in a horn is through an air column rather than through solid material, the high-frequency response of the horn is much better than that of the cone. The overall efficiency at all frequencies is better.

In spite of these advantages, straight (unfolded) horns are not commonly used for general-frequency coverage systems or low-frequency units in homes because of their bulk and their relatively high cost of manufacture. They do find wide use, however, in two forms, as follows:

1. In straight form. In dual speaker systems (to be discussed later) and in connection with the tweeter speakers, which reproduce only the high-frequency portion of the audio-frequency range.
2. In modified folded form. In special speaker enclosures (to be discussed later).

The high-fidelity sound enthusiast should therefore be familiar with some of the fundamentals of horns. The following basic information about horn design is included to clarify these fundamentals rather than to act as constructional information, although a few audiophiles have constructed their own horns with good results.

The most common type of horn design is the *exponential horn.*

$$\frac{A_x}{A_T} = e^{mx}$$

where,

A_x is the cross-sectional area at distance x from the throat, in square inches,
A_T is the cross-sectional area of the throat, in square inches,
e is the natural logarithm base, 2.7183,
m is the flare constant of the horn, in inverse inches,
x is the distance from the throat, in inches.

This equation is of importance primarily in defining the flare constant, m. The greater the flare constant is, the faster the diameter of the horn increases.

The flare constant determines how long a horn with a given mouth-to-throat area ratio must be, but it is of much greater importance in another connection. Each horn has a cutoff frequency below which no sound energy can be coupled through it. Below the cutoff frequency, the throat area offers to the driver a pure acoustic retance and no resistance; thus, no power can be transmitted. The cutoff frequency is dependent solely on the flare constant for a given horn in air of a given temperature and humidity. The relation between them is as follows:

$$f_c = m\frac{V}{4\pi}$$

where,

f_c is the cutoff frequency, in hertz,
m is the flare constant, in inverse inches,
V is 13,500 inches per second, the velocity of sound in air at 20°C.

Linear measuring units must be consistent throughout and in this case are inches because they are the units most appropriate to horn structures of practical size.

We are primarily interested in the type of flare required for a given cutoff frequency. A convenient and commonly used method of specifying flares is that of stating the distance along the axis of the horn over which the cross-sectional area doubles. Doubling the area is the same as multiplying the diameter by 1.414 (the square root of 2); consequently, the shape of the desired horn can be laid out if a series of diameters at the proper distances from the throat are plotted.

The relation between the area-doubling distance (x_D) and the cutoff frequency for exponential horns is plotted in Fig. 6-15. Notice

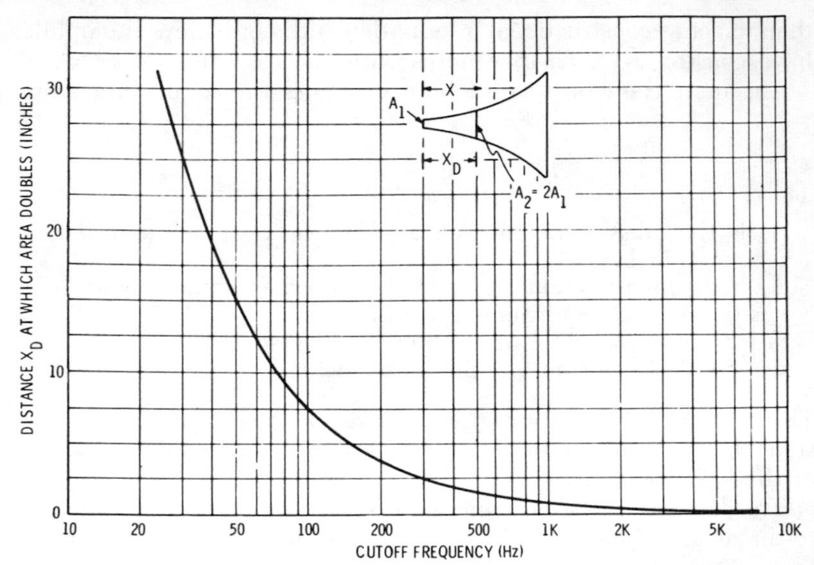

Fig. 6-15. Relation between cutoff frequency and distance in which cross-sectional area doubles for exponential horns.

that a horn must have a doubling distance of 25 inches for a 30-Hz cutoff frequency. Another horn, which need only be used for frequencies above 3000 Hz, can double its area each quarter inch! This indicates why tweeter horns are relatively short and flare out very rapidly, especially if designed for a high-frequency crossover such as 3000 Hz.

If the cutoff frequency were the only consideration, the throat and mouth diameters could be made so nearly the same that even with a small flare (large x_D) for low frequencies the horn could be made short. For the best frequency response, however, there are important reasons for keeping the throat small and the mouth large; they are as follows:

1. When the horn is to be used for full-frequency coverage, the throat must be small to couple properly to a small diaphragm. If the diaphragm is not kept relatively small, it suffers attenuation at the high frequencies because of its mass, just like the cone. If the throat diameter is not nearly the same as the diaphragm diameter, there is loss of energy in the acoustic transfer from the diaphragm to the horn. When the horn is to be used only at low frequencies, such as for a woofer in a dual system, then the throat can be very large.

2. If the mouth is not made large enough, the sound tends to be reflected back toward the throat, and serious attenuation of the low-

frequency components takes place. For this reason, the diameter of the mouth should be kept to a minimum of a half-wavelength at the lowest frequency to be reproduced. A wavelength is equal to the distance the sound wave travels during the period of one cycle. It is thus equal to 13,500 inches per second (the velocity of sound) divided by the frequency. Since we are interested in a half-wavelength and this is our suggested minimum size for the mouth, the following relation is pertinent:

$$D_M = \frac{6750}{f_l}$$

where,

D_M is the diameter of the mouth, in inches,
f_l is the lowest frequency to be reproduced, in hertz.

For example, to maintain good operation down to 67.5 Hz, the mouth diameter should be 100 inches. This gives an idea of just how bulky horns with good low-frequency response could become.

All of the foregoing discussion of horns has implied the use of a circular cross section, but a square cross section can be used also. The same relation holds as far as cross-sectional area is concerned, except that the dimensions of the sides instead of the diameter are given.

When a low-frequency horn is designed, the cutoff frequency closely approximates the lowest frequency to be reproduced. However, a tweeter horn must have a cutoff frequency appreciably lower than the crossover frequency at which the tweeter stops operating. This is because crossover should be gradual rather than sharp, as explained further in a later discussion of crossover systems.

A horn may be driven by any of the previously described driver types. Tweeter horns sometimes feature crystal or capacitor drivers, but in general the moving-coil dynamic driver is most common. This arrangement is illustrated in Fig. 6-4, which shows the magnet, voice coil, and small diaphragm of a typical unit. As previously explained, the diaphragm is small enough to vibrate with efficiency at the highest frequencies in the reproduced range. If the diaphragm were open to the free air, it would have very poor efficiency at low frequencies because of the small area of contact; however, because of the transformer action of the horn, the large area of air at the mouth of the horn is effectively coupled to the small diaphragm area in the driver.

The diaphragm is closed in on all sides, except for the port which accommodates the throat of the horn. The space between the diaphragm and the throat of the horn is known as the *sound (air) chamber.* In air chambers of simple annular shape, attenuation at high frequencies is sometimes encountered. At high frequencies, the wavelength is small and the different portions of the diaphgram are at different distances from the mouth of the horn. This means that appreciable phase differences appear, and resultant cancellations occur between the high-frequency sound components coming from different parts of the diaphragm, as illustrated in Fig. 6-16A. To overcome this phase problem, special chamber designs like those of Fig. 6-16B are often used. These make it necessary for all the sound energy to flow through ports of roughly equal length to the horn mouth, thus minimizing high-frequency phase differences and cancellation. This path equalization is also aided by use of a curved diaphragm like that shown.

Up to this point, we have been discussing straight horns, that is, horns whose axis is a straight line. The advantages of horns can also be obtained by using the same flare and by curving or folding the length of the horn to have space. Folded horns in high-fidelity systems are most often employed as, or in conjunction with, speaker enclosures. They are discussed later in the chapter on that subject.

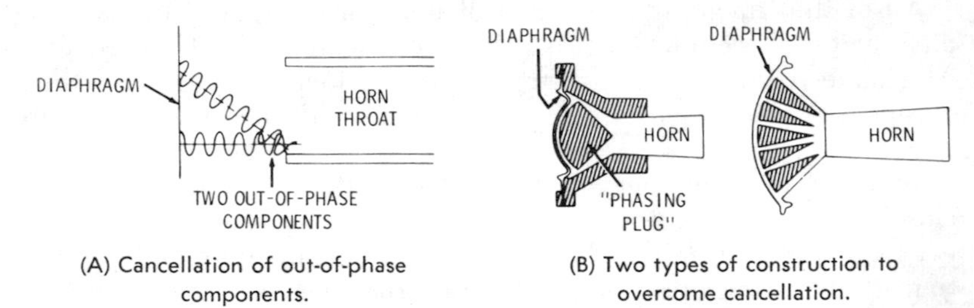

(A) Cancellation of out-of-phase components.

(B) Two types of construction to overcome cancellation.

Fig. 6-16. Methods for overcoming out-of-phase cancellation of high-frequency components in the air chamber of a horn driver.

DIRECTIVITY OF BASIC UNITS

It is desirable that listeners in any part of a listening area receive sound of the same quality. This ideal can be approached but never quite reached because speakers have a directivity characteristic. As could be expected, greater volume of sound is obtainable from the front of a cone or the mouth of a horn than from other parts of the radiator. However, overall volume loss with direction is not so important as long as there is a reasonably low level of distortion and good balance of frequency components. It is the change of directivity of a speaker with respect to frequency which constitutes an important problem.

Radiators of both the cone and the horn types tend to concentrate radiation of the high-frequency components of sound in a narrow cone about the axis of the radiator. The degree of directivity of a speaker is indicated by a directivity pattern, the basic function of which is indicated in Fig. 6-17. The axis of the radiator is considered the reference line with an angle of zero degrees. Directivity patterns are normally shown as a top view in a horizontal plane through the radiator axis. A cone or a circular or square horn in free space should have the same pattern in a vertical plane; but, of course, room reflections and speaker mounting may cause it to be different. The pattern line in Fig. 6-17 indicates the relative sound intensity radiated in any direction by its distance from reference point *O* in that direction. For example, line *OA* indicates by its length that the

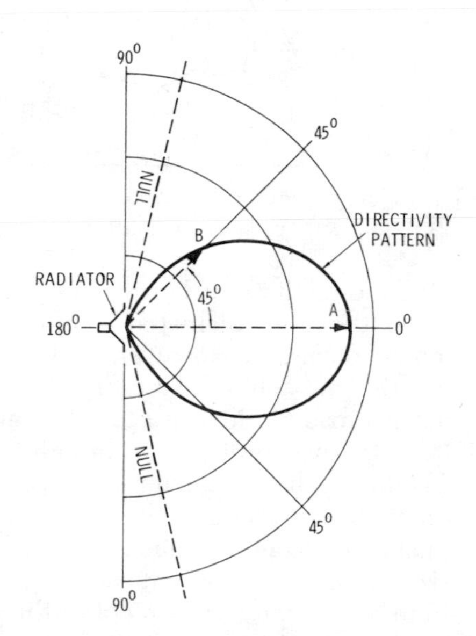

Fig. 6-17. Radiation pattern of a typical cone at one frequency.

sound radiated along it is a maximum compared with that in any other direction. At a 45-degree angle, line *OB* is a measure of the relative sound intensity in that direction. Since *OB* is only about half as long as *OA*, a listener along that line would hear only about half the volume that a person along *OA* at the same distance from *O* would hear. At angles near 90 degrees, the pattern indicates zero radiation; of course, in any practical setup, such a zero area would not exist because sound would reach there by reflection.

Because directivity normally varies considerably with frequency, a complete diagram must show separate patterns for each of at least several frequencies. Typical variation of directivity with frequency for a 12-inch cone is illustrated in Fig. 6-18. It is assumed that the speaker is mounted in an infinite baffle (baffles are discussed later). Notice how much narrower the radiation pattern is at highs than at lows.

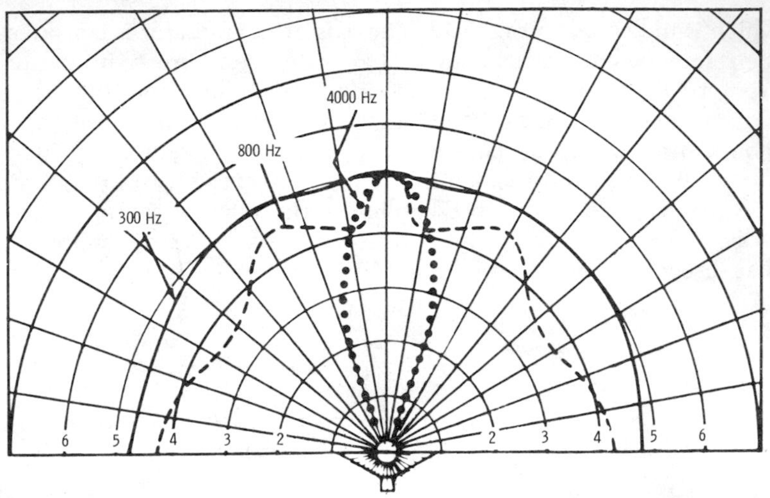

Fig. 6-18. Variation of directivity with frequency for a typical cone radiator.

The directivity of horns is not much different from that of cones when the cone diameter is approximately the same as the horn mouth diameter. However, at low frequencies at which the wavelength approaches the mouth diameter, the horn directivity becomes much broader. Contrary to what might be expected, directivity for these frequencies becomes broader as the mouth diameter decreases. Then at higher frequencies at which the mouth diameter is several times the wavelength, the directivity narrows slightly as the mouth diameter becomes smaller. All in all, for cones and horns of practical sizes, it may be said that the directivity of the cone is little different from that of the horn, except at the highest frequencies, at which the horn gives wider distribution.

The limited directivity of speaker radiators at high frequencies is a factor given considerable attention in speaker system designs. One approach is to use more than one speaker, pointing each of several units in a different direction. This is more frequently done with horns than with cones for the following two reasons:

1. Because of the small throats of the horns, they are much easier to mount at an angle and close together than the cones.
2. Horns suitable for high frequencies are more compact and give slightly better distribution.

A typical example of how a number of horns may be combined to give better high-frequency distribution than is possible with one horn is shown in Fig. 6-19. This type of structure is very popular for tweeters in dual systems and is frequently referred as to a *multicellular horn.* With such an arrangement, distribution of high-fre-

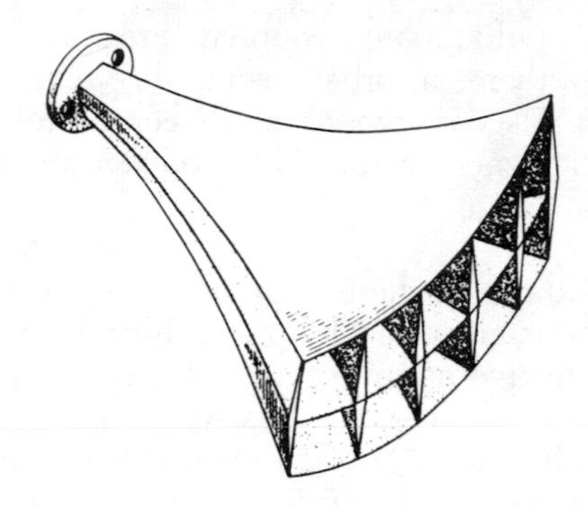

Fig. 6-19. A multicellular horn.

quency sound components can be made almost as broad as that of the low-frequency components. In most cases, the throats of all the horn units are fed by the same driver unit; in a few more elaborate installations, separate drivers are used.

DUAL AND MULTIPLE SPEAKERS AND SYSTEMS

It has previously been explained that a simple single-cone speaker has definite limitations as far as frequency range is concerned. For good low-frequency response, the cone should be relatively heavy, its suspension should be as soft as possible, and its area should be as great as possible. At the highest audio frequencies, these measures are all detrimental, and the cone should be as light and small as possible with a stiff suspension. Thus, we have the generally accepted conclusion that a cone of conventional design cannot produce acceptable high-fidelity response (from 60 to at least 12,000 Hz). This conventional cone, if designed for reasonable low-frequency response, ordinarily becomes unacceptable at about 8000 Hz. In some cases, the fall-off at the high end is such that it can be at least partially compensated for by treble boost in the amplifier, but, in many speakers, "hitting the highs harder" leads to annoying distortion.

All speaker designers agree about this limitation, but they do not agree about the best way to overcome it. There are three main approaches to the problem of extension of frequency response of the conventional speaker to satisfy high-fidelity requirements:

1. Special design of a single cone to extend its response.
2. Combination of two radiators, one for high frequencies and one for low frequencies, into one physical assembly, or closely attached to each other usually along a common axis (exemplified by the coaxial type of construction). Sometimes three radiators are used in the same arrangement. In some units, the

radiators are coupled through mechanical compliance between them; in others, separate voice coils are employed.

3. Use of two or more completely separate speakers, each designed to reproduce only a specified portion of the frequency range.

No one of these approaches is universally recognized as best. The proponents of each approach present convincing arguments, but the subjective nature of any final test has prevented any obviously conclusive choice. It is this which helps lend fascination to the pursuit of high fidelity, and the reader can expect to enjoy many hours of speculation concerning his own choice of a speaker system. Rather than favor one method over another, we present the most common arguments for each system, and this should equip the reader to form his own opinions.

Single-Cone Radiators

This type is favored by its proponents not only because of its relative simplicity but also because it is claimed that separate tweeters (high-frequency radiators) have a tendency to become "fuzzy" because of a phenomenon called *rim resonance.* The contention is that the rim of the tweeter cone or horn resonates at some high frequency and oscillates at that frequency, causing interference when high frequencies are being reproduced. It is claimed that a single cone can be so designed that a high-frequency portion in the center will operate independently at high frequencies and that it will be loaded by the outer portion of the cone to prevent rim resonance. It is also claimed that no dual arrangement can make the high- and low-frequency sounds appear to the listener to be coming from the same source, and that the difference in construction of the high- and low-frequency portions causes detectable quality differences (coloration) to the trained ear. In the single-cone arrangement, both high- and low-frequency sounds emanate from the same cone; this is supposed to eliminate the problem of duality.

Coaxial Arrangement

Proponents of this type point to the limitations of response of a single-cone unit because of the conflicting requirements for size, mass, and suspension at the two ends of the audible frequency range. This means that some additional unit must be introduced to divide the frequency range into two or more parts; and with each part designed for optimum operation in its own range, the best response and minimum distortion are obtainable. The coaxial enthusiast, although favoring a dual arrangement, eliminates the completely separate systems because of the danger of phase differences and the tendency claimed that the high- and low-frequency sounds seem to come from different sources. (This is a similar argument to that used by single-cone proponents.) To minimize such spatial distortion, this design has the tweeter radiator right inside the low-frequency (woofer) cone and coaxial with it; therefore, the apparent source of both frequency-range components is the same.

Separate Woofer-Tweeter Arrangement

Those who favor the woofer-tweeter arrangement argue that interaction and loading resulting from the placement of the woofer and tweeter together in the coaxial arrangement cause distortion not present in separate arrangements. They state that there is also a rather narrow distribution of the high-frequency components of reproduced sound from the coaxial type, and this distribution is due at least partly to the action of the woofer cone as a wide-mouth horn at the high frequencies. This can be overcome by separating the tweeter so that its energy distribution will not be influenced by the woofer.

As can be concluded from the review of pros and cons, each approach has inherent potential advantages and weaknesses. However, the designers and manufacturers of the better speakers of each type have taken measures to minmize each weakness, and speakers of high quality can be obtained in any of the three categories. We reiterate that the choice is the buyer's and should be exercised only after careful consideration of all claims, plus his own application problem. Even better, listen to each type under as well-controlled conditions as possible; but unfortunately it is seldom practical to find these conditions.

CONSTRUCTION FEATURES OF SPEAKERS

To help in a study of the various models of speakers in all categories, let us now consider some of the constructional features which are used to ensure high-fidelity performance.

Extended-Range, Single-Cone Type

Some of the measures taken to extend the response of a single cone have already been mentioned. One of the most important is the division of the cone into two parts: (1) one part which resembles a small cone and which is the center portion of the main cone, and (2) the second part, which is coupled to the first (the remainder of the main cone) by a compliance which extends this second part to its full dimension. This is illustrated in Fig. 6-20. The high-frequency portion of the cone is connected to the remainder of the cone through a mechanical compliance which is a ring of softer material than the cone. This compliance material allows the center high-frequency portion to operate as a separate unit but transmits low frequencies to the remainder of the cone with a blending action in midrange. The whole cone assembly acts together as a low-frequency radiator.

The cone, especially its center portion, of this type of unit is made with a curved flare. As previously explained, this helps high-frequency response. Frequently, the center portion of the cone is also made of harder material than the outer portion; this helps the center portion to operate independently at the high frequencies. An example of another type of extended-range speaker is shown in Fig. 6-21.

Coaxial Type

The coaxial principle is probably exemplified in more different commercial models than any other. We cannot review all combinations and types, but a few representative ones will give the reader sufficient general information to recognize the others. First, consider the generalized diagrams of the two main types of dual coaxial units shown in Fig. 6-22. Fig. 6-22A shows the type employing a single

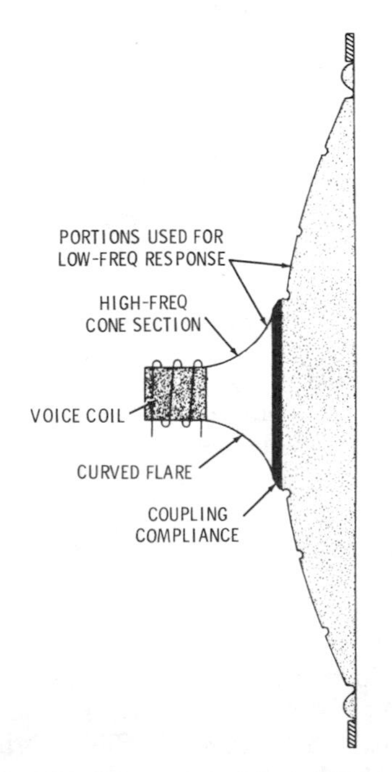

Fig. 6-20. Construction of a typical single-cone, extended-range speaker.

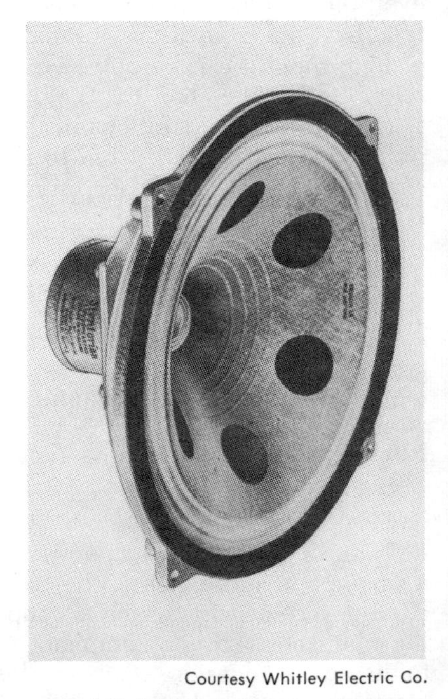

Courtesy Whitley Electric Co.

Fig. 6-21. Typical single-cone, extended-range speaker.

voice coil for both high and low frequencies. The voice-coil form is fastened rigidly to the tweeter radiator so that the high-frequency components will be efficiently transmitted to it. It is fastened to the woofer cone through a soft mechanical compliance which transmits low-frequency components but which tends to reject high-frequency components and keep them out of the low-frequency radiator. In this

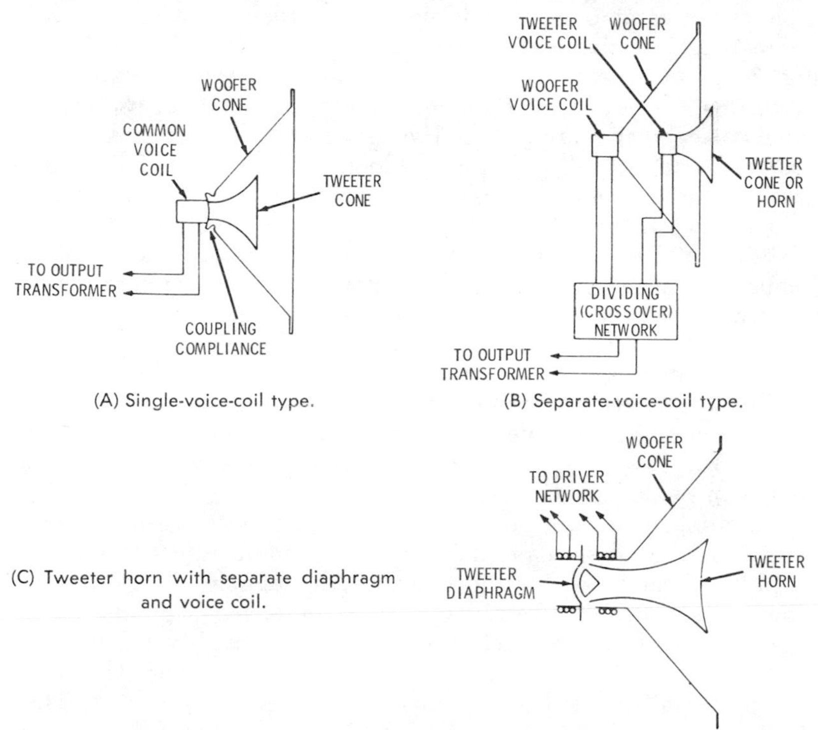

(A) Single-voice-coil type.

(B) Separate-voice-coil type.

(C) Tweeter horn with separate diaphragm and voice coil.

Fig. 6-22. Basic arrangements employed in coaxial speakers.

way, the physical construction causes the unit to function as a mechanical divider network which automatically separates high- and low-frequency components. It will be noted that actually the extended-range, single-cone speakers described previously are similar to the coaxial type because the tweeter section operates separately in its range, and because the total cone and the high-frequency portion are mounted coaxially with each other.

The other general type of coaxial unit is shown in Fig. 6-22B. In this type, each radiator has its own voice coil. Because the voice coils are separate, an electrical dividing network must be employed; therefore, only the high-frequency af currents are fed to the tweeter voice coil, and only low-frequency currents are fed to the low-frequency voice coil. If appreciable signal power at frequencies outside its intended range is applied to either voice coil, distortion and overloading will result. Divider networks, also called crossover networks, are discussed later.

When a horn is used for the tweeter, a separate diaphragm and voice coil must be used to excite it, as illustrated in Fig. 6-22C.

Some representative coaxial speakers as they appear in models commercially available are illustrated in Fig. 6-23. The speaker in Fig. 6-23A may be considered a transition between the extended-range single-cone and the duocone (single voice-coil) coaxial types. Small conical domes are fastened to a special corrugated single-cone structure, and a small radiator is added in the center. The domes break up the surface of the cone for the high-frequency components, reducing losses and helping to distribute radiation better.

The type of coaxial unit in Fig. 6-23B employs a separate high-frequency cone coupled to the common voice coil through a mechanical connection. Besides its function as a high-frequency radiator, the small cone is said to improve low-frequency response by addition of its mass to that of the large cone, and to act as a diffuser for the large cone in the middle range of frequencies.

A coaxial unit employing a separate cone-type tweeter is shown in Fig. 6-23C. The tweeter is mounted on brackets fastened to the metal rim-support frame of the woofer.

The problem of proper distribution of high-frequency sound components in coaxial speakers has led some manufacturers to the use of multicellular horns for the tweeter. An example of this type is shown in Fig. 6-23D. The tweeter has a completely separate driver of the type required for excitation of a horn. This driver fits inside the woofer, which has a voice coil 3 inches in diameter. The horn is a single unit divided into sections by baffles at the mouth. This division into sections directs the high-frequency components over a wider radiation angle than would be obtainable without such division. In the unit illustrated, the flare cut-off of the horn (which is exponential) is 1800 Hz, which is far enough below the 3000-Hz crossover frequency to ensure smoothness in the transition between woofer and tweeter. Fig. 6-24 shows another method of achieving better high-frequency distribution with coaxial design and the use of an acoustical lens in the high-frequency horn.

Separate Woofer-Tweeter

Systems employing physically separated woofers and tweeters are commonly custom-built or at least are composed of units by different manufacturers. A few systems are sold by one manufacturer as integrated units, but we shall concentrate on the separate woofer and tweeter units and how they are combined.

Courtesy RCA

(A) RCA Duocone.

Courtesy North American Philips Corp.

(B) Philips Type 971 OM.

Courtesy Quam-Nichols Co.

(C) Quam coaxial speaker.

Courtesy Altec Lansing

(D) Altec Type 604C.

Fig. 6-23. Various types of coaxial speakers.

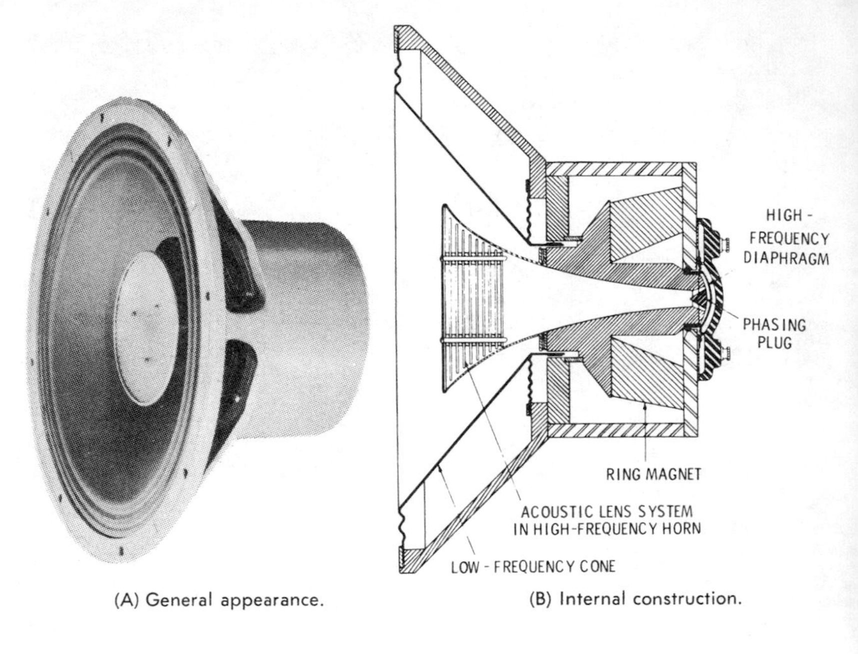

(A) General appearance.

(B) Internal construction.

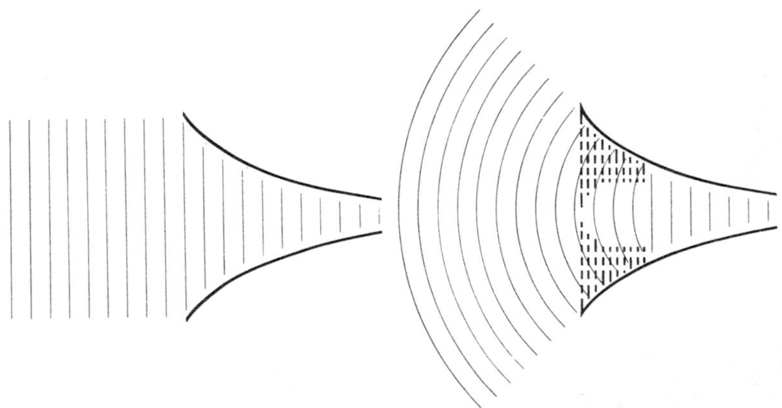

(C) Distributions with and without acoustic lens.

Fig. 6-24. Coaxial speaker with horn-type tweeter and acoustic lens.

Because the tweeter takes over above the crossover frequency, the woofer does not have to have extended range in the high-frequency direction as is the case with full-coverage, single-cone speakers. Full-coverage, single-cone speakers are frequently good woofers but are overly expensive for the purpose because the special design effort expended in improvement of their high-frequency response is wasted in this application. On the other hand, just any good-sized cone speaker is not necessarily a good woofer because, in spite of their large size, single-cone speakers do not always have good low-frequency response. Low-frequency response is doubly important in a woofer: (1) because this is the primary function of the woofer, and (2) because reproduction does not seem balanced if high frequencies are well reproduced by the tweeter without the low frequencies, or vice versa. The woofer should therefore have a resonant frequency as low as possible, with 50 Hz or lower being a rough guide. As will presently be explained, the resonant frequency of the low-frequency system (as opposed to the speaker unit itself) depends on the type of baffle or enclosure used. However, the speaker should be designed and constructed so as to have as low a self-resonance as possible in any event.

Although any good conventional speaker with the aforementioned features will do as a woofer, greatest economy can be effected in the purchase of a unit expressly designed as a woofer. The economy arises from the fact that most conventional speakers when designed for better-than-average low-frequency response are also designed with some improvement in high-frequency response in mind. The latter is, of course, not needed in the woofer application; however, the woofer must have good response well beyond the crossover frequency to ensure good transitional operation. This is no problem at all when low crossover frequencies of 300 to 1000 Hz are used, but some speakers may start to show some drop-off or other poor response characteristics when the crossover frequency is as high as 3500 Hz (in which case the woofer should have good response to about 4500 Hz).

As mentioned, speakers designed expressly and solely as woofers in dual or multiple systems are few and far between, so no representative types are illustrated here. What many audiophiles do is to build up their speaker systems gradually. For example, a first step could be a full-range, single-cone speaker with carefully checked low-frequency resonance. This unit can be used alone for very good high-fidelity reproduction until a tweeter and suitable divider network can be purchased. With the addition of the latter, the speaker system is extended to the dual type.

When a speaker is to be used only for reproduction of low-frequency components, in other words as a woofer, it is sometimes mounted in back of a slot instead of a circular opening. The slot is usually rather narrow, and its length is less than the diameter of the speaker cone. (Such a slot is shown schematically in Fig. 7-2C in the next chapter). Such a slot acts as an acoustic low-pass filter. It attenuates all sound components above a certain frequency. It also improves the loading and impedance match to the speaker cone at low frequencies.

Unlike woofers, tweeters cannot normally be used alone. All types therefore have the same special purpose—to reproduce only the higher-frequency portion of the audio-frequency range. One of the most important characteristics of a tweeter is its low-frequency cutoff point. The minimum frequency of operation of the unit must be below the crossover frequency of the system so that it will overlap the range of the woofer. If a low cutoff frequency (300 to 1000 Hz) is employed, the range of the tweeter from the crossover frequency to the limit of audibility for full-range fidelity in the system makes it difficult to obtain a unit with uniform response, minimum distortion, and wide-angle distribution over that range. On the other hand, relatively economical and simple driver and radiator arrangements will handle the range necessary with a high crossover frequency such as 3000 or 3500 Hz.

The construction of separate tweeter units is much like that of the tweeter portions of the examples of coaxial units given in the previous section. Horns are more popular than cones because of their greater efficiency and greater potential frequency range. A typical tweeter assembly is shown in Fig. 6-25.

Standard stock models of cone-type speakers can sometimes be used as tweeters. However, it must be emphasized that just because a speaker is small it is not necessarily a good tweeter. In fact, most ordinary small speakers are not good tweeters and often have no better high-frequency response than an average woofer. Obviously, such a speaker would provide no improvement at all over just a woofer and would be likely to add considerable distortion resulting from application of high-frequency signal components (from the

Courtesy Electro-Voice, Inc.

Fig. 6-25. Typical tweeter-unit construction.

divider network) to which it does not properly respond. If you are able to hear tones up to 15,000 Hz very well,* it is advisable to test a prospective tweeter speaker by applying a signal from an audio signal generator or oscillator and by noting how well the output holds up to the limit of your own audibility. In making the test, it is important to remember that as frequency rises, directivity sharpens; so, be sure to stay directly in front of the cone when listening for the upper-limit signal.

Multiple Combinations

Some audio engineers believe that the audio-frequency spectrum should be divided into more than two parts with speakers for each part for proper full-range reproduction. For example, systems employing three or four speaker radiators are commercially available—some in the coaxial form, others in the separate form. The use of three or more ranges will reduce the width of each range so that uniformity of coverage for each unit is much more easily obtained. For example, the woofer need cover only from the low limit, say about 30 to 60 Hz, to about 500 or 1000 Hz. For this range, it is certain that the woofer cone will operate as a whole and that no attenuation due to breakup will occur. Another separate radiator is employed for the middle range of frequencies, from 500 to 1000 Hz to about 3000 to 8000 Hz, with the exact limits depending on individual design. This middle-range speaker unit is sometimes referred to as the *squawker,* and the three-speaker system is known as a *woofer-squawker-tweeter* combination. Because of the extended range of the squawker over the normal top frequency of a woofer, the tweeter can start at a relatively high frequency, and its design requirements are not so rigid as for a tweeter in a two-way system. An example of a three-way coaxial or triaxial speaker is shown in Fig. 6-26. The woofer and squawker cones are mechanically connected (or divided) in a duocone arrangement. The tweeter, with its own separate driver, is mounted inside the squawker. An electrical crossover network is used to divide the amplifier output between the tweeter and the woofer-squawker unit.

Commercial units are also available for systems employing four sections. Typical frequency ranges are: low-bass section, 35 Hz to 200 Hz; mid-bass section, 200 Hz to 600 Hz; treble section, 600 Hz to 3500 Hz; very-high-frequency range, 3500 Hz and above.

Multiple systems have the advantage of allowing relatively simple driver-radiator units to be used for each range. Since the unit for

---

*Children and adolescents can often hear sounds of frequencies up to 18,000 Hz, but each person's limit of audibility decreases with age and often drops to 10,000 Hz or lower.

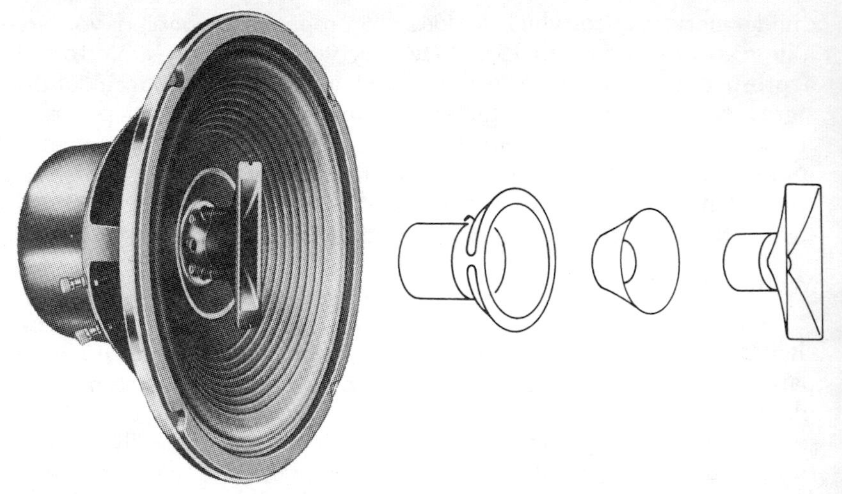

Fig. 6-26. A three-way coaxial speaker.

each range does not have to work as hard as in single-radiator systems, uniform frequency response should be and usually is easily obtained. However, in the rush to a dual or multiple system, the reader should not forget that there are other things of equal and sometimes even greater importance. The changes in directivity and in fine shadings of harmonic content between different components in a multiple system sometimes cause a consciousness in the listener of the takeover from one unit to the other. Some critics claim the apparent change of source location, particularly between the high and low notes of a single instrument, is evident in some systems. Then, in addition, if there is even a slight slip-up in the design or fabrication of the crossover network or the assumed takeover frequencies, severe distortion may result. In multiple systems, there are more parts which can self-resonate and cause trouble. These things are brought up not as pure criticism but to give the reader as much perspective as possible. While the dual or multiple system may be the complete answer for some enthusiasts, it is by no means a panacea in general. It should be remembered that any system—single, dual, or multiple—can give excellent results only when carefully designed and installed.

ELECTRICAL DIVIDER (CROSSOVER) NETWORKS

In dual or multiple speaker systems, the audio-frequency energy from the amplifier must be divided so that only the appropriate frequency components are fed to each unit of the system. In most cases, the individual parts of the system, although designed for optimum operation in their specified respective portions of the frequency spectrum, are subject to distortion and sometimes even overheating if they are driven to full power rating at frequencies outside their normal range. This is an important factor adding to the more obvious one: overall efficiency is substantially reduced by feeding too much low-frequency energy to a high-frequency unit and high-frequency energy to a low-frequency unit.

In dual systems employing a common voice coil, the mechanical compliance between the high- and low-frequency radiator sections of the cone divides the energy after it has been converted to mechanical motion of the voice-coil form. The compliance acts as a low-pass filter, eliminating most of the high-frequency components from the woofer, or larger section of the cone. Low-frequency energy is fed to the high-frequency portion of the radiator, but not to it alone, because at low frequencies it acts only as part of the total mass composed of both the low- and high-frequency portions.

When each of the units in a multiple system has it own voice coil or at least two have separate voice coils, the division of low- and high-frequency energy must be done electrically by divider networks.

The simplest type of divider network consists merely of a single capacitor, as illustrated in Fig. 6-27. The fact that the reactance of a capacitor is inversely proportional to frequency is employed to distribute the audio signal. In the arrangement of Fig. 6-27A, the tweeter and woofer voice coils are connected in series, and a capacitor is connected across the woofer. The value of capacitance is made such that at frequencies above the desired range of the woofer the reactance of C becomes so low that it shunts the woofer (C acts as a bypass caapcitor). Low-frequency components can be kept out of the tweeter if a parallel connection of the voice coils is used with a capacitor in series with the tweeter circuit, as illustrated in Fig. 6-27B.

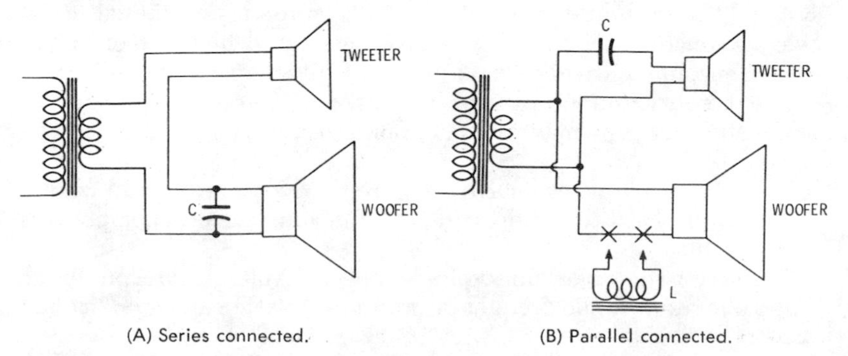

(A) Series connected. (B) Parallel connected.

Fig. 6-27. Simple divider circuits employing single reactances.

Inductance can be used with capacitances to make the divider network more complete. For example, in Fig. 6-27B the inductance (L) can be connected in series with the woofer as shown. The inductance, the reactance of which increases with frequency, chokes the high-frequency components out of the woofer, and the capacitor (C) blocks low-frequency components out the tweeter. The values of C and L must be such that the reactance in each case is about equal to or a little lower than the voice-coil impedance in the frequency range to be attenuated.

A capacitor or inductor provides gradual attenuation with frequency, as the range of undesired frequency components is approached. Although the crossover range should not be too narrow, simple reactance circuits as in Fig. 6-27 are ordinarily too broad in the changeover region. Instead, a combination low-pass (for the woofer) and high-pass (for the tweeter) filter circuit is usually employed. With this type of circuit, much more rapid attenuation can be made near the crossover frequency than is possible with simple capacitor and inductor arrangements as illustrated in Fig. 6-27. Attenuation of about 12 dB per octave* is considered proper in most applications. Gradual crossover arrangements attenuate at about 6 dB per octave. Filters with sharper cutoff than this can be constructed by use of additional components, but power losses in the filter become excessive, and the additional sharpness is not necesary anyway. A typical divider-network response graph is shown in Fig. 6-28. The curve of woofer output crosses the curve of tweeter output response at the crossover frequency. This intersection is also at the −3-dB, or half-power, level; at the crossover frequency, half the output power is being fed to each unit. From this, it can be seen why the respective individual response characteristics of the woofer and tweeter units must overlap substantially. If the crossover level were lower, there would be a lessening of total output in the crossover region, and this would result in frequency distortion in the system. The dash-line curve of Fig. 6-28 represents a gradual crossover attenuation of 6 dB per octave, compared to the more commonly encountered solid-curve value of 12 dB per octave.

The construction of divider networks is not as simple as the schematic diagrams may indicate. Some of the reasons are:

1. The capacitors ordinarily require fairly accurate odd values that are hard to obtain without connecting several components together.
2. There is no polarizing voltage, applied voltages are purely ac voltages at audio frequency, and electrolytic capacitors cannot

*An *octave* is a range in which the frequency doubles.

be used. At the values necessary, other types of capacitors are relatively bulky and expensive.

3. Current at low impedance is appreciable; therefore, fine wire cannot be used for the coils, which have hundreds of turns for the values required. Again, the values are odd, so standard units are not applicable.

To illustrate the problem, the schematic diagram of one of the more popular types of divider networks is shown in Fig. 6-29, with formulas for calculating the values required for any crossover frequency f_c and speaker impedance Z and an example of its use in a

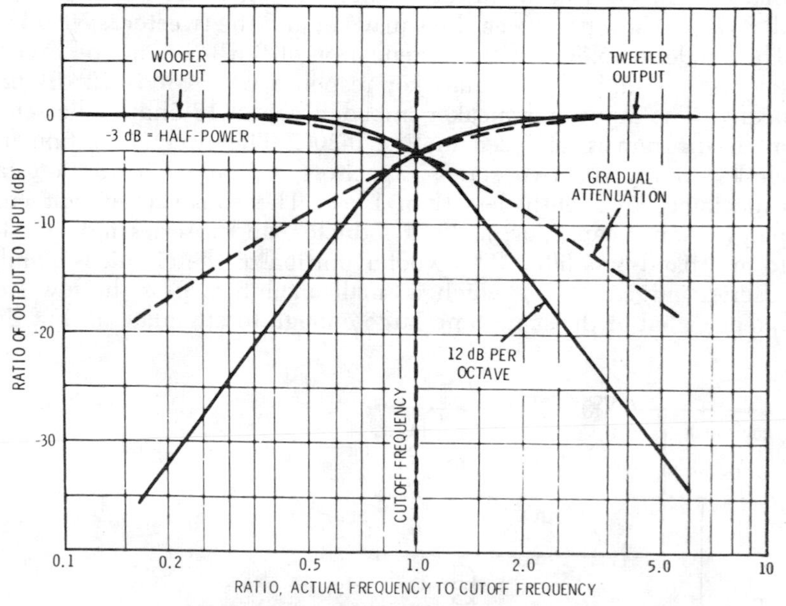

Fig. 6-28. Divider-network response curves.

practical application.* The values obtained in the solution of this example illustrate the previous statements about the odd values of the capacitances involved. The circuit shown in Fig. 6-29 can be expected to give an attenuation of about 12 dB per octave.

Although the difficulties mentioned must be taken into consideration, they are not insurmountable, and some audiophiles prefer to construct their own divider networks. If the reader wishes to do so, it is suggested that he consult some handbook for data as to diam-

*For those familiar with filter theory, these are m-derived half sections with equations simplified for $m = 0.6$.

eter, number of turns, etc., for his calculated inductance values. Although the number of turns and overall size of the coils can be reduced by the use of iron cores, this is not recommended, because the later introduce nonlinearity. The capacitors can be of the oil-filled variety and can be made up from standard sizes connected in series or parallel.

For those who prefer to buy their crossover networks ready-made, the latter are available from a number of manufacturers of audio-frequency equipment. The physical construction, schematic diagram, and response curves of a commercial three-way network are shown in Fig. 6-30. Note from the response curves (Fig. 6-30B) that the crossover between the woofer and squawker is 500 Hz and that the crossover between the squawker and the tweeter is 5000 Hz. This model employs gradual attenuation of 6 dB per octave, except for the low end of the squawker response where it is 12 dB per octave. The input and speaker impedances are 16 ohms. The schematic diagram is analyzed in Fig. 6-30C. The tweeter portion includes simply the two series capacitors, C2 and C3, for a total capacitance of slightly less than 1 μF. The squawker circuit employs a series capacitor, a shunt inductor, and a series inductor to form a band-pass filter. The woofer portion of the circuit is simply a series inductor (L1) which is small enough to pass the low frequencies but at the same time large enough to attenuate all but the

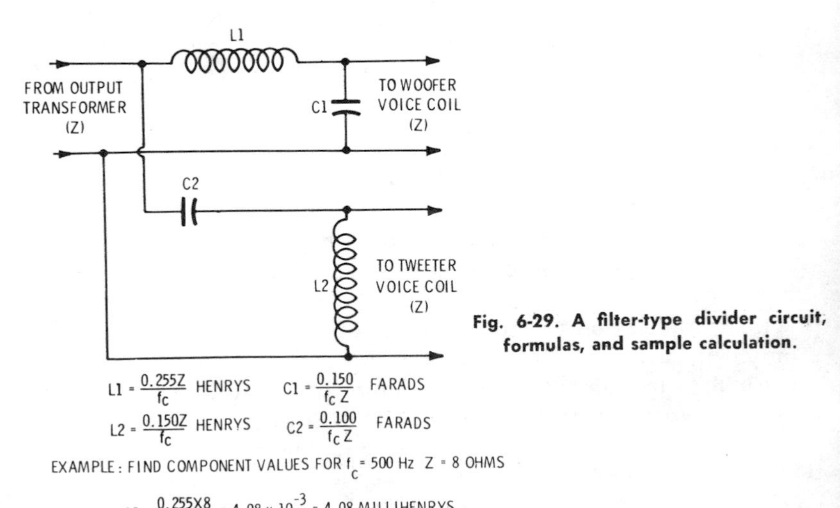

Fig. 6-29. A filter-type divider circuit, formulas, and sample calculation.

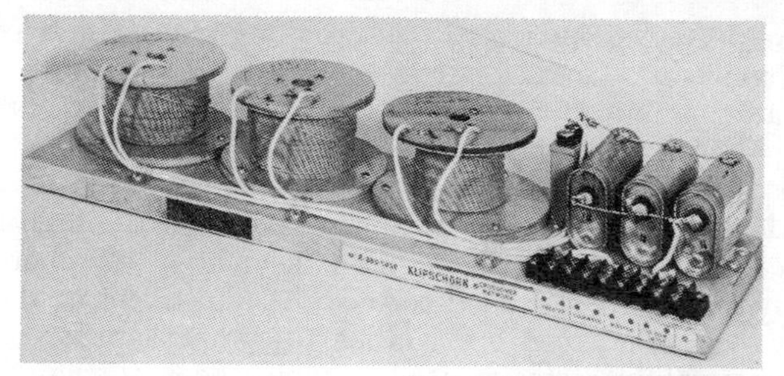

Courtesy Klipsch & Associates, Inc.

(A) Construction.

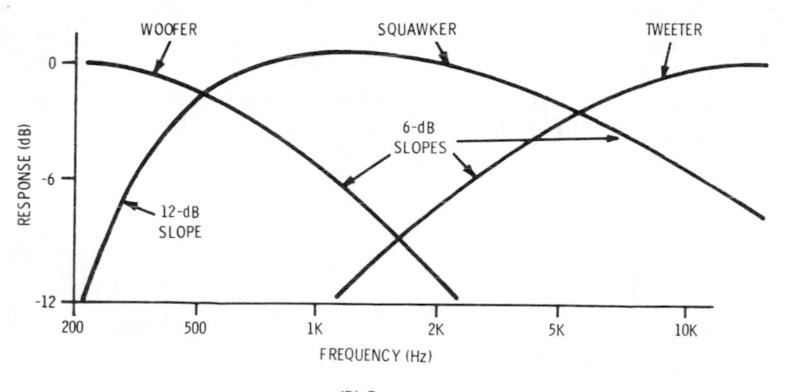

(B) Response.

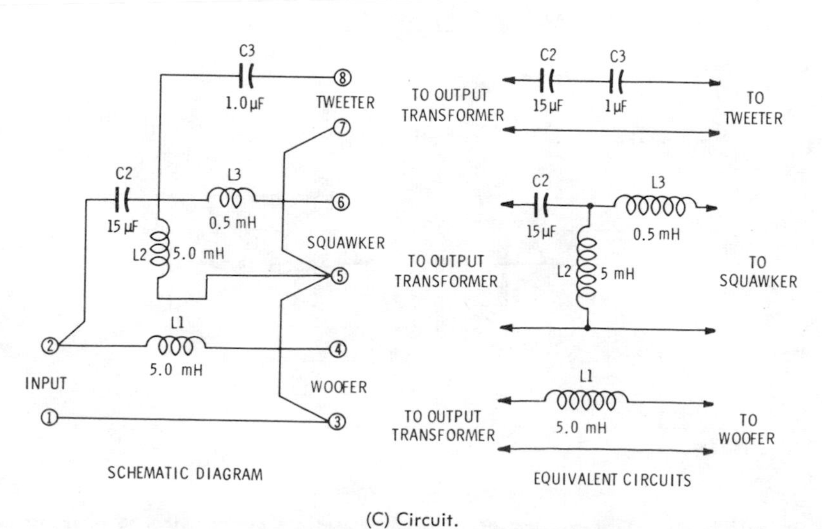

(C) Circuit.

Fig. 6-30. Design features of a commercial three-way divider network.

lowest frequency components in the woofer circuit. The tweeter and woofer circuits are of the single-reactance type and cause the slow rate of attenuation (6 dB per octave). The squawker circuit has more rapid attenuation because it is of the composite filter type.

When a horn-type tweeter is used with a cone-type woofer, there is a tendency toward energy unbalance between the high and low frequencies because of the higher efficiency of the horn. Some divider networks are designed to provide adjustment of the output of either section, relative to the other, to compensate for the difference in reproduced efficiencies. A typical commercial crossover network of this type is illustrated in Fig. 6-31A. The response characteristic of the network is shown in Fig. 6-31B. The response shown applies for the 0-dB position. Note that the tweeter response can be reduced either 2 dB or 4 dB with respect to the woofer response by moving the tap.

Fig. 6-32A shows response ranges for an eleven-speaker system arrangement developed by McIntosh (shown in Figs. 6-32B and

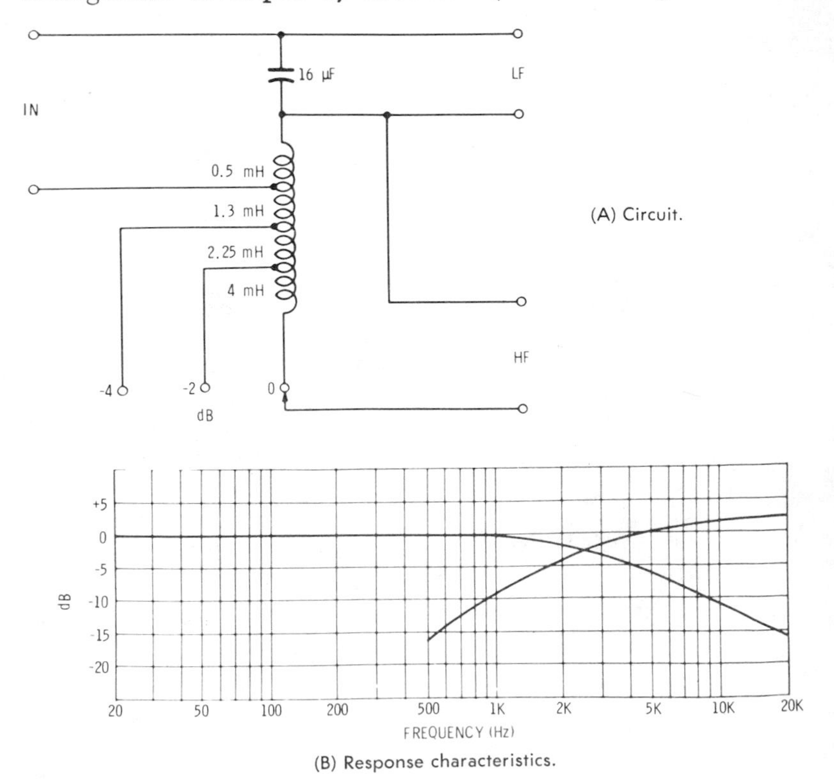

(B) Response characteristics.

Fig. 6-31. A divider network that provides adjustment of tweeter output to compensate for tweeter-horn efficiency relative to woofer efficiency.

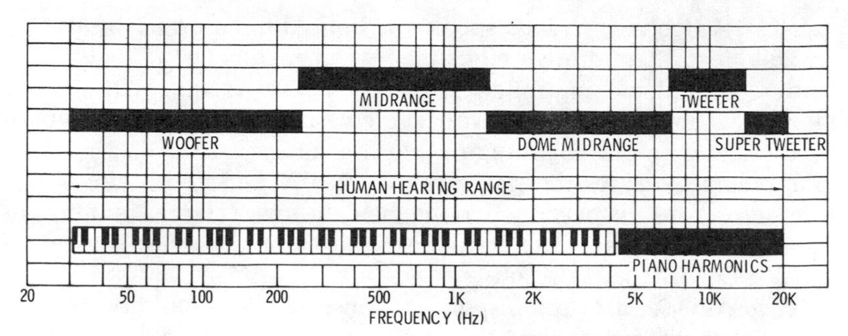

Courtesy McIntosh Laboratory Inc.

(A) Response ranges.

(B) External appearance.

Courtesy McIntosh Laboratory Inc.

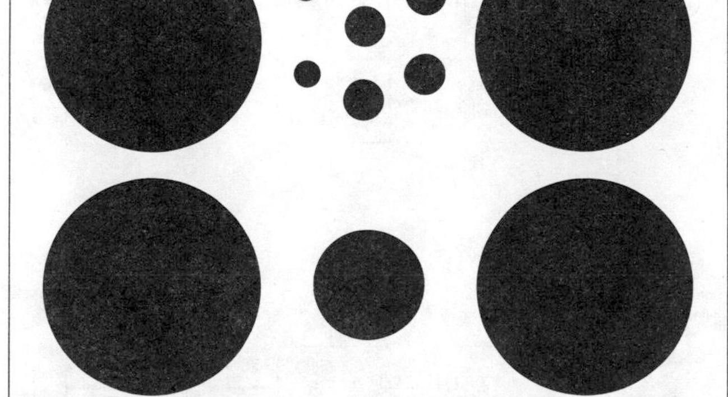

4-10" RADIATORS (12" LOUDSPEAKERS)
1-5" RADIATOR (8" LOUDSPEAKER)
4-1 1/2" DOME RADIATORS
2-COAXIAL SUPER RADIATORS

(C) Speaker arrangement.

Fig. 6-32. Eleven-speaker system.

6-32C). A twelve-inch bass speaker is limited to an upper frequency of 250 Hertz. A midrange eight-inch speaker radiates to 1500 hertz, and several dome midrange speakers radiate from 1500 to 7000 hertz. A compound coaxial speaker continues the radiation up to 14,000 hertz on one diaphragm, and the super-tweeter on the other diaphragm increases the range to over 20,000 hertz with almost flat response. This dispersion arrangement using a large number of speakers allows the listener to hear the complete audio range regardless of listener position.

When divider-network response is considered, it should be remembered that the actual acoustic attenuation at the crossover frequencies may be much greater than that indicated by the electrical circuit. This is because, even when separate voice coils are employed, there may be between units mechanical compliance which acts as an attenuator.

The divider networks described thus far are connected between the output transformer and the speaker units. Because the power level is high and the impedance low, the inductors must be capable of handling fairly high current and are therefore bulky. Because of the low impedance, the capacitors must also be large, as has been explained. For these reasons, some systems provide frequency division in the amplifier ahead of the power-output stage. At that point, the power level is so low and the impedance so high that simple resistance-capacitance–type filters can separate the high-frequency components from the low-frequency components. From this dividing point to the speakers, two separate channels are provided. There are separate output amplifiers and separate output transformers, one for each channel, as illustrated in Fig. 6-33. This is not as expensive an arrangement as it might seem at first because, since each channel

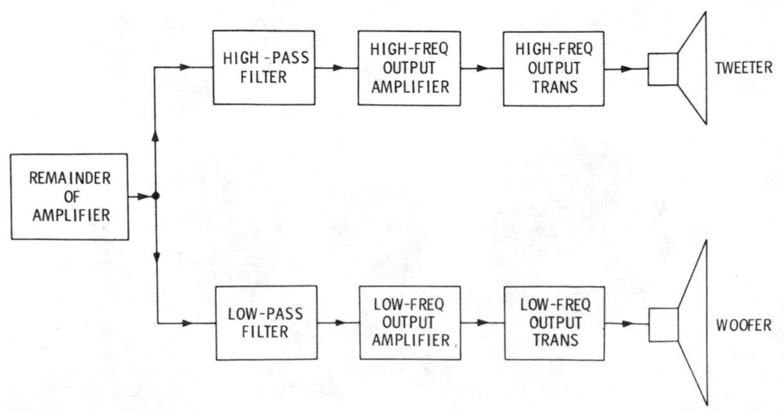

Fig. 6-33. Dual-speaker system in which frequency components are separated at low power level and fed through separate output stages to speakers.

Courtesy Altec Lansing

Fig. 6-34. Altec Lansing 771B Electronic Crossover Biamplifier.

need have only a limited frequency response, the output transformers can be relatively inexpensive.

Fig. 6-34 shows the Altec Lansing Electronic Crossover Biamplifier, which provides built-in crossover circuitry prior to the power amplification with separate low-frequency and high-frequency amplifiers in a single compact unit. The electronic crossover circuit divides the input signal into two signals, one feeding the low-frequency amplifier (bass) and the other feeding the high-frequency

Fig. 6-35. Typical dual-voice-coil unit for use as middle speaker in stereo arrangement.

Courtesy University Sound

amplifier (treble). The low-frequency amplifier provides 60 watts to drive the bass speaker, and the high-frequency amplifier provides 30 watts to drive the treble speakers. In this way, more effective use of the full power output makes available more speaker drive where it is most needed (bass).

In addition, extreme low-frequency power demands do not affect high-frequency reproduction because separate amplifiers drive separate bass and treble speakers. Separate external crossovers are not needed. The crossover frequency is switchable to 500 Hz, 800 Hz, or 1500 Hz with 12 dB/octave slope to adapt to several combinations of speakers. This unit may be mounted inside the speaker cabinet and driven directly from a tuner.

CENTER SPEAKER FOR STEREO

Individual speaker requirements for stereo reproduction are substantially the same as for monophonic reproduction. The same low-distortion and wide-frequency-range requirements must be met in each unit of the stereo system.

One special feature developed especially for stereo is the dual voice coil. A typical speaker of this type is illustrated in Fig. 6-35. Two voice coils are wound on the same form on the cone structure of the speaker. This arrangement forms a convenient method for mixing the left and right stereo signals for the center speaker. Since the center speaker is usually required only to reproduce the lower frequencies, which are not as directional, the dual-voice-coil speaker is usually a woofer. It is normally used with additional left and right speakers, which must reproduce only the frequency components above approximately 200 Hz.

7

Speaker Baffles and Enclosures

In our discussion to this point, we have considered cone-type radiators as though all the sound energy were released from the front of the cone. This is essentially what we meant in an earlier statement that direct-radiator speakers would be considered as being mounted in an infinite baffle. Now that we have considered the design features of speakers, we shall proceed to describe the effect of their accessory equipment.

BAFFLES

Actually, sound energy is released from both sides of a cone. This is natural because there is air on both sides of the cone and the cone moves as a unit; however, when the cone moves forward, the air in front of it is compressed and the air in back of it is rarefied. The sound released from the rear of the cone is of opposite phase to that released from the front of the cone. If the sound from the rear is allowed to flow so that it meets the sound from the front, cancellation takes place, and the response of the speaker drops off sharply. Such cancellation is substantial only when the paths to the meeting place are short compared to a wavelength and maximum when the total path length from the front of the cone to the back is exactly equal to zero or one wavelength. Sound waves from the rear change 360 degrees in phase in one wavelength and therefore oppose front waves. The wavelength of sound becomes longer as the frequency decreases; consequently, front-to-back interference is worst at the lowest frequencies and ordinarily marks the cutoff frequency of the speaker mounting. Such interference is not appreciable at higher frequencies at which the wavelength is small compared to the path length between the front and back of the cone.

At these frequencies, the compressions and rarefactions are so closely spaced that there is no definite general cancellation action as at the low frequencies.

The longer the path length between front and rear, the lower is the frequency at which interference can take place. By extending the edges of the cone with some rigid flat material, we make it necessary for sound waves from the rear to travel out to the edges of the material before they can meet the sound waves from the front and interfere with them. The added material is called a *baffle,* and its principle is illustrated in Fig. 7-1. With the speaker alone (Fig. 7-1A), the front and rear waves must travel only along one side of the cone to meet at the edge. This path is so short that a speaker alone without baffle will usually not reproduce much below about 350 Hz. In Fig. 7-1B is shown the situation when a baffle is added. The length of the interference path is increased by the width of baffle material on each side. The reader can clearly demonstrate this effect by operating a speaker connected to a radio receiver or record player. First, listen to the speaker alone; then place it against a temporary, improvised baffle. The latter can be a large piece of cardboard or a corrugated carton with a hole cut in it. The increase of low-frequency response will be very clearly noticeable.

Baffles should be made of good sound-insulating material and should be soft enough to prevent rattle. Soft woods are satisfactory, but material like *Celotex* is more appropriate. The speaker must be securely fastened to the baffle, and the baffle must be rigidly mounted to prevent rattle.

If a baffle is to be of limited size, the speaker should not be mounted in the center. The center is a bad position because the

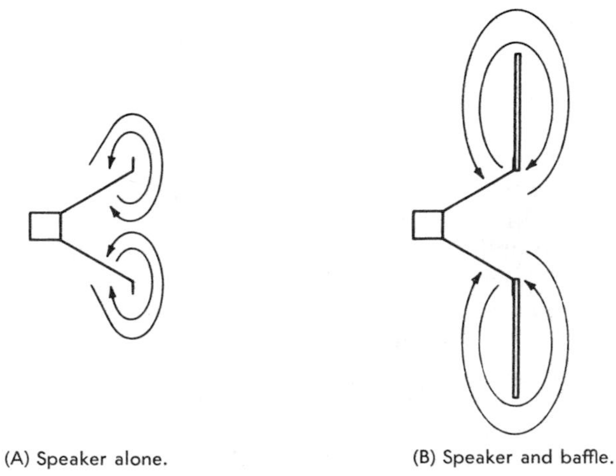

(A) Speaker alone. (B) Speaker and baffle.

Fig. 7-1. Increase of front-to-back interference path length by addition of baffle.

path lengths to all four edges are the same, and the frequency components at which the path length (one side) is one wavelength are severely attenuated. The center position is illustrated in Fig. 7-2A. If the speaker is moved toward one corner, as in Fig. 7-2B, the uniformity of response is much better because the path length to each edge is different, and the interference attenuation is distributed. The interference does not have to take place around the edge of the baffle but somtimes is purposely made to take place through a port, as illustrated in Fig. 7-2C. The design principle of such an arrangement is to equalize the response peak due to speaker resonance by the sharp attenuation around that frequency by spacing the port so that the sound travels a half wavelength. Some audiophiles adjust the size and shape of the port until it balances out the resonant peak of the speaker.

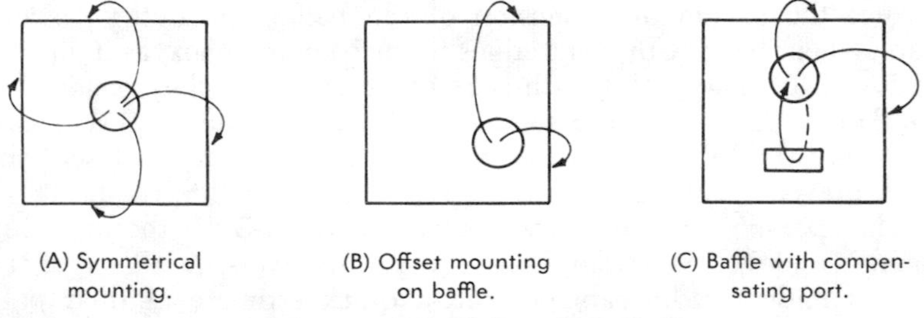

(A) Symmetrical mounting.

(B) Offset mounting on baffle.

(C) Baffle with compensating port.

Fig. 7-2. Mounting a speaker on a flat baffle.

For a theoretically perfect speaker, the ideal baffle is one which has infinite lateral dimensions. The interference path lengths are then infinite, and no matter how low the frequencies of the sound components, cancellation or reinforcement due to baffle limitations cannot take place. Obviously, an infinite baffle in the fullest sense cannot be realized. However, if the baffle dimensions are sufficiently large that the frequency at which the path length is a half wavelength is far below the range of frequencies to be used, the baffle is referred to as *infinite*. For example, a speaker mounted in a hole in the wall of a house and placed at least 6 feet from the nearest opening, with its back opening on one room and its front on the adjacent room or outside is, for all practical purposes, mounted on an infinite baffle. For use with an infinite baffle, one or two identical wide-range speakers with essentially flat response over the desired frequency range are recommended to be placed at ear level or directed toward ear level. Speakers for this application should have very low resonance characteristics because there will be no provision for eliminating such defects. In the previous discussion of speakers, it was stated that each direct-radiator speaker was assumed to be mounted in an infinite baffle because this removes the effect of front-to-rear interference and allows us to consider the inherent effects of the speaker.

SIMPLE ENCLOSURES

An infinite baffle, or an approximation of it, is one of the best speaker mountings. However, its size is a disadvantage in an ordinary home. For example, to reduce the frequency of interference to 50 Hz, the baffle must be at least 12 feet square! Because of this size problem, various arrangements have been developed in an attempt to get the same effect without the use of so much space. This has led to the design of speaker enclosures.

The evolution from a flat baffle to a simple enclosure is illustrated in Fig. 7-3. The simple flat baffle is symbolized in Fig. 7-3A. To reduce the maximum dimension of the baffle, the outer portions can be bent back at the four edges to form an open box, as indicated in Fig. 7-3B. The path length is as great as for the flat version, but the lateral dimensions are smaller. In Fig. 7-3C the process is carried one step further, and the back of the enclosure is closed; the back prevents any sound from the rear from getting to the front.

The open-back cabinet arrangement of Fig. 7-3B is the one commonly employed for radio and television receivers. A glance at the size of midget radio cabinets and a quick estimate of their path lengths will quickly show why low-frequency response is lacking

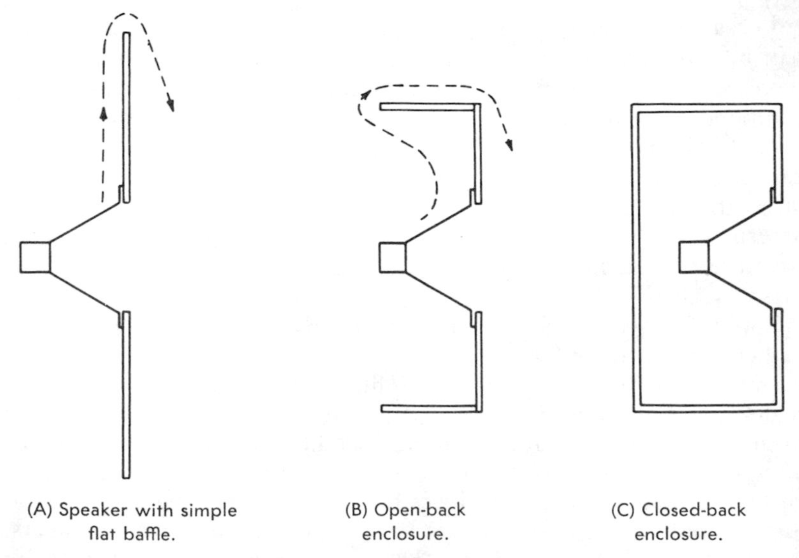

(A) Speaker with simple flat baffle.

(B) Open-back enclosure.

(C) Closed-back enclosure.

Fig. 7-3. Evolution from a flat baffle to a simple enclosure.

in this type of receiver. The small speaker used is also a low-frequency limiting factor, but the cabinet is usually the important limitation. Large console models employ larger speakers and larger cabinets, but the path length is seldom sufficient to allow reproduction as low as 150 Hz, unless some special cabinet design other than the simple open-back box is used.

Another disadvantage of the open-back box is the fact that it acts as a resonant tube at some frequency well within the operating range, unless it is very large. Sounds of frequency near resonance are reproduced with annoyingly excessive volume relative to other sound components. Any sudden sound peaks of any frequency or sounds of low frequency can shock-excite the box into oscillation at the resonant frequency. Low-frequency sounds which are not attenuated by the interference path all seem to sound the same, because of shock excitation at the resonant frequency. This accounts for the fact that many console radio receivers and some of the earlier juke boxes emitted a constant booming during reproduction of music.

It would seem, then, that simply closing the box as in Fig. 7-3C would be the answer; and the box could be as small as desired as long as it holds the speaker, because the sides and back would block the rear-to-front interference path. Unfortunately, this is not so. As soon as the box is closed up tight, as in Fig. 7-3C, the air in it is no longer free to move in the open as in the open-back cabinet. Instead, the action of the cone causes pressure changes in the cabinet rather than a combination of pressure and velocity. This means that the springiness or compliance of the air is an important factor. Compliance is acoustic capacitance and combines with the compliance of the cone suspension in such a way as to raise the resonant frequency of the system to a value higher than that of the speaker alone. The reason for this effect can be noted from the equivalent circuit for the closed box, given in Fig. 7-4. The mass (in-

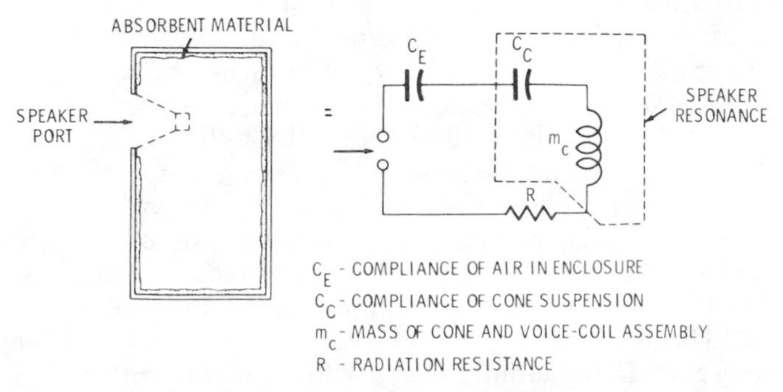

Fig. 7-4. Equivalent circuit of closed-box enclosure.

ductive effect) of the cone, the compliance (capacitive effect) of the cone suspension, and the compliance (capacitive effect) of the air in the box are all effectively in series with each other. The smaller the box, the smaller is the acoustic capacitance it simulates. Use of a small box lowers the capacitance connected in series with the series-resonant circuit of the speaker and thus raises the overall resonant frequency.

An important feature of the closed box is the fact that, because there is no motion of air in and out of it, there is no inertance or inductive effect. There is only compliance, or capacitive effect. Accordingly, the box does not resonate itself as do some other types which we shall consider later. All it does is enter the resonant circuit of the speaker unit and raise the resonant frequency of the system above that of the speaker.

Compliance and thus the capacitive effect of a closed box increases with size if the box can be made large enough so that the equivalent air compliance capacitance is large compared with the equivalent speaker-suspension compliance capacitance. Then the box will raise the resonant frequency only a negligible amount above the resonant frequency of the speaker. On the other hand, because the only effect of the box is to raise the resonant frequency, a speaker with a very low self-resonance could be put into a relatively small box. Then the system resonance would still be low enough for good results, even though it is raised above the resonant frequency of the speaker alone. Generally, a closed box of a given volume will raise the resonant frequency of the system a given percentage above the resonant frequency of the speaker.

Because the closed box keeps back radiation from getting around to the front and interfering, it is frequently referred to as the *infinite-baffle enclosure* and sometimes even as *infinite baffle*. As should be clear from the preceding discussion, this closed-box enclosure is not at all equivalent to a true infinite baffle unless it is so large that its effect on the resonant frequency is negligible.

Although the primary consideration in connection with a closed box is its effect on system resonance, this is not by any means the whole story. Conditions are such that as frequency increases above the resonance value, the system goes through a series of minor resonances which, if not counteracted, cause substantial irregularities in the response characteristic. There is also an effect due to reflection of back-radiated energy from the inside of the back and walls of the enclosure, with irregularities resulting from phase variations between cone vibrations and such reflected energy reaching the cone. These effects are minimized in practice by padding the inner surfaces of the cabinet walls with sound-absorbent material such as heavy felt, glass fiber, rock wool, or cellulose. This padding material acts as an acoustic resistance, lowering the Q of the box at high frequencies and tending to smooth out the response.

The *boffle* is a development of the foregoing, fitted with a system of screens which act as two-stage acoustic filters damping out cabinet and air resonances.

The compliance of the box increases with the size of the speaker cone. Therefore, the larger the cone, the greater is the percentage increase of speaker resonance caused by the enclosure. The resonant frequency of larger speakers is ordinarily so much lower than the resonant frequency of smaller speakers that the compliance increase is overcome, and the system resonance is lower with a larger speaker. Practical designs are such as to limit the increase of resonance due to the cabinet to 20 percent over the speaker resonant frequency. Authorities do not agree on any standard cabinet volume necessary for a given size of speaker, probably because of the wide variations of speaker resonance and suspension compliance in different models of speakers of the same cone size. Data from different sources specify that from 400 to 700 cubic inches of cabinet volume per inch of nominal speaker diameter should be used as a minimum. Since it is safer to stay on the large side, let us assume that 600 cubic inches per inch of nominal speaker diameter is a good minimum value. On this basis, some typical speaker sizes and minimum required enclosure volumes would be as given in Table 7-1.

Table 7-1. Minimum Enclosure Volume for Typical Speakers

Speaker Size (inches)	Volume	
	(cu in)	(cu ft)
6	4800	2.8
10	6000	3.5
12	7200	4.2
15	9000	5.2

One significant design feature in many of these infinite-baffle types of enclosures is the breakup of the rear wall to minimize internal reflection effects which become severe at higher frequencies. These breakup arrangements vary from two-surface double panels to complicated assemblies of small flat surfaces all at different angles. Another method of minimizing these reflections is to make the whole rear portion of the enclosure semispherical, to break up reflections into an infinite number of angles. Representative of this type is the "kettledrum," which looks like the instrument after which it is named, with the speaker unit mounted in the diaphragm portion of the drum.

One manufacturer makes use of the infinite-baffle box but makes the rear wall with a flexible mounting to the remainder of the box. This makes the cabinet give under sound-pressure waves, reducing the stiffness of the enclosure acoustically. The acoustic impedance of the box thus becomes equivalent to that of a much larger box. This enclosure is used with an extended-range, single-cone speaker to give excellent performance in a relatively small space.

In the infinite-baffle enclosure, all radiation from the rear of the speaker is eliminated. This minimizes interference but also makes systems of this type less efficient than others which utilize the sound from the rear.

One way to overcome this loss of efficiency is to use two or more identical speakers in the enclosure. When two or more speakers operate on the same audio-frequency signal and are mounted close to each other, they interact. Sound energy from one speaker reinforces the vibration of the cone of the other speaker, and vice versa, in what is known as mutual coupling. The action is similar to that in a multielement antenna array. The result is that the efficiency of the combination is much better than the efficiency of either speaker alone. Another advantage arises from the fact that even speakers of the same model are seldom exactly alike, because of tolerances in manufacture. These tolerances are actually an advantage because the difference in the resonant frequencies of the respective speakers tends to distribute the speaker resonance effect and make low-frequency response more uniform.

This idea is frequently used in infinite-baffle enclosures but is just as applicable to other types of enclosures. An example is given under the topic of bass-reflex enclosures.

BASS-REFLEX ENCLOSURES

The infinite flat baffle and the closed box eliminate back interference by isolating the rear of the cone from the front. In another, more popular arrangement, radiation from the rear of the cone is used to reinforce the front radiation rather than cancel it. This arrangement is illustrated in Fig. 7-5 and is known as the *bass-reflex enclosure.* It is the same as the closed box except that an opening is cut below the speaker. This opening, or port, allows air to flow in and out of the box as the speaker cone moves back and forth. The basic design principle is that the back radiation is fed through a path, including the box as a whole and the port, such that the back radiation emerges and reinforces radiation from the front.

Addition of the port adds acoustic inductance (inertance) to the load offered to the speaker by the cabinet, because of the motion of air through the port. Although the stiffness of the air in the enclosure is not so great as in the closed box, it is still appreciable. The flow of air through the port is also accompanied by friction, which is an acoustic resistance effect. As illustrated in the equivalent circuit in Fig. 7-5, the port inertance, friction, and enclosure compliance form a parallel-resonant circuit. This circuit is connected effectively in series with the series-resonant circuit of the suspension and mass of the cone of the speaker.

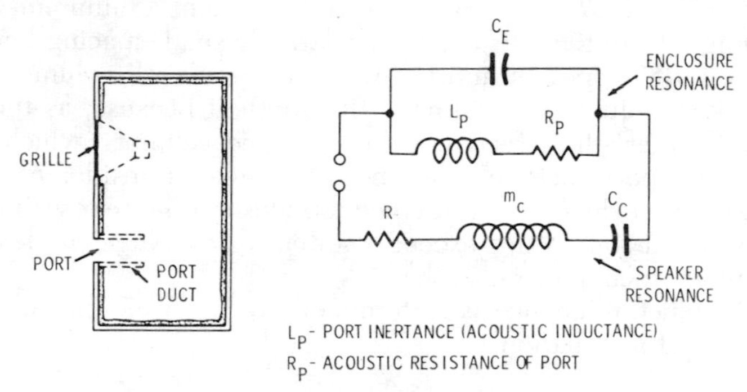

Fig. 7-5. The bass-reflex enclosure and its equivalent circuit.

Because of this more complex acoustic-impedance pattern, the design of a good bass-reflex enclosure involves a little more than simple path length. The effective length of the path of the sound from the rear of the cone is primarily dependent on the nature of the impedance offered by the enclosure itself, as expressed in the equivalent circuit. The design is simply a matter of keeping the resonant frequency of the enclosure in proper relation to the resonant frequency of the speaker. Because the enclosure offers parallel resonance, and the speaker series resonance, one of these resonances can be made to offset the other. Accordingly, one of the most popular methods of design places cabinet resonance right at the speaker resonance frequency. Others believe in making the enclosure resonate slightly lower in frequency than the speaker, feeling that, although this does not completely compensate for speaker resonance, it extends the overall response to a lower frequency.

Placing cabinet resonance exactly at the speaker resonant frequency does not eliminate the bump in frequency response due to speaker resonance, unless the *Q*'s of the two resonant systems are approximately equalized. Such equalization is accomplished by loading the enclosure with acoustic resistance until the overall response is smoothest. Acoustic resistance is supplied by absorbent material on the inner walls but more efficiently by placing grille cloth or other meshlike material over the port. When air is forced through small holes, friction, which is acoustic resistance, is developed.

The port cannot be placed too close to the speaker because then the path from the rear of the cone starts to become direct, like that in a small flat baffle, rather than including the impedance of the box as a whole, as it should. To keep the overall cabinet size to a minimum, some designs add a duct in back of the port, as shown by the dash lines in Fig. 7-5. This allows maintenance of a minimum direct path length from the rear even for relatively small spacing between the port and the speaker and keeps down the overall volume. There is a limit to volume reduction by this method because, as the port becomes larger, the effective volume of the cabinet (which must not include port-duct volume) becomes smaller inside. Also, the end of the port duct must not come too close to the rear wall of the enclosure and should not exceed one-tenth of a wavelength at the resonant frequency.

The cabinet resonance is determined by the well-known Helmholtz resonator equation:

$$f = 2070 \sqrt[4]{\frac{A}{V^2}}$$

where,

f is the resonant frequency of the cabinet, in hertz,
A is the area of the port, in square inches,
V is the volume of the cabinet, in cubic inches.

Calculations of the design parameters of the bass-reflex enclosure are beyond the scope of this book, and our approach has been to explain which factors are important and why. Dimensional information has been carefully worked out and put into graphical form, an example of which is shown in Fig. 7-6. The cabinet volume must have a definite value and should not be as large as possible, as occurs with the flat baffle and the closed box.

Because of variations in speakers and tolerances in cabinet construction, the actual cabinet and speaker resonant frequencies seldom coincide exactly, even though designed to do so. Because the area of the port influences the cabinet resonant frequency, this area may be varied after construction and assembly are complete, to trim the cabinet resonance to match that of the speaker. The port can be covered in part by a book or other flat, rigid device, and the open area can be varied.

An audio-frequency signal generator (audio oscillator) of reasonably constant output is useful in checking for proper port-area adjustment. The front of the speaker cone mounted in the enclosure is watched as the generator output frequency is passed through frequencies from 30 Hz upward. As the resonant frequency of the

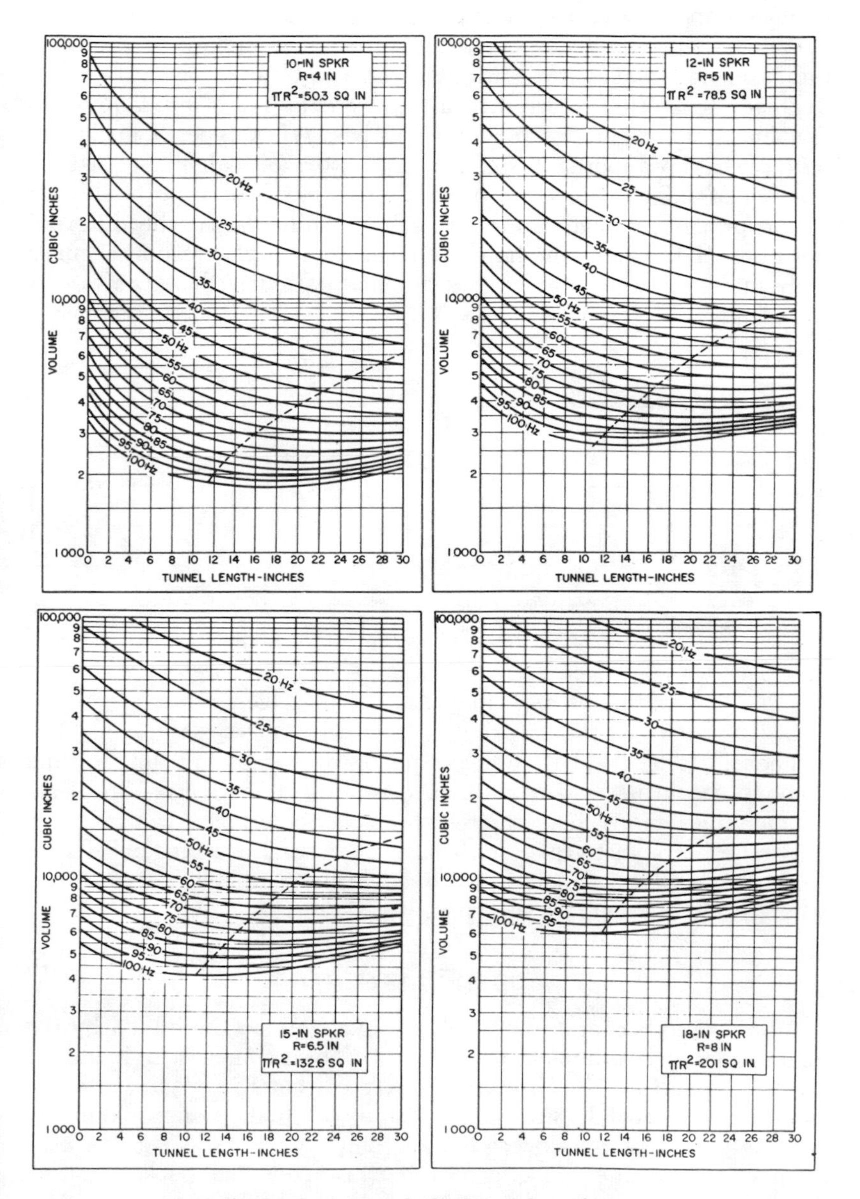

Fig. 7-6. Design graphs for bass-reflex cabinets.

speaker is reached, cone excursion increases greatly to a maximum and subsides to smaller deflections at higher frequencies. The audio signal is adjusted for resonance. Then the port area is varied until cone excursion becomes minimum. The speaker resonance is then being damped by cabinet antiresonance.

Because the bass-reflex enclosure reverses the phase of the back radiation and makes it additive to the front radiation at low frequencies, it is often referred to as an *acoustic phase inverter.*

Some designers prefer to make the bass-reflex type of enclosure with distributed ports. Instead of one port of calculated size, a number of smaller ports are used. The areas of these ports add up to the proper calculated port area. It is claimed that the different positions of the different holes prevent dips in the response curve at frequencies at which the interference path length becomes one wavelength.

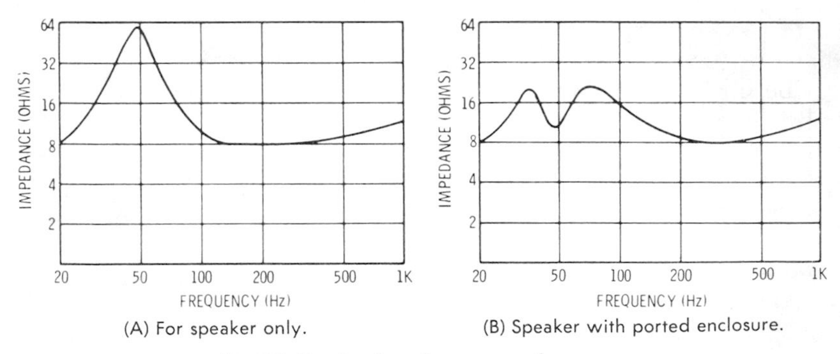

(A) For speaker only.

(B) Speaker with ported enclosure.

Fig. 7-7. Speaker impedance versus frequency.

The effect of the bass-reflex type of enclosure in damping the response peak of speaker resonance is illustrated in Fig. 7-7.

Sometimes more than one woofer or tweeter unit is used in a bass-reflex cabinet to take advantage of mutual-coupling efficiency. An example of a small bass-reflex enclosure employing two 6-inch woofers and a tweeter is shown in Fig. 7-8.

ACOUSTIC LABYRINTH

Another method of acoustic phase inversion is exemplified by the acoustic labyrinth, depicted in Fig. 7-9. The cabinet is divided into parts by a series of baffles in such a way that the spaces between the baffles form a lengthened passage, or duct, of approximately constant cross section between the back of the speaker and the front of the cabinet. The labyrinth thus feeds the back radiation around to the front.

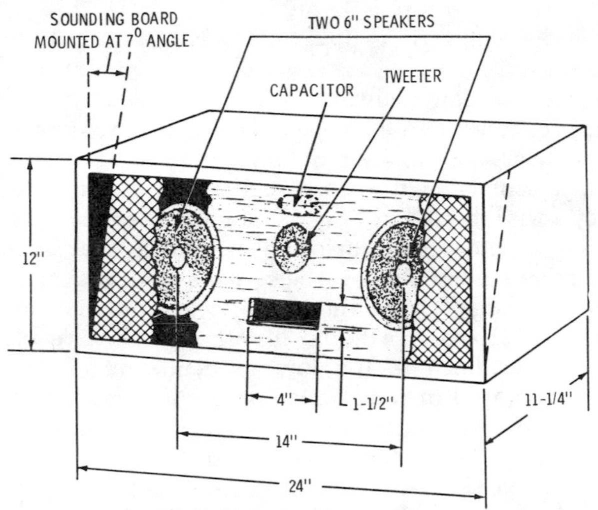

Fig. 7-8. Multiple-speaker arrangement.

Besides the acoustic inversion effect, the labyrinth has another very important design consideration. The pipe or tube simulated by the space between the baffles acts as a tuned line when it is exactly a quarter-wavelength long at the resonant frequency of the speaker. It simulates a parallel-resonant circuit which equalizes the series resonance of the speaker unit in the same manner as in the bass-reflex cabinet. The labyrinth has two main beneficial actions: (1) It equalizes the resonant bump in the speaker response and spreads out low-frequency response, and (2) it provides reinforcement of the sound in the range near the frequency at which the duct length is a half-wavelength (twice the resonant frequency of the speaker).

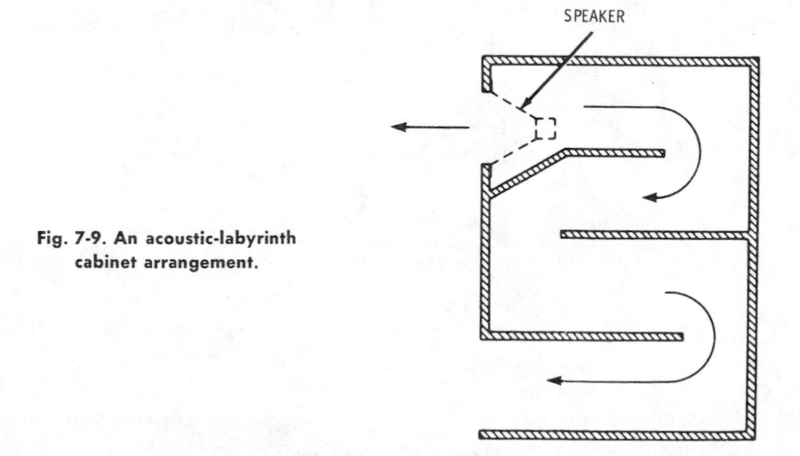

Fig. 7-9. An acoustic-labyrinth cabinet arrangement.

HORN-TYPE ENCLOSURES

Audio engineers have always been attracted to horns because of their high efficiency and good frequency response above cutoff frequency; but, as was brought out in our previous basic discussion of horns, a conventional horn structure for a low enough cutoff frequency for high fidelity (say 50 Hz or lower) is so large at the mouth as to be prohibitive for ordinary use. To overcome this problem, a number of designs have been developed to simulate the performance of a large horn without the large dimensions necessary in the conventional type. In home installation, the use of horn-type enclosures is confined to the low-frequency range, and they are ordinarily employed in conjunction with a separate tweeter (which may itself have a separate horn), or they are driven by a coaxial or extended-range, single-cone driver. In the home, cone-type drivers are almost universally used to drive the horns at low frequencies.

One of the first measures taken to reduce size is the folding of the horn. Folded or rolled exponential horns have been used for centuries in musical instruments, and these look like that shown in Fig. 7-10A. Of course, this type has a small throat and in electrical sound systems would be used with a diaphragm-type driver. In high-fidelity systems, horns are driven by cone-type speakers (for low frequencies) and must therefore have large throat diameters.

A low-frequency horn application more typical of high-fidelity equipment is shown in Fig. 7-10B. The front of the cone faces away

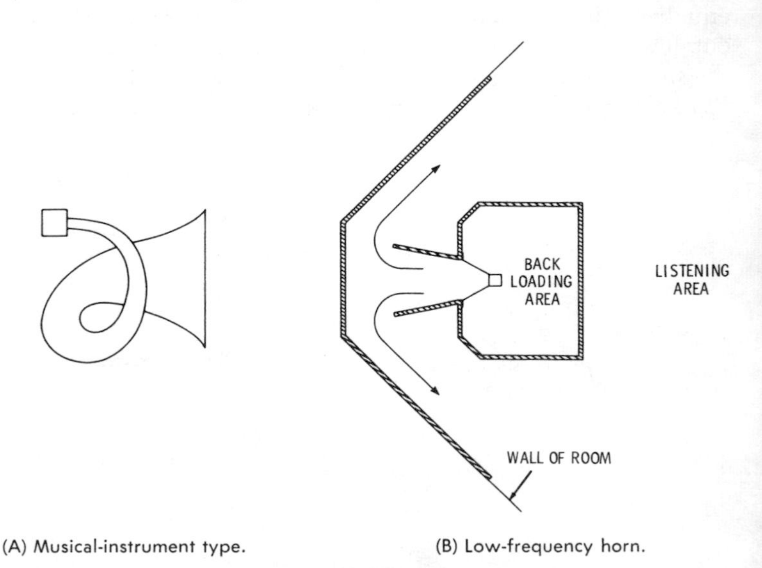

(A) Musical-instrument type.

(B) Low-frequency horn.

Fig. 7-10. Folded horns.

from the listener, but its output is directed around between the baffles toward the listening area. The baffles are so constructed and mounted that the closed area leading from the speaker cone to the room roughly approximates an exponential horn of the low-frequency cutoff desired. This type of horn in practical design cannot in itself have the full mouth area necessary for theoretical cutoff in the desired range of approximately 50 Hz. The structure is normally designed to be placed in the corner of a room, and the walls of the room become continuations of the sides of the horn, as illustrated.

Notice the back-loading area in the structure of Fig. 7-10B. This is considered necessary for cone drivers because without it the loading on the cone is not symmetrical, and distortion results. This is why in practically all low-frequency horn structures some balanced loading is provided. The back loading is not always a closed area, but in some cases constitutes slots or ports.

One of the best known of the better reproducing systems is the Klipschorn arrangement. A layout of the interior of a typical commercial type of the Klipschorn is illustrated in the diagrams of Fig.

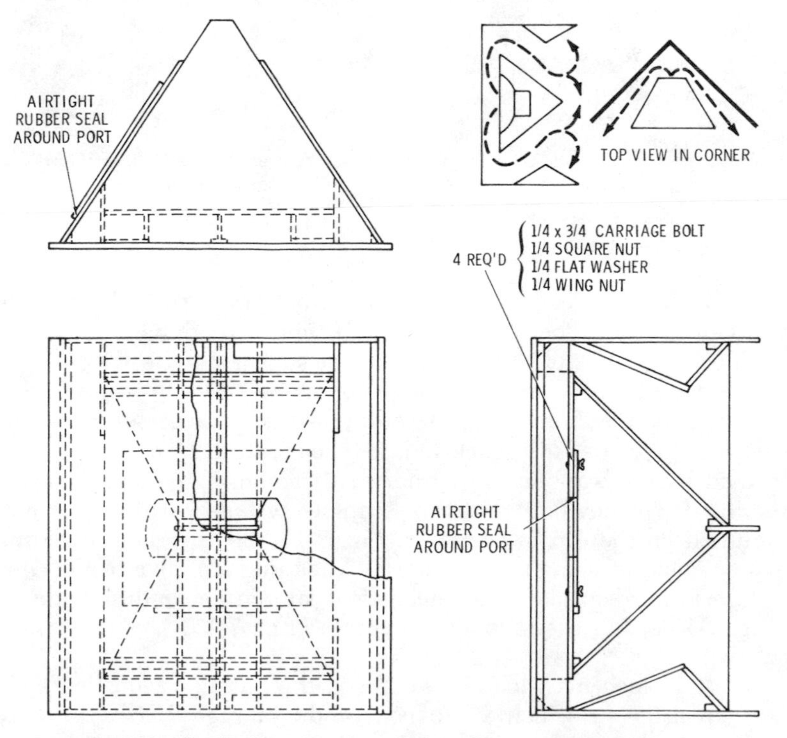

Fig. 7-11. Klipschorn arrangement used in Electro-Voice assemblage.

7-11. The principle is the same as that of Fig. 7-10B. A cutaway view of the original Klipschorn is shown in Fig. 7-12. An enclosure in the center of the cabinet back-loads the woofer. The front of the woofer cone opens into a narrow front channel, which is the throat of the horn. From there, the woofer output is led around the center enclosure through ducts of a gradually increasing cross-sectional area. These ducts open out to the back of the cabinet. A vertical

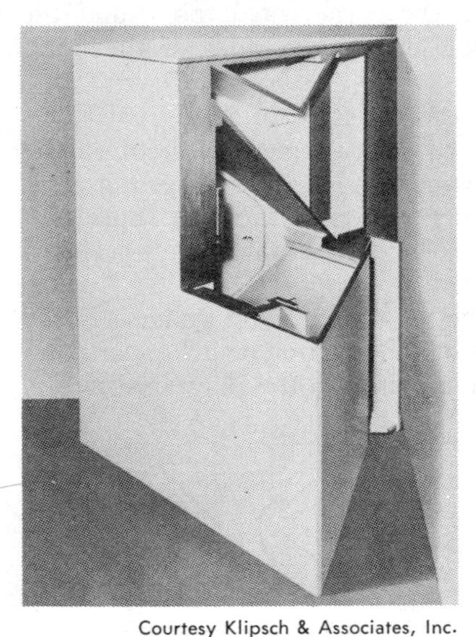

Courtesy Klipsch & Associates, Inc.

Fig. 7-12. Cutaway view of Klipschorn woofer horn.

Courtesy Klipsch & Associates, Inc.

Fig. 7-13. Rear view of Klipschorn with tweeter on top.

baffle at the back leads the sound to emanate from the sides through vertical openings between the baffle and the remainder of the cabinet. The sides of the cabinet are at such an angle with respect to the walls that the space between them and the walls forms a continuation of the horn. As the sound waves pass the front of the cabinet, they are still enclosed in the continuation of the horn formed by the walls in the corner of the room. A cutaway view of the low-frequency horn and the complete system, with tweeter horn mounted on the top, is shown in Fig. 7-13. Another horn-enclosure arrangement utilizing a labyrinth back-loading is shown in Fig. 7-14.

The Empire speaker with acoustical suspension and adjustments (Fig. 7-15A) has the following features (Fig. 7-15B):

1. A 12-inch mass-loaded woofer with floating (acoustical) suspension, a 4-inch voice coil, and a large (18 lbs) speaker ceramic structure.

2. Sound-absorbent rear loading.
3. Die-cast midfrequency/high-frequency full-dispersion acoustic lens.
4. Imported marble top.

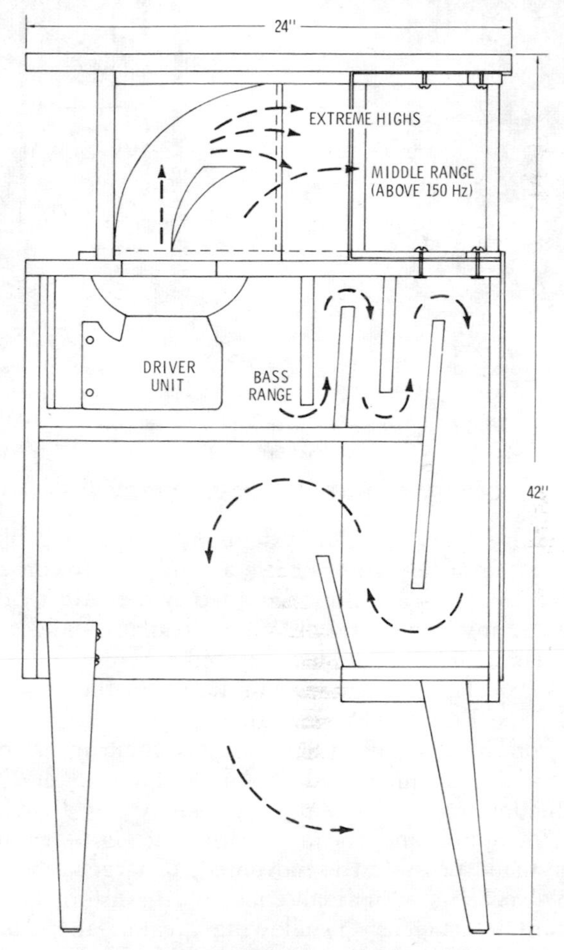

Fig. 7-14. Cross section of acoustic-labyrinth horn.

5. Ultrasonic domed tweeter.
6. Full-presence midrange direct radiator.
7. Totally damped acoustic fiber enclosure.
8. Exclusive dynamic reflex stop system for enriched bass response; can be adjusted to suit room acoustics.
9. Front-loaded horn with 360-degree-aperture throat.
10. Terminals concealed underneath.

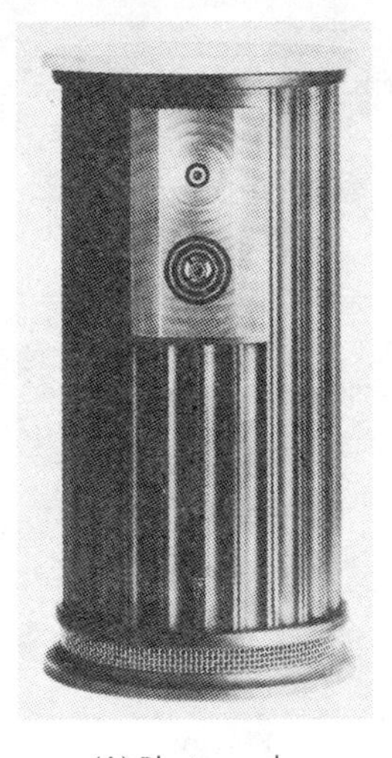

(A) Photograph.

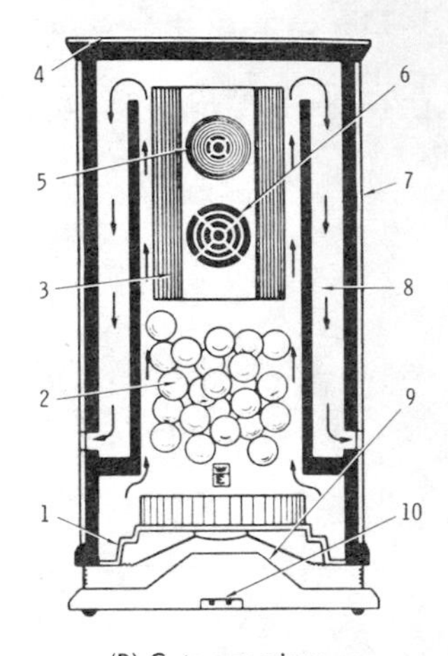

(B) Cutaway view.

Courtesy Empire Scientific Corp.

Fig. 7-15. The Empire Grenadier 800P speaker.

COMPACT HI-FI SPEAKER SYSTEMS

Compact high-fidelity speaker systems in decorator cabinets suitable for placing on a bookshelf or use as end tables, consolettes, or other decorative pieces are now designed and made to reproduce as accurately as any other type of system available—and at reduced prices over larger systems. These systems reproduce accurately from 40 to 20,000 Hz and can be purchased in the price range of $50 to $1000. See Figs. 7-16A through 7-16Q.

Exponents of larger speakers and systems point out that low bass response requires movement of large amounts of air and therefore hi-fi reproduction requires a large speaker cone. However, new designs and materials and the application of the natural laws of physics have made possible the movement of larger amounts of air with multiple speakers with small cones by the use of loose mountings (in acoustical suspension) allowing greater lateral movement, and by adding to the effective mass of the moving material. The increased travel of the cone moves the same amount of air as a larger cone with normal travel, and it provides more accurate control and reproduction. The latter is accomplished by the use of heavier and stronger magnets.

Acoustical suspension is provided by the reduction of forces on the cone at the mounting ring by the use of soft materials such as polyvinyl chloride (PVC) painstakingly applied to provide greater stretch with dynamic concentric alignment and negligible resistance to wide lateral movement. The force necessary to damp the movement of the cone is provided by the elastic air sealed inside the enclosure; thus it is said that the cone is in acoustical suspension or suspended on pressurized air and the elastic mount.

In these designs, the compression and decompression operate in a nearly linear manner, reducing much of the distortion inherent in larger speaker designs. In addition, tweeters and midrange reproducers have been improved in quality and width of dispersion, creating overall improvements that widen the utility and application of lower-cost high-fidelity systems.

For output measurements of Acoustic Research speaker AR-3, see Fig. 7-17. In this figure, the response curves of each radiator are shown separately so that interference patterns—which reflect the particular position of the recording microphone rather than reproducing quality—are eliminated. The speaker provides level controls for independent adjustment of the midrange unit and the super-tweeter section.

The speaker system shown in Fig. 7-18 is a small unit containing nine small (4-inch) speakers, all producing the full audio range. There are no crossovers, and therefore no distortion or loss is introduced on this count. The speakers are long-excursion, high-compliance (acoustic-suspension) speakers that can move large amounts of air. By acting in unison (mutual coupling of speakers), the group provides the same effect as one larger speaker with a cone area approaching the sum of the cone areas of the nine speakers. In addition, an electrical equalizer (Fig. 7-18C) is used to compensate for loss of bass response and to adjust overall performance to suit the room and the listener's taste.

The advantages of this system are: (1) The response of each speaker in the group is different from that of each of the others, causing the resonant frequency of each speaker to be different from that of every other speaker; therefore the resonance of each speaker is minute when compared to the output of the acoustically coupled group operating in unison. As a result, individual resonances and other distortions are negligible in comparison to the total output. (2) The speaker enclosure is pentagonal in shape, thus avoiding standing-wave resonance. There are no sides parallel to any baffle on which any of the nine speakers are mounted. (3) The omnidirectional effect of the speakers is increased by placement of speakers so that only 11 percent of the sound (one speaker) is direct-radiated toward the listener and 89 percent (eight speakers) of the sound is toward the rear to provide reflected sound in all directions from the wall back into the room. This provides more uniform dispersion at high frequencies throughout the room. (4) An equalizer is part of the system; it provides accurate compensation

Courtesy Acoustic Research, Inc.

(A) Acoustic Research AR2ax with front cover removed.

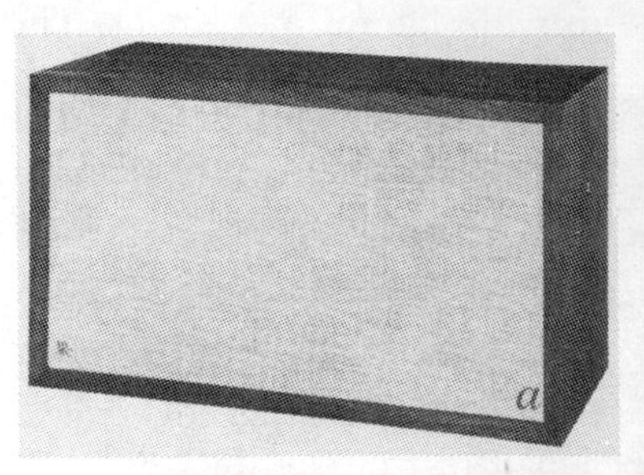

Courtesy Acoustic Research, Inc.

(B) Acoustic Research AR2ax with front cover in place.

Courtesy Electro-Voice, Inc.

(C) Electro-Voice compact speaker system.

Courtesy Heath Co.

(D) Heathkit Model AS-48 speaker system.

Courtesy Empire Scientific Corp.

(E) Empire Model 8500 wall-mount speaker system.

Courtesy University Sound

(F) University Mediterranean-style end-table system.

Fig. 7-16. Compact

Courtesy Fisher Radio

(G) Fisher WS-50 omnidirectional system.

Courtesy Lafayette Radio Electronics Corp.

(H) Lafayette Criterion speaker system.

Courtesy Altec Lansing

(I) Altec Lansing speaker system.

Courtesy James B. Lansing Sound, Inc.

(J) JBL monitor-type speaker.

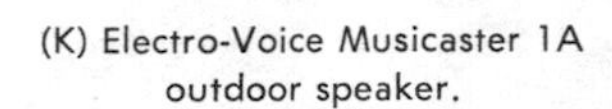

(K) Electro-Voice Musicaster 1A
outdoor speaker.

Courtesy Electro-Voice, Inc.

(Fig. 7-16 continued on next page.)

speaker systems.

Courtesy Bogen Div., Lear Siegler, Inc.

(L) Bogen "Row 10" speakers.

Courtesy Fisher Radio

(M) Fisher XP-16 consolette speakers.

Courtesy JVC America, Inc.

(N) JVC decorator speaker group.

Fig. 7-16. Compact

Courtesy Bose Corp.

(O) Bose 501 speaker system.

Courtesy McIntosh Laboratory Inc.

(P) McIntosh speaker system.

Courtesy Electro-Voice, Inc.

(Q) Electro-Voice consolette speaker.

speaker systems (Cont).

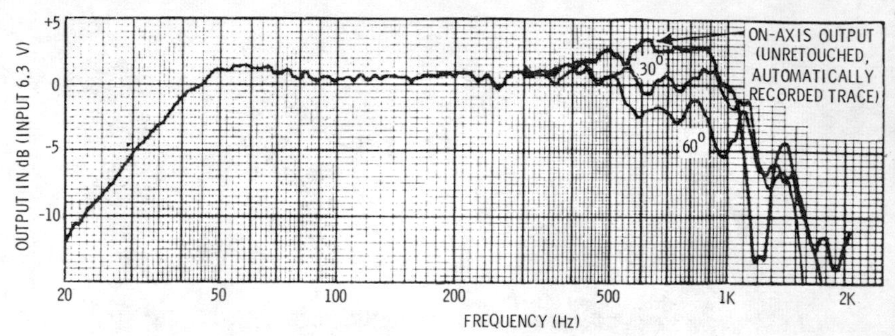

(A) Woofer frequency response.

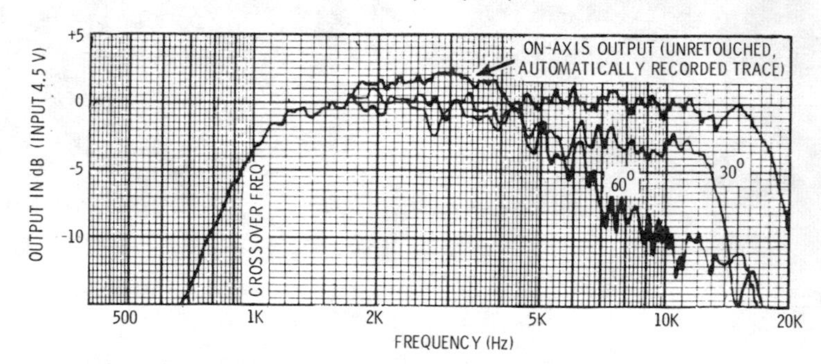

(B) Midrange response in test baffle.

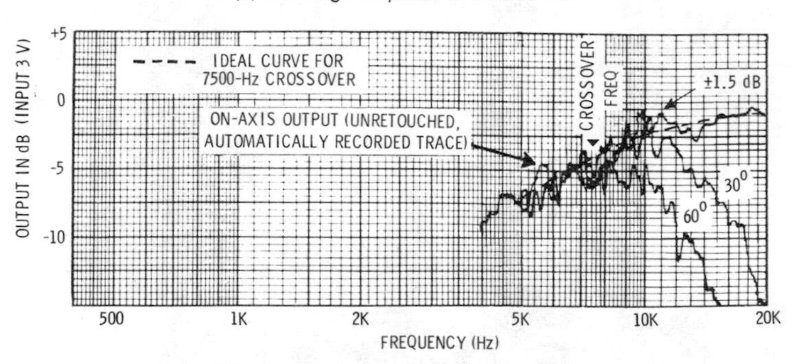

(C) Super-tweeter response in test baffle.

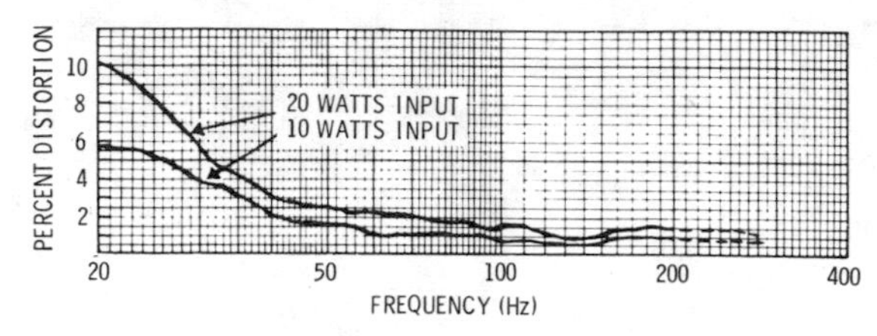

(D) Woofer harmonic distortion.

Fig. 7-17. Output characteristics of Acoustic Research AR-3 speaker.

(A) Front views.

(B) Rear views.

(C) Active equalizer.

Courtesy Bose Corp.

Fig. 7-18. Bose 901 Direct/Reflecting speaker system.

for the bass rolloff in acoustic-suspension speakers, and for the effects of radiation impedances, speaker characteristics, enclosure dimensions—and even for the grille cloth effects. There is negligible distortion introduced by this unit. The equalization is provided in the system before the amplifier, thereby eliminating bulky components and providing greater accuracy.

ELECTRICAL SPEAKER EQUALIZERS

Electrical equalizers for compact speakers are now available for use instead of mechanical baffles and horns to compensate for bass-response rolloff, and to provide for variation of effective bass power output requirements due to location of a speaker in a room and other room audio characteristics. (Examples of these units are shown in Figs. 7-18C, 7-20, and 7-21.)

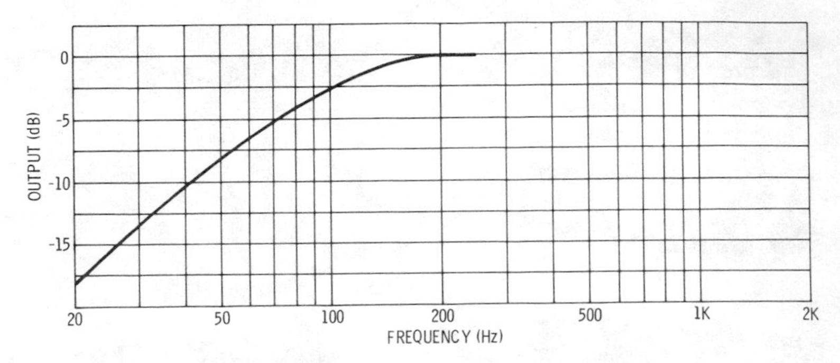

Fig. 7-19. Reduction in bass response in acoustic-suspension speakers.

The equalizer is primarily designed to compensate for the rapid falloff of bass output that occurs in high-quality acoustic-suspension (air-spring) compact speakers. In compact acoustical-suspension speakers without horn enclosures and/or other mechanical reinforcement, additional power is required to provide the large amounts of air movement necessary to produce a proper reproduction of base notes below 200 hertz. The baffle and horn can provide the needed reinforcement in larger units. The bass output of the acoustical-suspension speaker drops off as shown in Fig. 7-19. This sharp dropoff of response is within the short range of 20 to 200 hertz.

Most preamplifier sections of modern hi-fi systems do not provide for compensation of this large dropoff in such a short span of frequencies. If the bass boost in the average preamp (which usually has a fixed slope from 20 to 1000 instead of 20 to 200 Hz) is used to increase the bass response to compensate for the rolloff in the speaker, this boost will increase the response in a different proportion (slope) than is required, thereby introducing more distortion.

The equalizer shown in Fig. 7-20 is specifically designed to provide compensation for bass rolloff of acoustical-suspension speak-

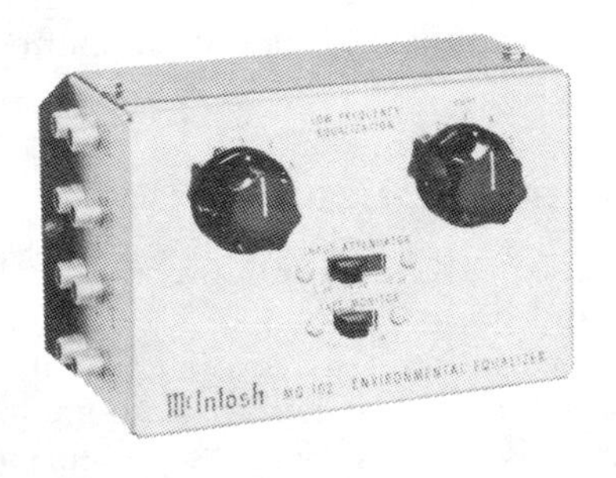

Courtesy McIntosh Laboratory Inc.

Fig. 7-20. McIntosh MQ 102 environmental equalizer.

ers. Another advantage of this equalizer is in compensating for room effects on the bass output of the speaker system in any particular configuration that may be used. The room in which a loudspeaker system is placed can substantially alter its sound balance. An equalizer may be designed to compensate for room conditions and for different positions of a speaker in any room.

Fig. 7-21 shows an equalizer that provides bass correction plus compensation for variations in speaker output requirements due to speaker placement or other factors. The power output of a bass speaker varies over an 8 to 1 range (9 dB) depending on where the speaker is placed in the room (Fig. 7-22). For example, if a speaker is suspended in the center of a large sound-absorbing room, a 20-hertz signal will radiate equally in all directions from the speaker. It will be radiating into a sphere. If the speaker is lowered to the floor in the center of the room, it will be radiating into a hemisphere. The signal striking the floor would be reflected upward and in effect would almost double the loudness. If the speaker is then moved along the floor to the center of a wall, the power will double

Courtesy McIntosh Laboratory Inc.

Fig. 7-21. McIntosh MQ 101 environmental equalizer.

again. The speaker is now radiating into a quarter of a sphere. However, if you move the speaker into the corner of the room, the radiation is concentrated into one eighth of a sphere, and the power doubles again to 8 times the power of the first position.

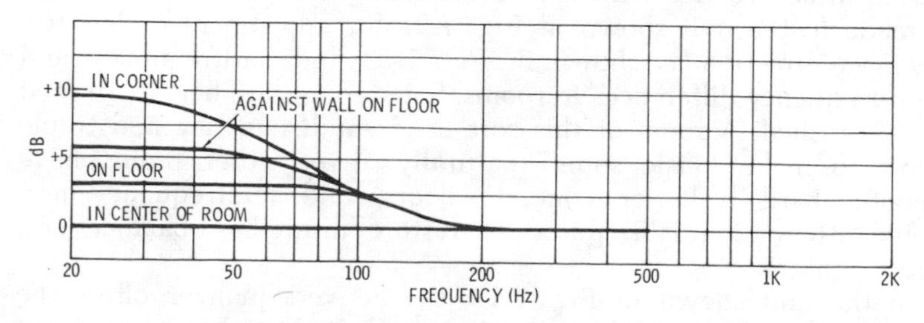

Fig. 7-22. Effects of room gain for different speaker placements.

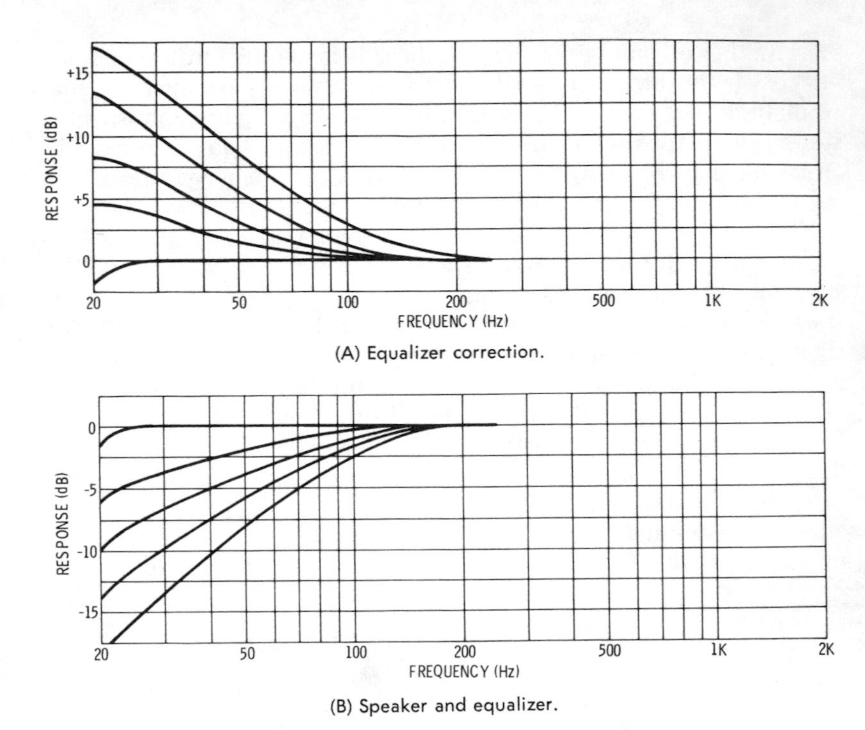

(A) Equalizer correction.

(B) Speaker and equalizer.

Fig. 7-23. Response curves for McIntosh equalizer.

To restore the system to a flat response under any of these conditions, while preserving the nearly perfect transient response of certain acoustic-suspension speakers, the electrical equalizer shown in Fig. 7-21 may be adjusted to provide a signal having a curve equal but opposite to the reduction or increase that occurs as a result of relocating the speaker. The equalizer can be adjusted to provide for room equalization for any location or condition of the room.

In the unit shown in Fig. 7-20, the electrical correction can be varied in accordance with the curves shown in Fig. 7-23. Provision is made in the unit shown in Fig. 7-21 for adjustment of low-frequency correction as shown in Fig. 7-23 and midfrequency and high-frequency differences in rooms. Some rooms are heavily draped and furnished. A room of this sort needs midfrequency and treble boost to make music sound naturally alive. Other rooms have smooth, hard walls or concrete floors, and midfrequency and treble attenuation is required to restore music-hall balance and effect.

In the unit shown in Fig. 7-18, the active equalizer offers the choice of 19 additional contours that can be selected from the front

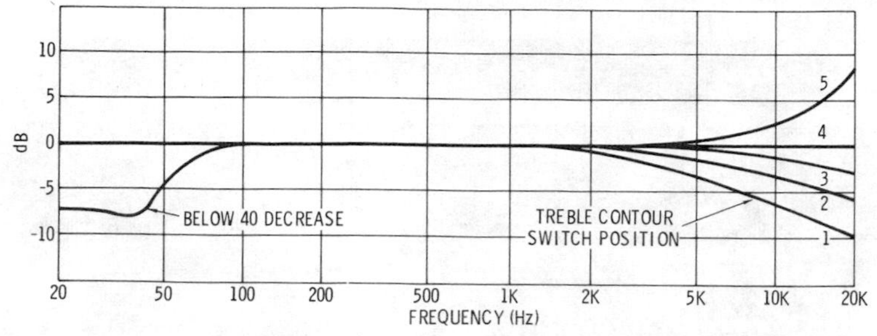

(A) Treble-level switch in normal position.

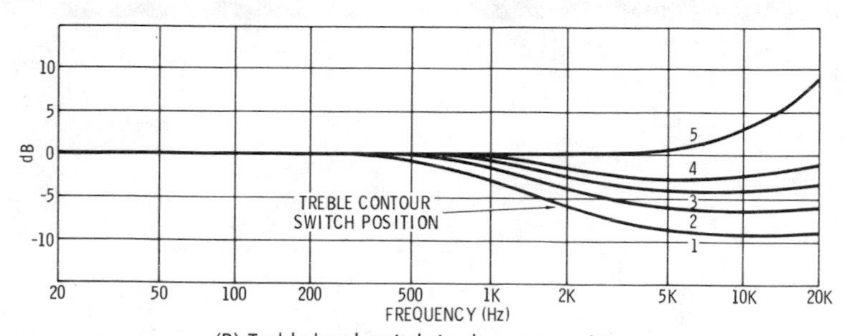

(B) Treble-level switch in decrease position.

Fig. 7-24. Response curves for Bose equalizer.

panel. This gives the listener flexibility to exercise his own taste in compensating for room and other variables. Various contours available with this unit are shown in Fig. 7-24.

The Altec Acousta-Voicette stereo equalizer shown in Fig. 7-25A is designed to adjust the stereo system output to the room coupling so that a flat acoustical response at the listener's position is allowed, as shown in Figs 7-25B and 7-25C. This device can correct the usual narrow-band room acoustic faults by adjustment of two sets (for stereo operation) of 24 filters spaced at center-frequency intervals of ⅓ octave from 63 Hz to 12.5 kHz. There is a broadband insertion control on each channel. Tape monitor, equalizer in, and equalizer out switching also is provided.

This unit can minimize room resonances and increase spatial perception by increasing the levels at high threshold areas for those with hearing deficiencies. It can provide for tuning the stereo system to the room to any listener's taste.

STEREO SPEAKER SYSTEMS

The basic design principles of speaker enclosures are the same for stereo as for monophonic reproduction. However, the fact that two

Courtesy Altec Lansing

(A) Front view of unit.

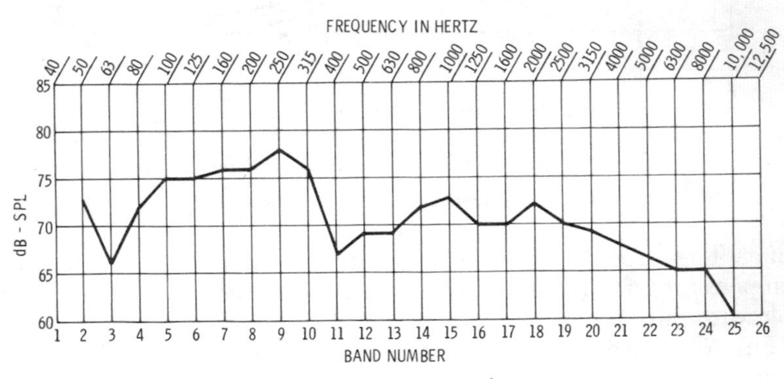

(B) Room curve before equalization.

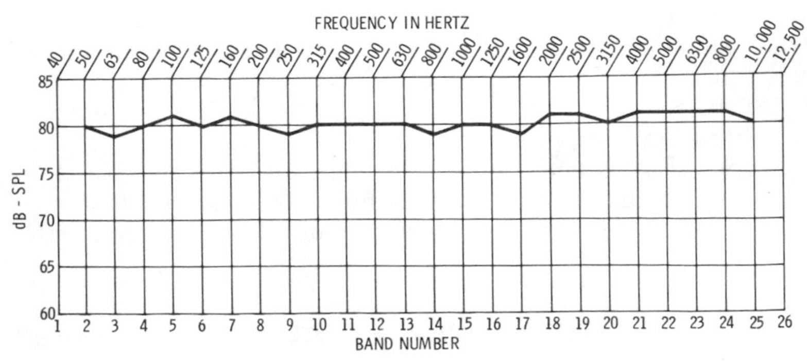

(C) Room curve after equalization.

Fig. 7-25. Altec Acousta-Voicette stereo equalizer.

or more sound reproducing sources must be used in stereo has led to variations in how these enclosures are arranged.

The simplest approach is two or more complete reproducing systems, one to handle the left signal and one for the right signal, and possibly two more for left-rear and right-rear signals. Stereo speaker-enclosure systems have evolved along the lines indicated by the arrangements illustrated in block form in Fig. 7-26. Many arrangements take advantage of the fact that the stereo effect is obtained predominantly with higher-frequency components of the sound, because at low frequencies the wavelength is so great that phase differences and depth distinctions between spaced objects are small. There is no definite frequency at which transition takes place, but there is a gradual change effect. Authorities place the dividing line for practical purposes between 150 Hz and 400 Hz. Frequency components in this range and below may come from a single source since little stereo effect can be produced under normal conditions.

In Fig. 7-26A, the existing system has been expanded by the addition of a limited-range (not much low-frequency response) compact enclosure to the full-range system previously used for monophonic reproduction. The added speaker provides all but the lowest frequency components and serves as the right-channel speaker for the stereo effect. The full-range system provides the left stereo speaker and reproduces all of the lowest-frequency nondirectional components.

Two compact wide-range enclosures and a low-frequency center speaker are illustrated in Fig. 7-26B. The center reproducer need only be a woofer, responsive to frequencies up to 500 Hz. A dual–voice-coil speaker, like that shown in Fig. 6-35, would be a logical choice for the center speaker. However, if the system is being evolved from a former mono system, the center speaker could be the full-range reproducer from that system.

The advent of stereo stimulated the trend toward smaller enclosures. Therefore, special techniques have been developed to reduce the size and price of full low-frequency response units. A number of compact single-channel reproducers have been developed with excellent characteristics for arrangements like those of Figs. 7-26A and 7-26B. They are available in all finishes and styles. Some are floor models; others are placed on a bookshelf or table. Some contain single speakers and have response from about 50 Hz up, so they can be used with a center woofer with no low-frequency loss; others are full-range arrangements, with claimed frequency response extending from 40 to 20,000 Hz. The latter could be used without center speaker, to provide left and right sources for a two-source arrangement. Most of the compact enclosures do not exceed 2 feet in their maximum dimension. Typical complete arrangements using compact enclosures are shown in Fig. 7-27.

There are still a large number of people who would prefer to have the complete stereo system enclosed in one cabinet to simplify furniture arrangement. Because the separation between left and

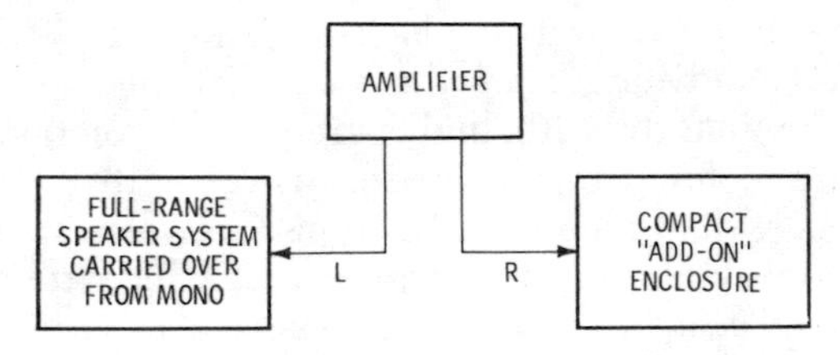

(A) Addition of compact second-speaker enclosure to existing monophonic system.

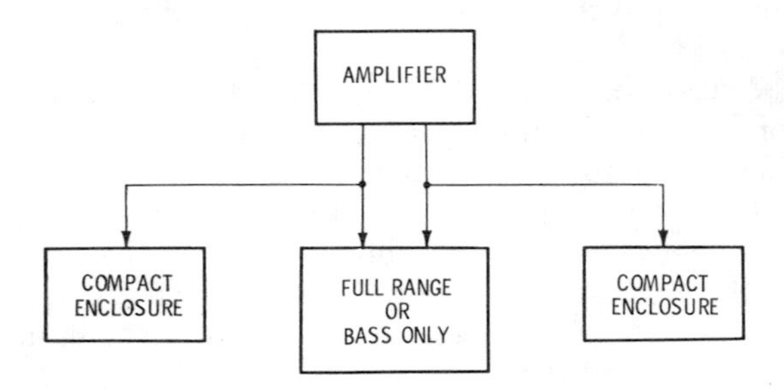

(B) Use of monophonic system as center woofer, compact enclosures at left and right.

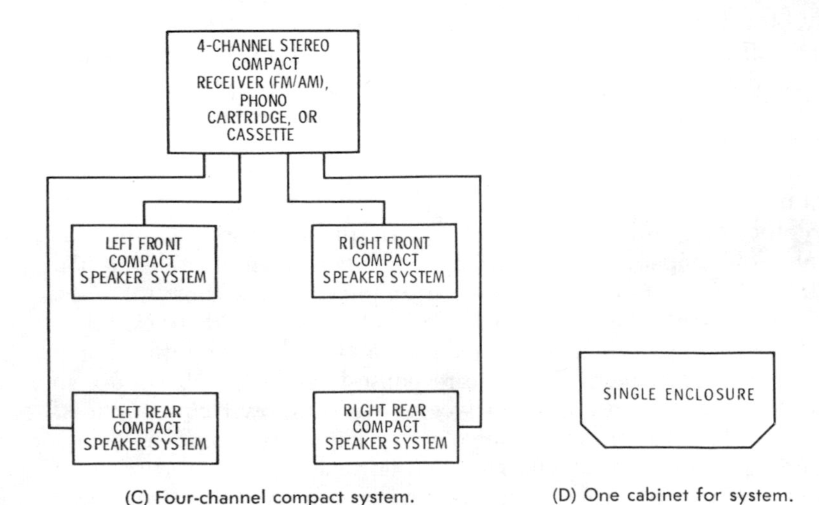

(C) Four-channel compact system.

(D) One cabinet for system.

Fig. 7-26. Enclosure arrangements for stereo.

right reproducers must be at least 5 feet (effectively), this means that the cabinet must be either 5 feet long or must have some arrangement by which the sound can be projected so as to simulate the proper spacing. Two examples of single-cabinet stereo-system reproducer enclosures are shown in Fig. 7-28.

A very elaborate stereo reproducing system is pictured in Fig. 7-29. The design of the reproducer is illustrated in Fig. 7-30. Basically it consists of two three-way (woofer, midrange, and tweeter) speaker systems, one mounted on each side of the enclosure structure. Their sound outputs are directed against a large round surface which reflects them to the listening area in such a way as to distribute the sound while maintaining its essential directivity for stereo effect.

Stereo Headphones

A good pair of stereo headphones is a worthwhile accessory to any high-quality music system. An increasing number of amplifiers and receivers include a stereo-headphone jack on their front panels, and, if no jack is available, connection to the speaker terminals is a simple matter. If a tape recorder is part of the music system, any serious attempt at live recording, as well as sound-on-sound recording techniques, requires headphone monitoring for proper balance and level adjustment.

A type of headset suitable for home listening is shown in Fig. 7-31. Each earpiece contains a 1-inch miniature dynamic speaker and is filled with plastic foam for resonance damping. The front of the driver cone is also loaded with plastic foam, and its enclosed rear is vented to the interior of the earpiece through two small holes. These design features are intended to give a wide, smooth frequency response with a minimum of peaks and holes.

The liquid-filled vinyl cushions mold themselves firmly but gently around the wearer's ear. External sounds are almost totally excluded, and it is said that the padded headband and soft ear cushions permit the listener to wear the headphones for hours without experiencing discomfort or fatigue.

The 8-foot plastic-covered cord has four conductors so that the two earphones can be electrically isolated if desired. They are fitted with a three-circuit phone plug that fits the stereo phone jacks of most amplifiers.

The frequency response of the phones shown in Fig. 7-31 is rated at 30 to 20,000 Hz. While stereo headphones provide a very different listening impression than do speakers, earphones have a wide-range response, smooth and with low distortion. High-frequency hiss is somewhat noticeable. Fig. 7-32 shows a record changer and built-in amplifier that is specifically designed for headset listening.

Courtesy Bogen Div., Lear Siegler, Inc.

(A) Bogen stereo system.

Courtesy Electro-Voice, Inc.

(B) Electro-Voice system with EV4 speakers.

Fig. 7-27. Examples of use

Courtesy Heath Co.

(C) Heathkit stereo center.

Courtesy NuTone, Inc.

(D) NuTone system with in-wall speakers.

of compact enclosures.

(A) Magnavox Astro-Sonic radio phonograph.

(B) Magnavox Agean, with color television receiver built in.

Courtesy The Magnavox Co.

Fig. 7-28. Single-cabinet stereo systems.

CHOOSING A SPEAKER SYSTEM

The judgment of speaker-system performance is, to a great extent, subjective. A number of system designs with excellent characteristics are available. Differences in the performance of these systems is such that the listener's taste rather than known measurable technical factors provides the basis of a choice. This is partly because the listener does not always conform to the ideal of wanting reproduction exactly like that which he would hear when present at the location of the performance. Various factors have conditioned him to tend to like a few deviations from realism, although constant experience with reproduction that is nearly perfect can educate one. There are a few people who just do not have a hearing system adequate to hear more than a limited degree of the audio range and who, either physically or by training, cannot distinguish between the performance of a poor radio receiver and that of a well-

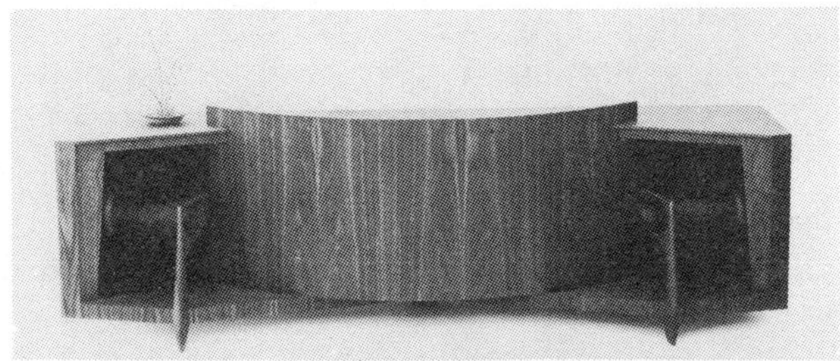

Courtesy James B. Lansing Sound, Inc.

Fig. 7-29. A full-range stereo horn enclosure.

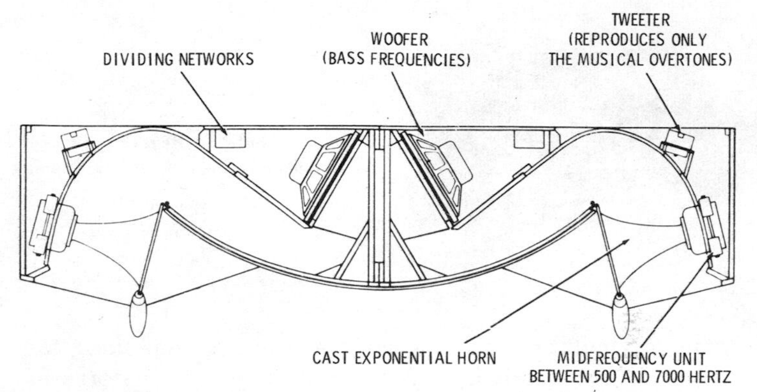

Fig. 7-30. Speaker arrangement in the enclosure of Fig. 7-29.

designed high-fidelity system. Then there are all the gradations in between. Keeping these facts in mind, we can note the following points in the selection of a high-fidelity speaker system.

1. There is no set formula for high quality, for reasons already explained. Each type of well-known system has its enthusiastic supporters, but the best audio engineers recognize both the strength and weakness of each system. Although they may favor one over the others for themselves, they realize that the type of listener and his situation influence things greatly. You cannot find the best system offhand; you must figure it out for yourself.

2. You cannot separate the choice of the speaker from the choice of the system as a whole. When you first decide you want high fidelity, you must also decide the optimum compromise between your ear and the cost. Optimum high fidelity costs money. Speakers cost the most of any part of it. The speaker system is thus very likely where the limit of cost will enter the choice. The first thing to do is

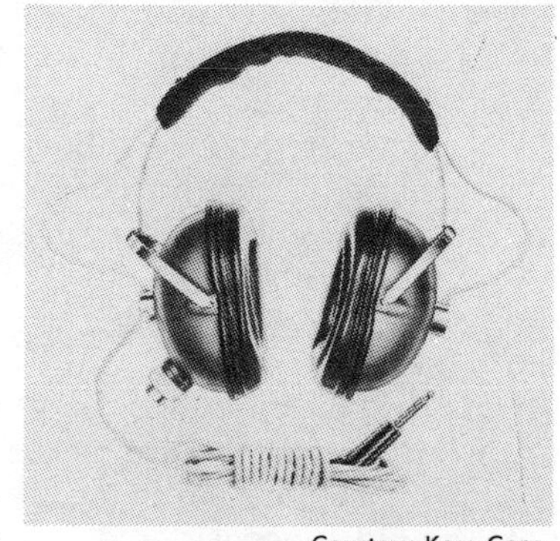

Fig. 7-31. Koss professional stereo phones.

Courtesy Koss Corp.

Courtesy Shure Brothers, Inc.

Fig. 7-32. Shure SA-10 record changer and amplifier for headset stereo listening.

determine the highest quality which really means something to you. The only way to do this is to shop around. Visit the stores where systems are sold, and, even better, become acquainted with people who have high-fidelity systems of various degrees of quality and different types of arrangements, and see which ones you like. More important, determine where you stop appreciating increments of increase in quality per unit of cost.

A vast majority of people are very happy with a system having response only to about 6 or 8 kHz, providing other distortions are low. A limit can almost be defined in terms of dollars because performance generally follows cost to some degree, assuming that parts of the system are properly matched. Once you have made your choice of quality level, the specifications of your whole system including input devices, preamplifier, as well as the speaker system, should be chosen to match. This is important because two of the easiest ways to waste money are: (1) to have major parts of a system not matched in performance and cost requirements, and (2) to pay the high cost of quality increments above your natural taste requirements.

3. Frequency-range limitation is not as annoying as other forms of distortion. When we speak of the response of a component or system, we assume (and thus probably take too much for granted) that the harmonic and intermodulation distortion over the entire range are low. It should be kept in mind that a limited frequency range in itself minimizes harmonic distortion because it cuts off components of harmonic frequencies above the range. Listener tests in general have definitely shown that a good, clean, low-distortion frequency response up to 4000 Hz is preferable to a frequency response that extends up to 10,000 or 15,000 Hz, but with slightly greater harmonic distortion. The prospective purchaser of a speaker system should, therefore, not place too much emphasis on nominal claims of frequency response but should assure himself as to how low all forms of distortion remain over the entire range.

4. The listener's ear is the final test. All specifications, charts, and data should be considered in their relationship to the sound to the

listener's ear, rather than to any technical tests; however, data from a valid technical test may be criteria as to whether the listener may some day acquire a more discriminating taste.

5. Although there is much to be said for the wide range of some of the single-cone, extended-range systems, the systems with the greatest range are generally those of the multiple type. The purchase of a multiple system does not in itself guarantee high fidelity. Such things as divided presence (consciousness of dual source locations), takeover, and fuzzy highs are common complaints. Many feel that a dual system is worthwhile in a higher-priced system but that the single-cone type of system is appropriate for lower-cost installations.

6. In separate-unit multiple systems, watch the phasing. The path length between the listener's ear and the source should be the same for both woofer and tweeter. If this requirement is not adhered to, the high- and low-frequency components join in the wrong phase relationship at many frequencies, and substantial distortion can result.

SPEAKER LISTENING TESTS

As has been explained, the final test of a speaker system is how it sounds. The listener who does not know what to look for in such tests is at a disadvantage. He may overlook factors which later, after he has purchased the system, may assume new importance. Experience has shown that proper attention to certain features will go a long way toward assuring continued satisfaction.

There are no standards of performance. Furthermore, most prospective purchasers of speaker systems do not have equipment to make the tests necessary to check a system against proper standards. Consequently, listener tests must be made as personal comparisons between different available systems.

The following suggestions should tell the reader what to look for in making listener tests on a speaker system. Each feature should be compared from one system to the next in an attempt to evaluate the one most pleasing. In any tests, every effort should be made to eliminate other components of the whole high-fidelity equipment by using the same or similar auxiliary equipment with each different speaker system. If the reader has not had experience in distinguishing between audio-frequency tones of different frequencies, it is suggested that he obtain an audio signal generator or test record made for this purpose, connect it to a speaker, and run through the audio range, familiarizing himself with the sound of audio tones of low, middle-range, and high frequencies. He will then be better prepared for the listening tests which follow.

1. Look for distinctions between various kinds of bass notes, such as those from different musical instruments. If a bass drum, plucked strings, or bass horn all sound alike, then there is probably excessive bass resonance effect in the speaker or the enclosure.

2. Compare performance on voice with performance on music. Voices should be crisp, natural, and highly intelligible with the same amplifier adjustments used for music from the same source.

3. Listen for sudden crescendos of percussion instruments and to pizzicato passages of music. Check for evidences of hangover transient distortion which would make it difficult to distinguish one pluck from another.

4. If possible, apply to the system a high-frequency tone of, let us say, about 10,000 Hz. If this is not possible, listen to needle scratching or interstation noise from an fm tuner. Then walk around in front of the reproducer, checking relative output directivity. The broader the output distribution at high frequency, the better is the system.

5. Tap the sides of the enclosure with your knuckles and note whether they are deadened properly or respond with undesirable vibration. All parts of an enclosure structure should be solid and dead.

6. Be sure that, for each speaker system checked, the amplifier boost controls are adjusted for the most pleasing performance. Also check the speaker balance control (if any) for best balance between woofer and tweeter output. This control is sometimes referred to as the *treble* adjustment of the speaker system. It is unfair to compare systems in which all controls are not adjusted for optimum performance.

8

Systems Design, Selection, and Installation

The assembly of a number of components to create a complete sound system is not difficult if some general principles of system design are observed. The major units which make up a sound system are record players, tuners, tape equipment, microphones, television receivers, preamplifiers, amplifiers, speakers, enclosures, baffles, horns, and other acoustical aids. Other parts, such as switching circuits, pads, microphone and speaker stands, cabling, and the like are also to be planned. In order to assemble a sound system which will perform in a desired manner, it is necessary to be familiar with all the components used, and what is required of each of them.

REQUIREMENTS

The most important physical and electrical design characteristics of a complete sound system involve size, power, fidelity, gain, compensation, filtering, number and type of controls, number and type of input (program) sources, and number and type of speakers. The design objective and limiting factors are elements which determine these characteristics. Each must be carefully considered before the components of a sound system can be specified. All of the components which make up a complete system must be carefully coordinated if the system is to operate properly. No component should be chosen without regard to the other parts of the system with which it is to be used.

AUDIO POWER

Before the components of a system can be selected, the power necessary to supply the required sound volume must be determined.

The absolute unit of power measurement, as applied to physical sound, is the *acoustic watt.* The acoustic watt is a unit of power of the physical activity of air at sound frequencies. It is not convenient to think in terms of the acoustic watt when designing a sound system, because the actual requirements are related to too many variables and the net results are extremely difficult to measure.

The objective is to provide the listener with a close approximation to an original live performance. Achievement of this objective requires the subjective equivalents of sound pressure levels that approach those of a concert hall. Although the peak sound pressure level of a live performance is about 100 dB, the average listener prefers to operate an audio system at a peak sound pressure level of about 80 dB. The amplifier, however, should also accommodate listeners who desire higher-than-average levels, perhaps to peaks of 100 dB.

A sound pressure level of 100 dB corresponds to less than one watt of acoustic power for an average room of about 3000 cubic feet. If speaker efficiencies are considered to be in the order of 5 percent, a stereophonic amplifier, then, must be capable of delivering about 20 watts per channel. Higher power outputs are required for speakers of lower efficiency. The peak-to-average level for most program material is between 20 and 23 dB (100 dB peak less 80 dB average equals 20 dB peak-to-average). A system capable of providing a continuous level of 77 dB and peaks of 100 dB would satisfy the power requirements of nearly all listeners. Moreover, because sustained passages that are as much as 10 dB above the average may occur, the power-output capability of a top-caliber system should not be below the value required for a peak of 100 dB of acoustic power while delivering a continuous level of 87 dB of acoustic power.

The *audio watt* is more convenient to work with than the acoustic watt. The audio watt referred to in this respect is the unit of electrical power at audio frequencies as measured at the output terminals of an amplifier. The power requirement of a specific installation can be expressed in terms of the number of watts of audio power required from the amplifier to be used if the efficiency of the speaker system is taken into consideration.

The efficiency of good speakers used with sound systems runs from under 5 to over 15 percent. Assuming that the speakers used have an average efficiency of 5 percent, it is possible to estimate the power requirements of a particular installation in terms of watts of audio power.

Fig. 8-1 is a chart that can be used for rough estimating. It shows the minimum power required for coverage of rooms of various sizes. The size of a room is based on its volume in cubic feet.

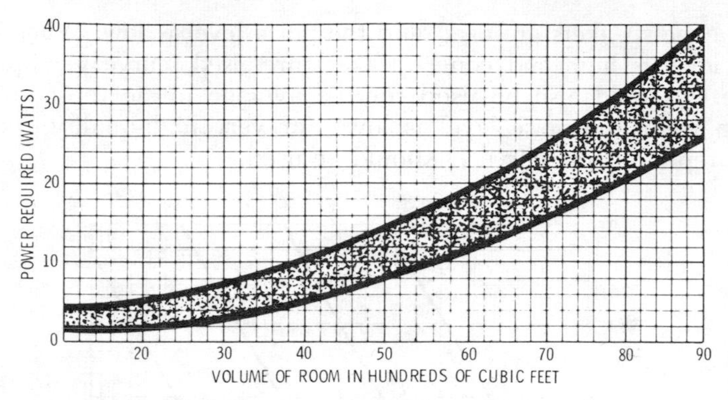

Fig. 8-1. Audio power required for various room sizes.

The shaded area indicates the approximate limits within which the power requirements fall. The range of selection is given to provide for various possible uses of the system. The wider the range of reproduction required, the greater is the power required.

The charts shown in Figs. 8-1 and 8-2 are based on use of efficient speakers with reinforced baffling. However, designers of speakers of the acoustic-suspension type, now commonly used in home hi-fi installations, generally trade off efficiency of speaker output for improved quality and reduction in size; the efficiency of such speakers may be as low as one percent. Therefore, when these charts are used to determine the optimum power requirements of acoustic-suspension speakers, it is recommended that the power requirements be multiplied by a factor of five.

Since a proportionate increase in power capability of an amplifier to drive a less efficient speaker is less expensive than the cost of a more efficient speaker of equal quality, the trade-off described above is usually in the best interest of the consumer.

Determining Power Required

The exact audio power required from the amplifier used with an indoor installation is governed by several factors. The most important of these are: the volume of the room or rooms to be served; the acoustic characteristics of the walls, ceiling, and floor; the average noise level prevailing; the frequency range of the system and of the material to be reproduced; and the efficiency of the speaker system.

The chart of Fig. 8-2 can be used to estimate detailed power requirements for rooms of various volumes. Curve A is for speech-reinforcement systems using high-efficiency horn-type speakers when the prevailing noise level is low. If the noise level is high, curve B applies, providing horn-type speakers are used. Curve B also applies when cone speakers are used and the noise level is low. When the noise level is high and cone speakers are used, curve C applies. Curve C should also be used for the average music reproducing system when the noise level is low. For very high-quality, wide-range reproduction, curve D should be used.

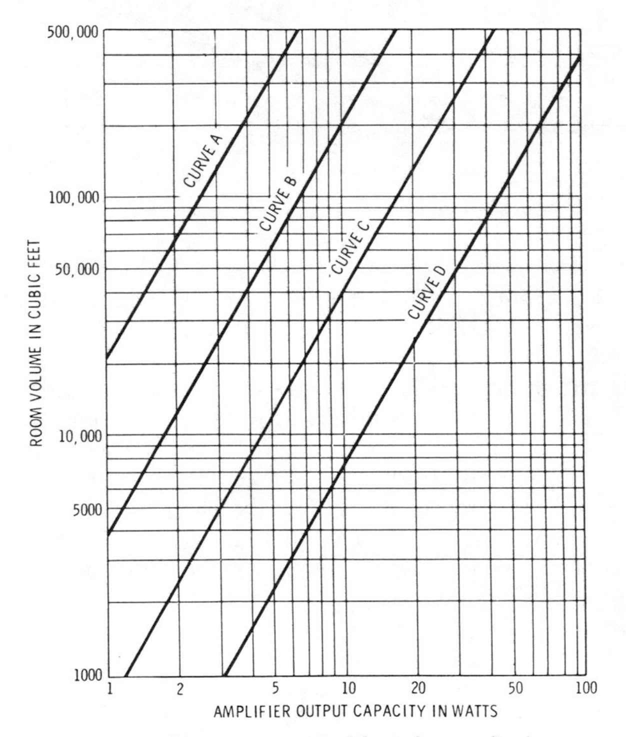

Fig. 8-2. Amplifier output required for indoor applications.

The curves of Fig. 8-3 give the amplifier power capacity for outdoor reproduction. The power required for outdoor installation depends on the distance between the speakers and the farthest point to be covered. Curve A is for a system using a single horn-type speaker covering an angle of 30 degrees. Curve B is for a system covering an angle of 60 degrees, and curve C is for a system covering an angle of 90 degrees.

In multiple-room installations, i.e., installations where several areas are to be served, it is necessary to determine the total power requirements of all the locations which are to be equipped with speakers, and then arrange the distribution of power to the various speakers in accordance with the requirements of the individual locations.

For top-quality installations, there are two reasons why it is advisable to select an amplifier with a higher wattage rating than is actually required. One reason is that the higher-power-rated amplifiers with the same ratings in regard to distortion and frequency

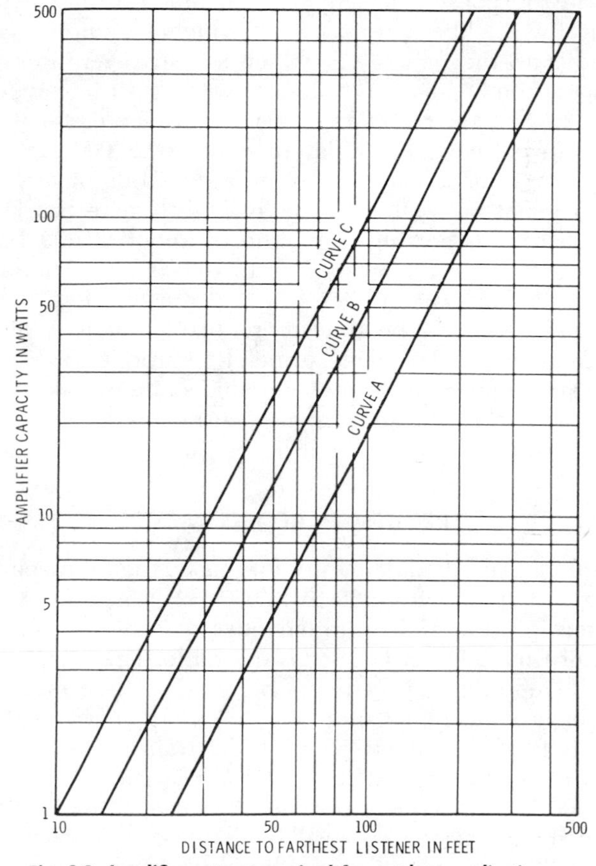

Fig. 8-3. Amplifier output required for outdoor applications.

response as the lower-power amplifiers will provide better operating characteristics at lower-power operation than the lower-rated units. The other reason is that greater dynamic range is available to handle extra loud crescendos, which may occasionally drive a smaller amplifier beyond its rated output, thereby causing noticeable distortion.

FIDELITY

The term "fidelity" includes a number of considerations. By definition, fidelity is the degree with which a system accurately reproduces without distortion at its output the essential characteristics of the signal that is impressed on its input. The fidelity requirements of a system depend largely on the type of material which is to be produced and on the listener's taste.

Frequency response is an important characteristic which helps to determine the fidelity of a system. Frequency response is a rating which indicates the range over which a system reproduces all frequencies uniformly; thus, a system may be said to have a flat response between 30 and 20,000 Hz, meaning that a curve of its output plotted against frequency is flat from 30 to 20,000 Hz. The term "flat" is generally understood to mean a deviation of less than ± 0.5 dB, which is not noticeable to the average human hearing system.

Although human hearing is limited to about 20,000 Hz at best, modern equipment is often rated up to 100,000 Hz. The implication and fact is that if an amplifier has *X* distortion at a higher-than-audible range, it should be able to produce less than *X* distortion at audible frequencies. The same principle is applied as that applied to the linear operation of class-A amplifiers: the smaller the operating portion of the overall response curve used the more nearly distortion-free the operation will be.

FREQUENCY REQUIREMENTS

For systems used predominantly for music, the frequency response should extend at least from 30 to 15,000 Hz. When a system is used for a direct pickup, and optimum fidelity is desired, a frequency response of from 30 to 20,000 Hz is desirable. This means that if the sound to be reproduced is being picked up directly from an orchestra or a singer, a wide range should be used. If, however, the source of sound is from recordings or electrical transcriptions, there is no necessity for an extremely wide range. A frequency response of from 30 to 15,000 Hz is adequate for such a system. This takes into consideration the average ear and all available components in the hi-fi chain.

Both the upper and lower limits of the frequency response of a system are important. It has been conclusively determined through subjective tests that the upper and lower limits are related, and that when the upper limit is raised, the lower limit should be lowered. It is a fairly well established rule of thumb among audio engineers that the product of the upper and lower frequency limits should be about 640,000 Hz^2.

This means that if the upper limit is 10,000 Hz, the lower limit should be $640{,}000 \div 10{,}000$, or 64 Hz. When the upper limit is 20,000 Hz, the lower limit should be around 32 Hz. The center point of the audible sound range is usually regarded as 800 Hz ($800^2 = 640{,}000$).

For best results, the response of a system should be the same number of octaves above 800 Hz as it is below. (An octave is the difference between two tones whose frequencies are related by a factor of 2. For example, one octave below 800 Hz is 800 ÷ 2, or 400 Hz. One octave above 800 Hz is 1600 Hz.)

The wider the frequency response of a sound reinforcement system, the more important it is to keep distortion content to the lowest possible percentage. Distortion can completely negate the advantage of a wide frequency-response characteristic.

If a sound installation is to be used for a-m radio reproduction, or if other limited-response programs (confined to frequencies between 100 and 8000 Hz) are to be used, designing a system whose upper limit is 8000 Hz and whose lower limit is 100 Hz will reduce to a considerable extent the power and response requirements without reducing the efficiency of the system. When it has been determined that a system is to be used for limited response only and reproduction of highest-fidelity components is not considered worth maximum investment, the system components should be selected so as to have a narrow frequency-response range throughout. This will save considerable money, and modern limited-range equipment will still sound good.

DISTORTION

Fidelity also includes the distortion characteristics of a system. For systems of the highest quality, the total cumulative distortion as fed to the speaker should not exceed 1 percent at normal operating level.

Distortion refers to the presence of components in the output which were not in the input, but were generated in the system itself, and which may be in the form of changes in waveshape or new components harmonically related to the input frequencies. For a detailed description of distortion see Chapter 1. If the equipment were to be used for speech reinforcement only, distortion content up to 3 percent might not be objectionable, and the highest frequency needed to be reproduced would be limited to approximately 7000 Hz. Overall system harmonic distortion becomes increasingly unpleasant to the ears as the upper frequency limit is raised.

GAIN

The overall gain of the amplifier parts of a system is important and must be carefully considered. The total gain of the amplifiers must be sufficient to drive the final stages to full output with the lowest-level input to be used. Gain is measured in *decibels,* or *dB.*

A decibel is an expression of a ratio of power or a ratio of voltage; thus, if we know the input signal voltage (E_i) of a preamplifier and the output voltage (E_o) of the same preamplifier, the overall gain requirements of this unit may be determined by using the following expression:

$$\text{Gain (in dB)} = 20 \log \left(\frac{E_o}{E_i}\right) + 10 \log \left(\frac{Z_{in}}{Z_{out}}\right)$$

where Z_{in} and Z_{out} are the input and output impedances, respectively.

If the input power (P_i) and the output power (P_o) are known, the gain may be determined by the following formula:

$$\text{Gain (in dB)} = 10 \log \left(\frac{P_o}{P_i}\right)$$

As a matter of convenience, specifications of hi-fi preamplifiers and amplifiers are usually given in overall ratings in a more practical manner. Specifications for basic amplifiers are described as a certain voltage input across a definite input load impedance necessary to produce full output according to power-output rating. Standard input requirements for basic amplifiers may vary from ½ volt to 3 volts of input across input load impedances varying from 100 to 500 kilohms.

Voltage input specifications must always be related to the load. A preamplifier or tuner rated to deliver 1 volt across 100,000 ohms will not drive a basic amplifier requiring 1 volt across 250,000 ohms to full output without more than rated preamplifier distortion. In some cases, input load conditions of basic amplifiers are not given, and ample drive should be provided to cover all contingencies. Also, basic amplifiers should have less than 2500-pf input capacitance to avoid distortion in the coupling network.

Preamplifiers are usually rated in dB of gain or input voltage level for full output. Fifty to sixty dB of gain will usually provide sufficient output (5 volts or more) across standard input impedances from the lowest-level program sources, such as dynamic phono cartridges or microphones. Outputs from these sources may be as low as 5 millivolts across 170 ohms. Input devices having lower outputs than this fall out of the practical application range, as noise becomes too great a factor.

If all units are rated in dB, decibels may be added or subtracted; an amplifier always adds a certain number of dB, and an attenuator subtracts a certain number of dB. Thus, if a preamplifier with a gain of 50 dB is used with a power amplifier with a gain of 60 dB, the total gain will be 50 + 60, or 110 dB.

The term "dBm" is used to indicate the volume level of a constant tone. It means that the level, or sound volume, of a constant tone is the specified number of dB above 1 milliwatt. These units are used in specifications of professional equipment and apply only to microphone levels as far as high fidelity is concerned.

The *volume unit* is another term commonly encountered in the use of pickup units. The volume unit is similar to the dBm in that it is used to indicate the level of a signal in dB above or below 1 milliwatt. The volume unit indicates that the measurement was made on average program material rather than on a constant tone as does the dBm. The volume unit is abbreviated VU.

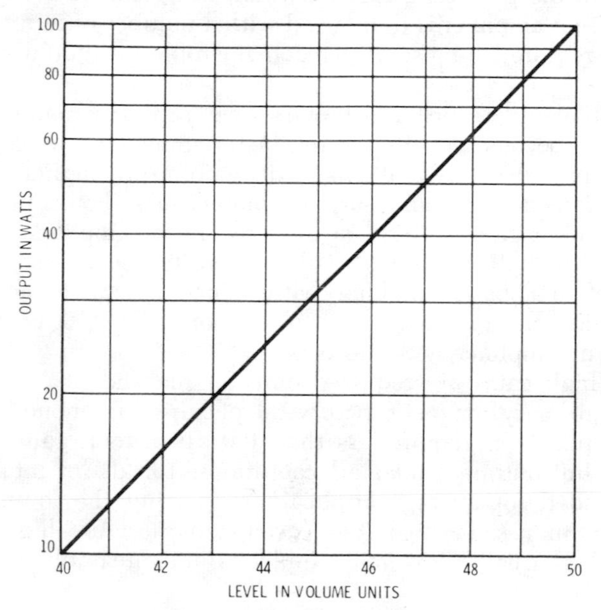

Fig. 8-4. Power versus volume units.

After the power output which will be required from the amplifier has been determined, and the microphones and other input sources have been selected, the gain which will be required from the amplifier can be determined. The first step in determining the gain is to find the volume level in VU to which the power output of the amplifier corresponds. The curve in Fig. 8-4 gives a number of typical values of output power and the volume levels in VU to which they correspond.

It is then necessary to determine the output level of the lowest-level microphone to be used. Since microphones are rated in a number of different ways, their ratings should be converted to dB below 1 mW/1 bar, as described in Chapter 4. The output level of microphones is usually a number of dB below 1 mW. Therefore, the output level must be added to the power output of the amplifier (in VU) to obtain the gain in dB required from the amplifier.

The following is an example of the procedure followed: Assume that the power required from the amplifier is 15 watts, which is equivalent to +42 VU (Fig. 8-4). If the microphone to be used with the amplifier has an output level of 53 dB below 1 mW, the gain required from the amplifier in this installation would be 42 + 53, or 95 dB.

In practice, it is usually wise to add a safety factor of about 10 percent. In the preceding case, it would bring the gain required to 105 dB. If an amplifier is to be used with a number of input sources, the gain required for use with each microphone should be determined.

If high-fidelity recording of one's own program material is desired, gain becomes an important factor. Since the average basic amplifier provides only a 40- to 75-dB gain, a preamplifier to cover all conditions of input and output combinations should provide for a distortionless gain of 70 dB or more. Many preamplifiers are rated at only 50 to 60 dB of gain; so, if hi-fi recording is a must, one must either obtain a basic amplifier with sufficient gain to add to the preamplifier gain to provide a total of more than 100, or one must obtain a preamplifier with more gain.

These high gains are required only for high-fidelity microphone pickup. Ordinary magnetic or crystal pickups for phono or mircophone applications require less than 100 dB of total gain for up to 60 watts full output. Under all conditions, 100 dB of total gain is recommended for average applications, barring the lowest-output hi-fi microphones, because it is easier and therefore less costly to produce high-quality low-gain amplifiers and preamplifiers.

BUILDING UP A HIGH-FIDELITY SYSTEM

Some of us may be fortunate enough to be able to go "all out" for high fidelity with our first system, but most of us cannot. In fact, the reader should not feel discouraged if his budget is limited, because he will probably learn more and derive greater enjoyment from building up his system in stages, starting with something relatively modest. Let us consider now some typical systems and also how an elaborate system can be evolved from a simpler one.

First, consider the simple breakdown illustrated by the block diagram of Fig. 8-5. It is convenient to think of a system as broken down into its three main parts: (1) a program source device, such as a record player or tuner, (2) an amplifier and controls, and (3) a speaker system. This arrangement is a minimum complement for either monophonic or stereophonic reproduction. The differences in stereo are only that provision must be made for a minimum of two channels (possibly four channels) and their related controls in the amplifier section, and that two or more speaker systems must be used. In addition, the source material must provide stereo signal outputs.

If you are starting from scratch, you should keep in mind that you will eventually want a stereo system. No modern high-fidelity system can be complete without stereo. However, we must consider here the possibility that you have a monophonic system now that is "left over" from pre-stereo days, or that you are starting from scratch and want to keep cost down in the first stage by limiting the system to mono. However, in either case, with a little careful planning, the monophonic system can be integrated into a stereo system later.

Fig. 8-5. Three main parts of a high-fidelity system.

There is one part of the high-fidelity system in which you can save by starting with mono, with no compromise in future use for stereo; that is the speaker system. A good wide-range speaker system will cost from $50 to $750, and of course you can spend considerably more. You will need to double this cost for two-channel stereo. If the rest of your system is designed for stereo, you can buy one speaker system first and operate with monophonic reproduction until you can afford the other speaker.

Actually, the cost of amplifiers, changers, and tuners is not much greater for stereo than for mono, if you insist on good high-fidelity reproduction. For example, a good stereo record player can be bought for approximately $75, and buying a record player of equal quality for monophonic reproduction will not save you anything. A good stereo amplifier, with two complete channels and all the controls you need, is not more than 30 percent higher in cost than a similar monophonic type.

At this point, individual needs must be analyzed—balanced design applied in relation to price and quality. The greatest attention should be given to the speaker. Fifty or one-hundred dollars more invested in this unit will give many times more return than the same amount spent in the amplifier. The next important unit where most return will be had per extra dollar spent is in the record player. Even a few dollars may make a great difference in this unit. Amplifiers are now so well developed that even the lower-cost combined units perform well. There is little noticeable difference in amplifiers until you double or triple the investment, and then the difference is small, compared with the differences between the lower- and higher-cost speakers.

Enclosures cannot be selected from a technical standpoint alone, as they must be selected to suit the overall room layout and associated furnishing of a room. In relation to investment, it is again recommended that emphasis should be placed on the speaker rather than the enclosure. A good speaker will sound good in a poor enclosure, but a poor speaker will still sound poor in a good enclosure. Furthermore, a good speaker will sound good with any reasonable baffling arrangement.

Fig. 8-5 shows a minimum-complexity arrangement which can still produce continuous-program hi-fi reproduction of excellent quality. With a record changer as the program source, this arrangement will be entirely suitable for playing medium-fidelity records and also give excellent performance on hi-fi records. Substituting a tuner (either a-m, fm, or both) provides another starter arrangement, designed for radio reception. If the tuner has high output, and if volume and compensation controls are provided in the tuner, only a basic amplifier need be provided. If the tuner does not possess a high output, then the preamplifier-amplifier unit will be required for equalization and volume control.

A good starter arrangement is given in the block diagram of Fig. 8-6A. When you purchase the speaker, you should keep in mind that you will want to add another speaker later, to provide for stereo reproduction. Therefore, the first speaker system might be something in a small enclosure, so that when the other speaker system is added the total space taken up will not be excessive.

The next step in the evolution is to add the second speaker system, as illustrated in Fig. 8-6B. Now, with this one addition, you have stereo performance, using the stereo record player and amplifier.

The next logical addition is a tuner, so you can hear stereo broadcasts (Fig. 8-6C). Your stereo amplifier should have inputs for both the record player and tuner, and a selector to choose either. You are now ready either to play records or listen to stereo broadcasts.

Perhaps you prefer a tape machine to either the record player or tuner; it can be purchased in place of either one, or in addition to them.

The diagram in Fig. 8-6C exemplifies a fairly complete stereo system, but you can still build from there. Fig. 8-7 illustrates a more advanced system. Here, the preamplifier is separate and is used with a basic amplifier unit. Record-player, tuner, and tape input devices are used. A third, "middle," speaker has been added, to minimize "hole-in-the-middle" effects. The same third-speaker circuit can be used to supply remote speakers. However, it should be noted that going from an integrated amplifier-control unit to a preamplifier and basic amplifier is a major change in system design because many of the control facilities of the original complete amplifier are duplicated in the preamplifier. The original integrated amplifier could still be used, but it might be economically desirable to replace it with a basic amplifier. The separate preamplifier will normally provide much greater flexibility than the original integrated amplifier.

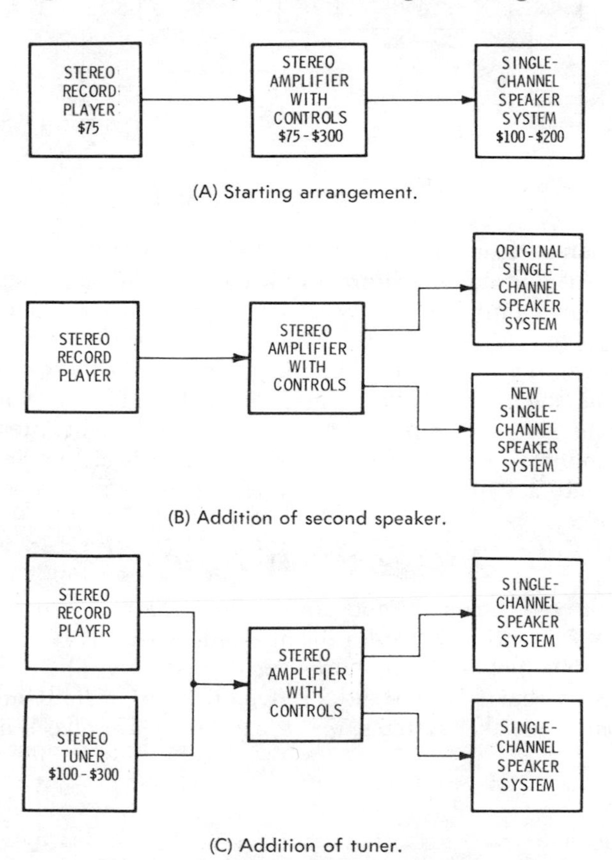

Fig. 8-6. Steps in acquiring a hi-fi system.

This system does not provide for control of the center channel because its origin in this arrangement is at the final output point of the other two channels. Control of the center channel is essential to true stereo effect. An arrangement that provides complete control of center-channel blending and volume is shown in Fig. 8-8. This particular arrangement utilizes a combination stereo preamplifier and power amplifier, with center-channel mixing controls and a low-level center-channel output. The center-channel output is fed

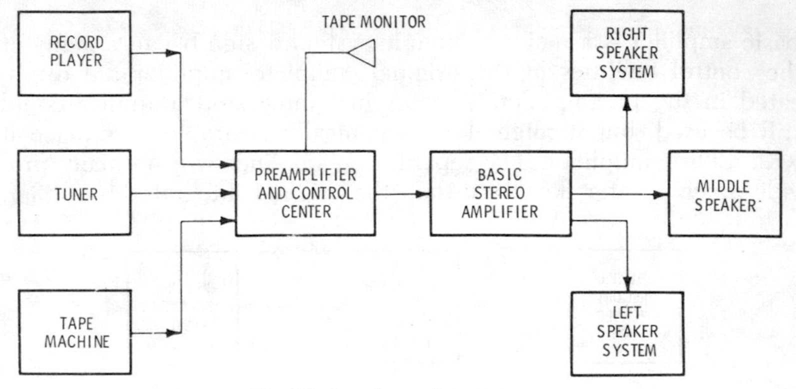

Fig. 8-7. An advanced stereo system.

to a separate single-channel amplifier which drives the center speaker. An additional feature of this arrangement is that an old single-channel amplifier may be utilized for the center channel; thus the stereo effect can be improved at low cost. The two-channel stereo power output is fed to the left and right speakers in the same manner as shown in previous arrangements. There are numerous combinations of preamplifier and control center or preamplifier, power amplifier, and control center that provide a low-level third-channel output that may be used in this arrangement.

FOUR-CHANNEL SYSTEMS

Four discrete channels and four speakers may be used to provide a four-directional surround-sound effect that is quite different from two-channel, two- or three-speaker stereo. The operational difference is that four instead of two different signals are fed to four separate speaker systems, and the speakers are placed in various

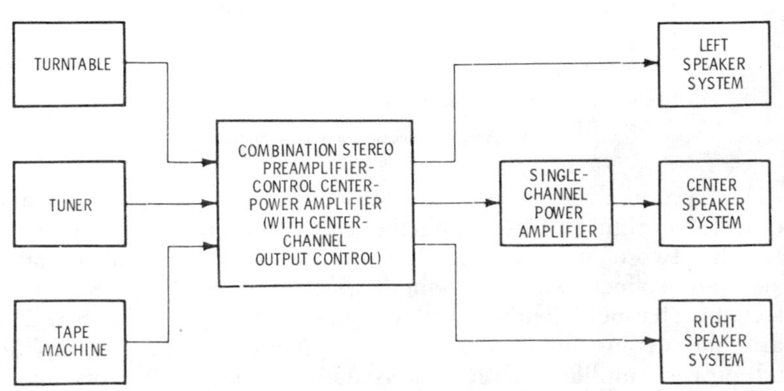

Fig. 8-8. Three-speaker output with controlled middle channel.

ways that, in effect, surround the listener to provide a concert-hall effect that can be quite dramatic compared with regular two-channel stereo.

There are three main techniques commonly in use to provide four-channel sound. They are operation with four discrete channels (sometimes called 4-4-4), the derived-sound technique (2-2-4), and the encoded matrix (4-2-4) technique. All of these systems are described from several points of view elsewhere in this book. The general theory of derived sound, four-discrete-channel operation, and matrixing operations is discussed in Chapter 2. Descriptions of available equipment to adapt for and/or provide four-channel sound are contained in Chapters 4 and 5.

Discrete Four-Channel Tape System (4-4-4)

The optimum method of providing four-channel sound is shown in Fig. 8-9. Signals from four-channel tape, either reel or cartridge, are played back through a four-channel integrated amplifier which contains four separate preamplifier-amplifier circuits with ganged controls and provision for balancing of all the channels. This arrangement is known as four-discrete-channel stereo (4-4-4). The equipment and tape in this system should maintain 30 to 40 dB of separation between channels and provide any degree of control and balance necessary to bring out the best of any four-channel material.

Four-channel open-reel equipment is expensive, and program material is scarce. This equipment is most suitable for professional use in developing program material in the studios. However, eight-track, four-channel cartridge tapes are available in fairly good supply, and excellent equipment is available, in the middle price

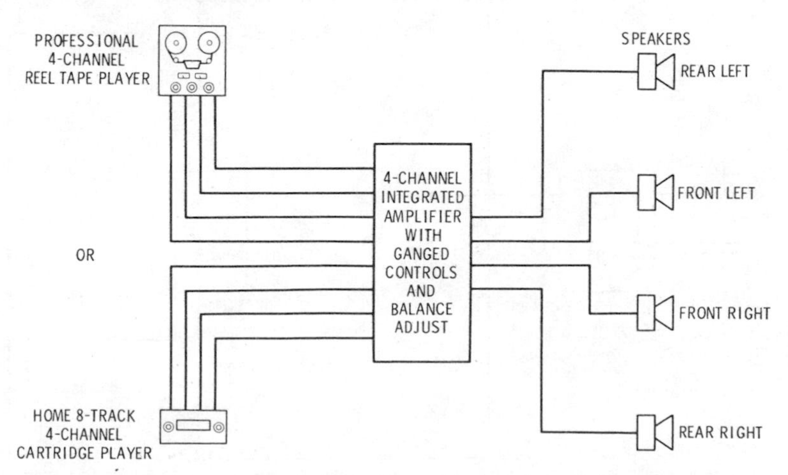

Fig. 8-9. System for playback of four discrete channels from tape.

range, that contains Dolbyized circuits and provision for use of chromium-dioxide tape that extends the useful audio range to provide excellent four-channel hi-fi sound for the home or car.

Derived Four-Channel System (2-2-4)

The simplest and lowest-cost method of creating the four-channel surround-sound effect is with the use of adapters and equipment for the derived-sound technique. This system fully utilizes the capacities of present two-channel systems by extracting parts of the signal content from the output power of a two-channel system and redirecting selected portions of these signals to two additional speakers (Fig. 8-10). This system obtains the additional channels of related but partially discrete information by means of a sum-and-difference extraction matrix to provide two additional signals from the power outputs of a two-channel stereo amplifier. A simplified connection diagram of the Dynaco Quadaptor is shown in Fig. 8-11.

When the original two output signals are fed through the matrix, the center front speaker is connected in series with the ground returns from the left and right speakers. The amplifier must have 6 dB of blending to maintain sufficient separation at the speakers. The 6 dB of blending is equal in amplitude and opposite in polarity to the cross talk which is introduced by the front-speaker connection, and therefore cancels out the cross talk and provides excellent separation of the speaker outputs.

The rear speaker is connected from the hot terminal of the left amplifier to the hot terminal of the right amplifier. No connection is made from the rear speaker to a ground terminal. The speaker

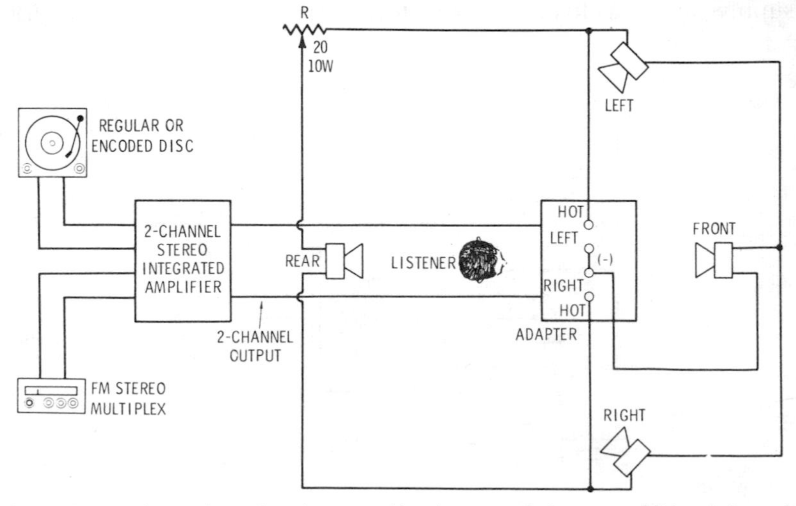

Fig. 8-10. Use of two-channel equipment with adapter to derive two additional channels.

impedance isolates the hot sides of the amplifier outputs, and the common or ground terminals in each stereo-channel output should be connected.

Derived-sound adapters may be used with nearly all component stereo amplifiers and receivers. A very few amplifiers (notably those with a "floating" output circuit) cannot be used with this arrangement. The reason is that this system requires a common ground reference between the two channels. Any restrictions against joining the common (ground) output terminals of a stereo amplifier or receiver should be pointed out in its instruction manual. This limitation is not related to the normal prohibition against paralleling the outputs of a transistor amplifier (direct connections between the "hot" or "high" terminals of the left and right channels).

If separate mono amplifiers are used, they should have similar circuits so that proper phase relationships are maintained between the two channels. It may be desirable to connect their ground terminals together.

When more speakers are added in the same room, the amplifier power requirements do not necessarily increase, provided that the

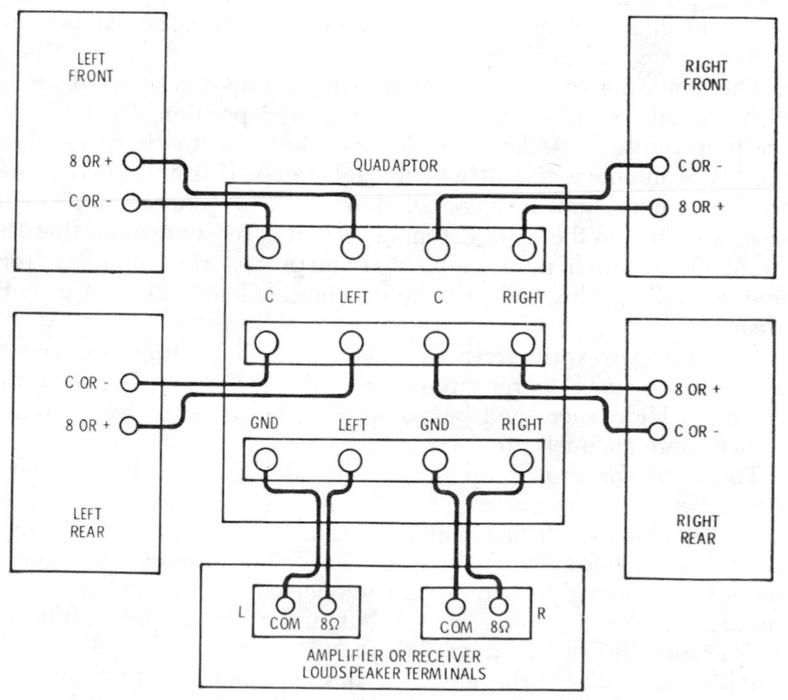

Courtesy Dynaco Inc.

Fig. 8-11. Simplified connection diagram for Dynaco Quadaptor.

sum of the power outputs of the two amplifiers exceeds the sum of the minimum power-input requirements of all speakers in the room; see Figs. 8-1, 8-2, and 8-3.

Fig. 5-60C shows the Dynaco SCA-80Q, which is a two-channel stereo preamp-amplifier with the Dynaquad circuits and controls built in and connected internally. This unit accepts two-channel input and provides properly phased and balanced derived-four-channel output. The input can be from standard two-channel stereo fm multiplex, disc, or tape. The advantage of the Dynaco system is its simplicity; only a standard stereo system with two additional speakers (making a total of four speakers) and the Quadaptor circuitry are needed.

The operation of the Quadaptor (shown in Figs. 5-60A and 5-60B) is as follows: The REAR LEVEL control simultaneously adjusts the volume level of the two rear speakers. It does not affect the loudness of the front speakers. In the fully clockwise position, with four similar speakers, the rear-speaker level is the same for a rear-signal source as is the front-speaker level for a front source. Because the listener will normally be seated closer to the rear speakers, this control provides attenuation as recommended (usually 3 to 7 dB). The control will likely be used at or near the ¾-clockwise position with four smiliar speakers.

The mode switch, a 3-position switch, provides derived "four-dimensional" sound in its normally centered position. In the lower position, the rear speakers are disconnected, and the front speakers are heard in conventional two-channel stereo. If it is desired to use another, remote pair of 8- or 16-ohm speakers simultaneously with some speakers in the main listening area, it is recommended that the Front Only switch position be used to protect the amplifier from undue loading. This will provide conventional stereo sound in both areas.

In the upper (spring-return) Balance Check position, the speakers are connected to reproduce only the difference between the channels. Here, there will be no output from a monophonic source (where both channels are identical).

To adjust for proper operation, this system must be balanced. The REAR LEVEL control is set to the maximum clockwise position, and the mode switch to Front Only. The desired stereo program is played at the normally used volume level. The amplifier or receiver should be switched to a monophonic operation mode, so that identical signals are appearing at the left and right amplifier outputs. This is usually identified as "Mono," "A + B," or "L + R" on the amplifier controls. If the amplifier has no mono switch provisions, a mono record or an a-m or mono fm radio program may be used instead.

When the Quadaptor switch is in the Balance Check position, little or no sound will be heard from any of the speakers. To obtain precise electrical balance and maximum separation, the balance control of the amplifier should be adjusted slowly for a null, or minimum sound output. If the amplifier has individual volume controls on each channel rather than separate volume and balance controls, the two volume controls must be adjusted for the null. If the amplifier provides independent left- and right-channel bass and treble controls, the two bass controls should be nulled, as well as the two treble controls. Normally this will be in their "flat" positions, or where the tone controls are not affecting the circuit. If it is preferred to have some tonal modification, the controls should be nulled with the amount of boost or cut which the listener normally employs.

Once a null has been established, release the Quadaptor switch, return the amplifier mode to normal stereo, and you should be listening to four-dimensional sound. Adjust the REAR LEVEL control so that the back speakers are just perceptible in the normal listening position.

This simple balancing procedure may be repeated whenever changes in the volume-control setting on the amplifier or differing program sources make it advisable.

Because the balance-control setting is critical to obtain optimum separation in this system, this control should not be used for shifting left-to-right emphasis when listening in the four-channel mode. Thus the listener is advised to shift his position to adjust balance inequities.

If there is reduced separation, lowered output level, and uncertain centering of soloists, or an ill-defined null, the cause is most likely to be reversed polarity (phase) in one of the connections to the Quadaptor.

The degree of increased realism which this derived system provides over conventional two-channel stereo varies with the program material. Some material is specially recorded to take advantage of the capabilities of this system. With existing two-channel material, the benefits to be derived are largely random and depend on the particular recording techniques employed. However, such benefits may be dramatic on some recordings, and it is a rare performance which does not show some improvement.

When the Quadaptor switch is placed in the Front Only position and then switched to four channels, additional audience-participation sound, including applause, should become apparent. In pop material, certain instrumentalists may appear to be located behind the listener, the result of unintentional microphone misphasing. Organ works frequently reveal added low-end power. On many classical recordings, the initial impression when switching from two to four channels may not seem so dramatic.

The benefits of four-dimensional sound will often be most apparent in smaller rooms, where space restrictions were previously a significant handicap in reproducing material such as organ-pedal fundamentals, or in creating any sense of "hall sound."

It should not be expected to hear four separate and distinct channels from the use of derived-sound techniques, as this would be in essence discrete four-channel sound or encoded four-channel sound. Realistic musical reproduction with derived sound will contain a similarity in relationship between all sound channels and significant overlap between them, and a large percentage of the program material will be included in common in all channels. The derived four-channel system attempts to take advantage of this principle to develop the full simulated reproduction of the two sound channels and of their phase and amplitude interrelationships in the four-channel mode. In effect, more information has always been on the record or tape than has been separated previously due to the masking effect of the louder signals (described elsewhere in this book). Recording engineers have long striven to find microphone-pickup techniques and performer placement in the studio or hall which could uncover more of the "flavor" of the live performance on playback.

The derived system provides for standard stereo reproduction from the front speakers without use of the rear speakers. If you switch off the rear speakers, you should hear the same left-to-right separation you always had. If a soloist was recorded in a central location, blended into the two channels, the solo will come from a virtual front center location between the front speakers. A monophonic program played through the derived system using only two speakers will also appear as a centered front source.

In operation, it is generally advisable to reduce the level of the rear-speaker output with a series resistor (R in Fig. 8-10). This could be a 20-ohm, 10-watt variable noninductive resistor. The rear speaker may give better results if it is raised to a level above the head of the listener and pointed directly at the listener.

Encoded Matrix Four-Channel System (4-2-4)

The 4-2-4 encoded matrix technique offers a compromise method that provides discrete four-channel stereo sound with sufficient separation to develop concert-hall realism, and at the same time is compatible with present disc reproduction techniques and fm multiplex broadcasting. Fig. 8-12 shows an arrangement for using this technique with a two-channel stereo system, an added adapter, an additional two-channel amplifier, and two additional speakers.

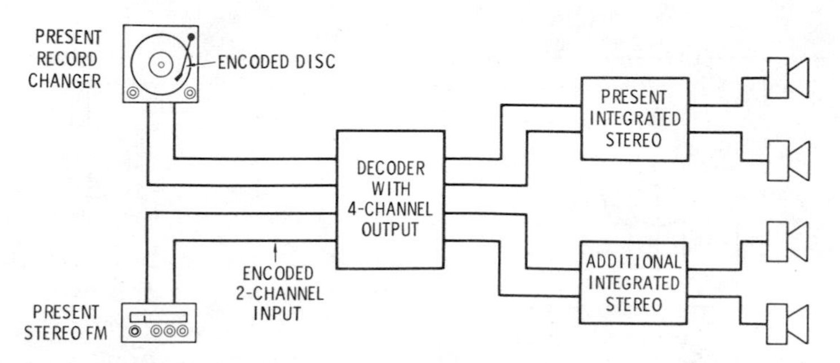

Fig. 8-12. Arrangement for reproduction from encoded discs, tape, or fm broadcasts.

The separation between channels is about half of that generally obtained with the arrangement shown in Fig. 8-9, but with propitious allocation of separation to favor the front speakers (where the effect to the listener is greatest) and to apply sufficient but reduced separation front to rear and rear left to rear right, this method provides an effect of concert-hall realism almost equal to that of the most expensive discrete four-channel system. Setup, connection, and operation of this system with the decoder shown in Fig. 5-62D connected as in Fig. 8-12 are as follows:

Four separate power amplifiers (two stereo amplifiers) and four speakers are required. A simple way to use the preamp and input-selector sections of most stereo amplifiers and receivers to feed the decoder, while the power-amplifier sections of that amplifier or receiver are used to power the left front and right front speakers of the four-channel system, is to operate the tape-monitor (also called tape-source) switch. This switch interrupts the normal signal path through the amplifier and introduces a new signal from outside the unit, independent of the input selector switch. The connections for this arrangement are shown in Fig. 8-13. If the unit has a separate control which permits selection of tape input without disturbing the normal input selector, you can easily connect the decoder.

The decoder can be inserted in the system anywhere ahead of the power amplifier. For example, the decoder can be inserted between separate preamp and power-amplifier components. However, some flexibility in system operation is lost with this method of connection, so the tape-monitor connection is recommended.

While the choice of equipment for a four-channel system is of concern to many, the criteria for good performance are less severe than for two-channel stereo. As is the case with two-channel stereo, good performance results when all four amplifier channels are the same and all four speakers are the same. However, the results are still totally satisfactory if the rear amplifier is of lower power and the rear speakers have narrower frequency range, because the rear

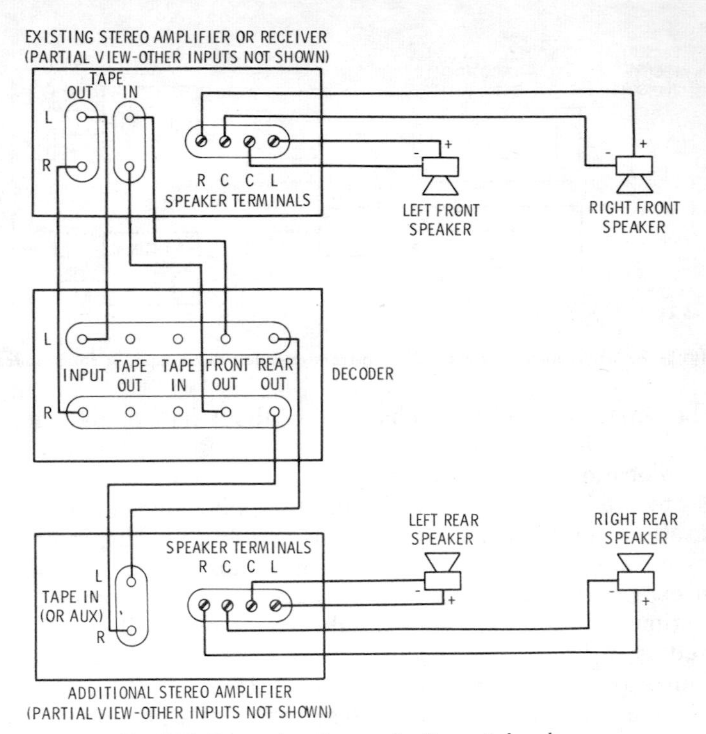

Fig. 8-13. Connection diagram for Stereo-4 decoder.

channels are enhanced substantially by the front channels, especially at low frequencies, and vice versa. Program material which has ambient information (hall sound) in the rear requires less power and frequency response in the rear channels than does the type of program material in which instruments are all around the listener. For best results, however, the two front speakers should be closely matched high-quality speakers.

With the decoder gain control set at maximum, the decoder output will be no higher than the input voltage from the tape jacks of the front amplifier. For this reason, the gain or input sensitivity of the rear amplifier must be high enough that the decoder output will drive it satisfactorily. If the sound level from the rear amplifier is sufficient when it is connected directly to the tape output jacks of the front amplifier, then the decoder will work with it as well.

Complete receivers (amplifier with tuner), integrated amplifiers, preamps and power amplifiers, integrated compact systems, consoles, or combinations of these may be utilized for four-channel equipment. The electronic unit used to drive the front pair of speakers will be called the front amplifier, and the electronic unit used to drive the rear speakers will be called the rear amplifier.

The unit to which program sources (phono, tuner, etc.) are connected should be considered the front amplifier. For this reason, a more complete unit (stereo receiver versus stereo amplifier) should be considered for the front unit. If two similar units are to be used (two amplifiers or two receivers), the better unit in terms of power output, distortion, etc., should be used as the front unit. If the two pairs of speakers are not identical, the better pair should be used for the front speakers.

When the equipment has been selected, refer to Fig. 8-13 and proceed as follows:

1. Using shielded audio cables of convenient length, connect the front-amplifier tape-output jacks to the decoder input jacks.
2. Connect the decoder front-output jacks to the tape-input jacks on the front amplifier. At this point, you should have cables going to and from the front amplifier, connected to the decoder input and front-output jacks.
3. Connect the decoder rear-output jacks to the rear-amplifier input. If the rear amplifier has tape-input jacks, use them. If not, use the auxiliary or other high-level inputs.
4. Connect the four speakers to the appropriate amplifier output terminals. The speaker locations are named with the observer facing the "front" of the four-channel system. Thus, the left rear speaker is connected to the left speaker terminals of the rear amplifier. Be sure to connect all speakers in phase, as outlined in the speaker instructions and elsewhere in this chapter.
5. If there is a tape recorder in the system, it may be connected to the tape-input and tape-output jacks on the rear of the decoder. These jacks duplicate the jacks previously used on the front amplifier, and, additionally, allow decoding of the output of the tape recorder into four channels.
6. Connect the ac line cord to a convenient outlet. If a switched ac outlet is available on one of the amplifiers, power will be applied when the amplifier is turned on, and the decoder master gain control can be left turned up all the time. In fact, if switched ac outlets are available on both the front and rear amplifiers, plugging the rear amplifier into the front amplifier and the decoder into the rear amplifier will permit the entire set of equipment to be turned on with the front-amplifier power switch.
7. Place the tape-monitor switches in the tape position to connect the output of the decoder into all four amplifier channels. Start with the amplifier volume controls turned all the way down, the decoder master gain control turned all the way up, and the decoder function switches in the source decode posi-

tion. Turning up the amplifier volume controls will provide the desired sound level in the room and also permit adjustment of front-to-back balance. If the volume controls are initially set for the loudest sound level you want in the room, with the front and rear sound levels about the same, then the master gain control on the decoder will reduce the sound from all four speakers equally as it is turned down.

After this initial adjustment, the rear-amplifier volume control can be used as the front-to-rear balance control. Adjust the rear sound level up and down slightly to move the balance point forward or backward in the room, and to adjust for best subjective effect. Enhancement of a two-channel classical work should provide a re-creation or simulation of the ambient sound as it originally existed in the hall where the recording was made. Normally this can be done by reducing the rear-amplifier volume somewhat, and perhaps turning down the rear treble control.

The left-right balance controls of both amplifiers work normally, although there will be less need to adjust the left-right balance in either front or rear.

8. To return to straight-through two-channel operation of the front speakers, (A) return the front-amplifier tape-monitor switch to the source position, and (B) on the rear amplifier, either switch back to source, or switch out of auxiliary depending on the hookup; or simply turn down the volume control.

   If a tape recorder is connected to the decoder as part of the system, it is possible to record conventional two-channel tapes, and play tapes through either the two-channel or four-channel system. The signals going to the tape recorder are identical to those coming from the tape-output jacks on the front receiver, where the recorder normally would be connected. If the decoder function switch is in the tape-monitor position, the output of the tape recorder will play back through the front amplifier in normal two-channel fashion. If the function switch is in the tape-decode position, the output of the two-channel tape recorder will be decoded into four channels, just as a record or fm broadcast is decoded in the source-decode position.

Improved control may be had from the arrangement in Fig. 8-11 by using the decoder shown in Fig. 5-62G. The operation technique is similar to that just described, but this adapter has more features to provide overall balance of the channels to each other with ganged controls. It also provides for monitoring visually and aurally, and it can provide "synthesis" for program material that is or is not

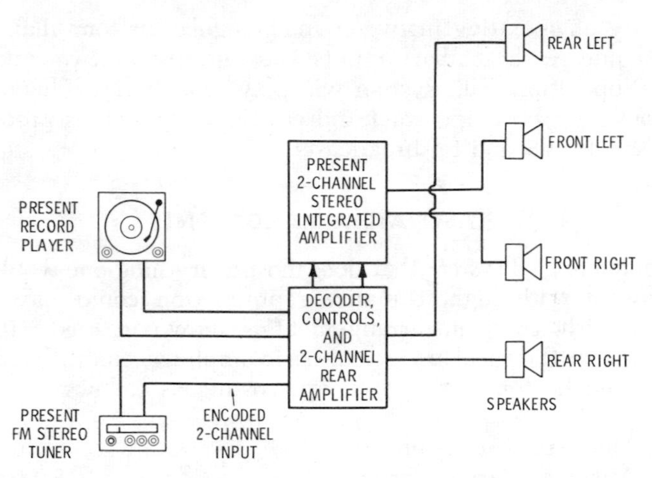

Fig. 8-14. Use of adapter-decoder-amplifier in 4-2-4 system.

encoded to develop delayed or phased outputs to the rear channels for simulated effects of four discrete channels.

Fig. 8-14 shows a similar system using a combination adapter–decoder–integrated-amplifier made especially to adapt two-channel systems to four channels with only one additional electronic unit and two additional speakers. Fig. 8-15 shows an integrated fm stereo receiver and four-channel amplifier with built-in decoder and the

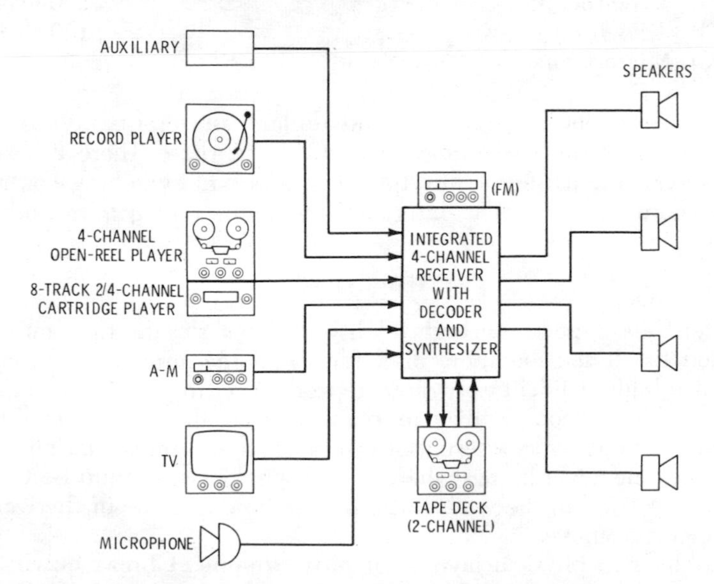

Fig. 8-15. System for two or four discrete channels or 4-2-4 encoded operation.

capability of operating in two-channel regular or four-channel decoded modes with auxiliary inputs and outputs for two- and four-channel operation. This system will play four-discrete-channel output from tape, both open-reel and cartridge, as well as process encoded 4-2-4 discs and fm broadcasts.

ELABORATE ARRANGEMENTS

A complete hi-fi system that does most everything one would ordinarily want with highest-quality reproduction could have many units. A fairly complete arrangement is shown in Fig. 8-16. It is expected that the speaker system, basic amplifier, control, and other units would be top quality, perhaps costing as follows:

(A)	Main speaker system	\$250 to \$700
(B)	Auxiliary speaker system (2)	\$100 to \$300
(C)	Speaker switcher and controls	\$20
(D)	Basic amplifiers	\$100 to \$200
(E)	Control center	\$100 to \$150
(F)	Record changer with reluctance cartridge and diamond stylus	\$85
(G)	Am/fm tuner	\$100 to \$150
(H)	Turntable with dynamic cartridge, diamond stylus, and hysteresis motor	\$150 to \$200
(I)	Tape recorder	\$200 to \$400
(J)	Television chassis	\$150
(K)	Microphone	\$50

One can spend several thousand dollars for top-quality components such as the preceding. In addition to these, there are such items available as: automatic time (on and off) switching controls, disc-recording equipment, intercom, telephone pickups, and others.

SYSTEMS LAYOUT

There are aspects of high-fidelity systems design that may be related to all the members of a family. Sometimes it is better to sacrifice highest fidelity of sound reproduction to gain better family relationship. A compromise in placement of the units may bring better harmony everywhere, both in music and family relations. For instance, one never knows where a wall speaker will sound best until the unit is tried in the wall, but once the hole is made in the wall it is difficult to move.

Problems in physical layouts of hi-fi equipment break down into three groups:

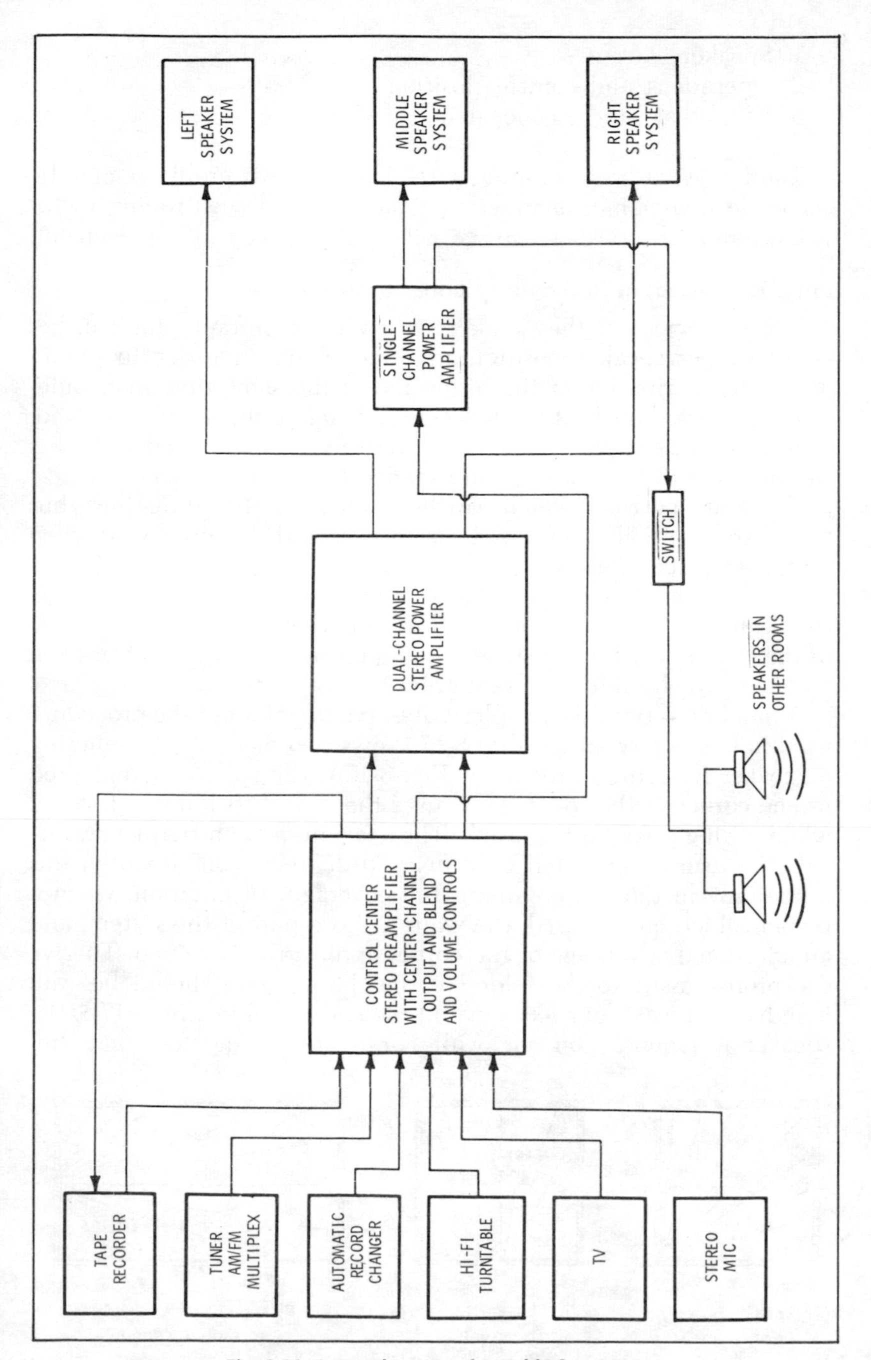

Fig. 8-16. A complete two-channel hi-fi system.

1. Speaker placement
2. Operations and control position
3. Units that can be concealed

The last is no problem anywhere. The first two problems may be solved in a combined manner or separately. The controlling factor is the room acoustics, and so we will begin with speaker placement.

Speaker Placement in Single-Channel Systems

The placement of the speakers is a very important problem, because proper speaker positioning is necessary in order to obtain adequate distribution of the highs and at the same time make adequate provision for bass reinforcement. If microphones are involved, improper speaker placement may cause feedback. Added to this is the problem of minimizing interference between speakers.

There are no rules which can be applied to all installations, but there are a number of points which should be observed when selecting speaker locations.

Speakers should never be placed at two ends of a room. As a rule, they can be mounted on one wall or in one corner of a room. Most of the sound heard by a listener should come from one point or from speakers which are equidistant from the listener.

A number of typical examples will serve to point out the procedure used in locating speakers. Fig. 8-17 shows two methods of mounting a speaker in a square room. In Fig. 8-17A, the speaker is mounted in one corner of the room. This gives the best distribution of sound when a single speaker is used. The maximum high-frequency distribution can be expected only over 90 degrees; consequently, this is the only method of obtaining full coverage. In addition, when a corner folded horn is used, the walls act as a part of the system, and an additional advantage of the corner application is gained. To give the proper assist to the folded corner horn, there should be wall lengths of at least four feet beyond the enclosure. In Fig. 8-17B, the speaker is mounted on one wall (or in it), giving less than full

(A) Speaker in corner of room.

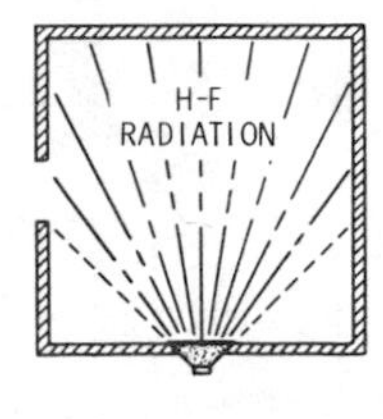

(B) Speaker in only one wall.

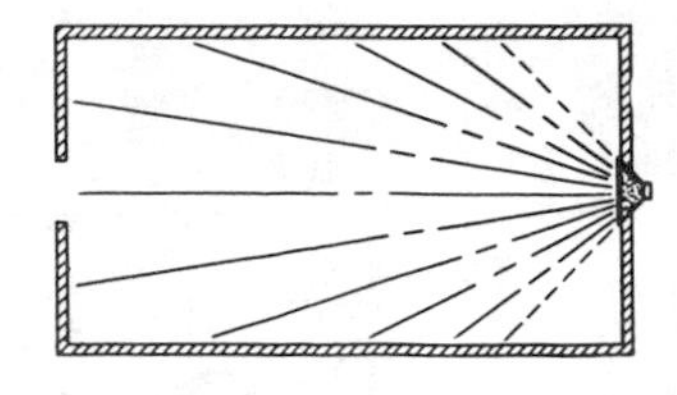

(C) Placement of speaker for a rectangular room.

Fig. 8-17. Speaker placement in square and rectangular rooms.

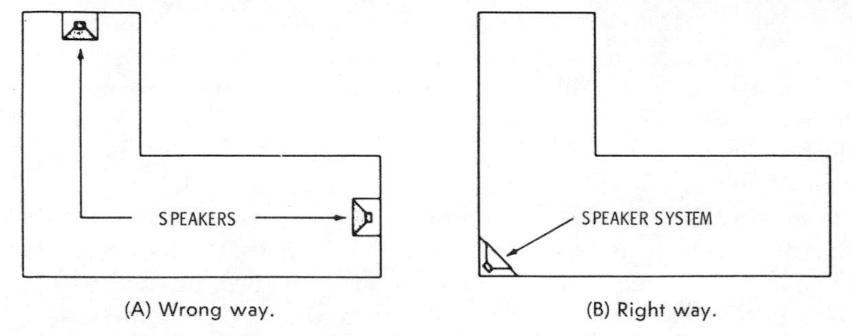

(A) Wrong way.

(B) Right way.

Fig. 8-18. Speaker placement in L-shaped room.

coverage of the highs. This same arrangement for rectangular rooms (Fig. 8-17C) will give better distribution because of reduction of reflection.

Fig. 8-18 shows the right and wrong ways to position speakers in an L-shaped room. Fig. 8-18A shows the wrong way, in which a speaker is mounted at the end of each leg of the L. In Fig. 8-18B, the correct position is shown. The speaker system is arranged at the junction of the legs of the L, bisecting the angle to obtain even distribution of sound.

Fig. 8-19 shows a two-room installation with adjoining rooms. A single speaker is mounted in each room. With this arrangement, the speaker lines are kept quite short, and if a listener is able to hear lows originating from two speakers, the speakers will be almost equidistant from him. This condition will exist near the openings between rooms.

Using an inside wall as an infinite-baffle arrangement may turn out in some instances to be an excellent low-cost arrangement, but there are many factors involved that may reduce the effectiveness. In an average home, a speaker may be mounted in a wall between

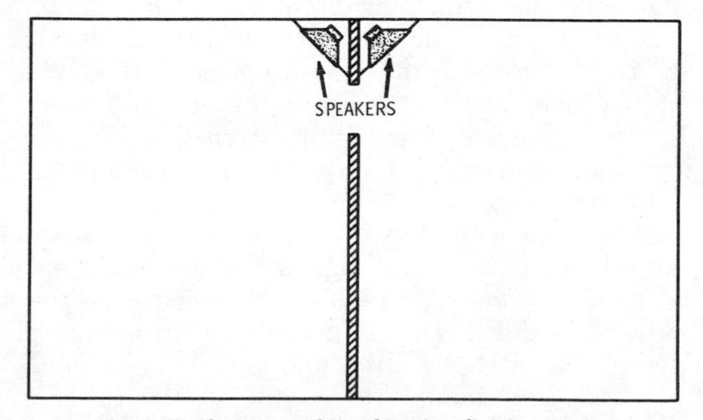

Fig. 8-19. Placement of speakers in adjoining rooms.

two rooms or between a room and a large closet. In any case, the baffle room must be in proper acoustic relation to the speaker to act as a first-class baffle. Room resonances in either room can be reduced by drapes on opposite walls, but this may not suit the decor. Severe reflections and other interference from the baffling room back to the speaker will cause distortion. A baffling room or closet must be acoustically treated to be effective. Chances are great that the acoustic nature of the room is such that a location where the infinite-baffling technique can be used to best advantage is not the optimum location for the speaker, and as a result the distribution of the highs only covers half or less of the room. Experience has shown that cutting up a house to provide for infinite baffling is not advisable as a practical matter on either a short- or a long-term basis. However, excellent results can be obtained if you persist and have a little luck.

In almost all indoor installations, the speakers are mounted from a normal sitting ear level on up to near the ceiling. This allows a clear path for the sound to a listener in any part of the room. Floors and ceilings are not recommended for speaker mounting, because the effects seem unnatural.

In outdoor installations, the speakers should be mounted 10 feet or more above the ground. All speakers should be located at the same point. This point does not necessarily have to be at the point of direct pickup, since in many outdoor installations sound from the point of pickup is a negligible factor.

Speaker Placement in Stereo Systems

As in the case of monophonic reproduction, the most critical link in the stereo reproducing chain is the speaker system. The requirements for stereo are even more rigid because proper speaker placement and balance are important for production of the stereo effect.

The use of all of the principles discussed in Chapters 6 and 7 for speakers and enclosures is more important in stereo systems than in mono systems. The reader is therefore referred to these chapters for fundamentals which concern the design of each individual speaker system. In stereo, we are primarily concerned with the use of at least two of these systems and how they are coordinated for best enjoyment of the stereo effect.

The exact requirements for placement, relative frequency response, and relative power-handling capabilities of the speaker systems to be used for stereo are matters on which even the most respected authorities differ. This is probably because the placement of the microphones at the source, the relative size of the source (orchestra, chorus, etc.), and the nature of the sounds involved differ widely in the different programs to be reproduced. One set of conditions cannot be optimum for all of them. So it is not possible to give positive single answers to most of the questions which the reader will have about acoustic reproduction for stereo. We shall therefore concentrate on reviewing some of the different ideas that have been set forth on the subject, aware that the listener is best qualified to make the final choice for himself.

Matching Speaker Systems—It seems generally agreed that the ideal situation is two or four completely matched full-range reproducing systems. Some have suggested that "reasonably good" stereo reproduction can be had with one elaborate system of speakers (for center, right, or left) and one or more cheaper systems which can have limited frequency response (high) and perhaps a small degree of distortion. However, this suggestion is usually made in connection with a transition from a monophonic system to a stereo system, and this arrangement is not considered a good substitute for two or four top-notch speaker systems.

In some compromises, cognizance is taken of the fact that the stereo effect is obtainable only with the higher-frequency signal components. Because of the large wavelengths of the lower frequencies when compared to the spacing between speakers and the distance to the listener, the low-frequency sounds seem to come from a wide area. (For example, at 50 Hz, the wavelength is over 20 feet.) Also, differences in phase due to sound-source location differences are not noticeable at low frequencies, and the stereo effect is lost (although presumably there is still an intensity difference). The fact that the stereo effect is not as prominent at low frequencies has led to the idea of concentrating the low-frequency components (which contain the most power) in one high-power woofer located in the center. The left and right speakers then need only reproduce the relatively low-power higher-frequency components. The left and right speakers should be carefully matched, but lower-priced speakers than are usually employed in a full-range system can be used. The use of a middle speaker also minimizes the "hole-in-the-middle" effect.

The left and right speaker systems must also be balanced as to phase. If the connections to one of the voice coils are reversed, the sound from one speaker tends to cancel that from the other, leading to distortion and lowered output. Correct phasing can be recognized as that resulting from the connection that gives the greater output. The polarity of each speaker is not important as long as both are phased the same way. Phase checking of speakers is discussed later in this chapter.

Placement—The ideal speaker placement is the same as that of the microphones with which the sound is picked up. Unfortunately, this cannot usually be arranged, because microphone spacing differs for different source material, and moving speakers around to match the program is not practical. Also, the layout and decor of the room must be considered, because, of course, the average home cannot be "designed around" the high-fidelity systems (although in isolated instances this has been done!). Probably the best general guide to placement is the triangle illustrated in Fig. 3-1. First determine approximately where the listeners will sit; then draw an angle of from 30 to 45 degrees between there and the wall where the speakers are to be located. The points where the two sides of the angle intersect the wall are the two approximate locations for the speakers.

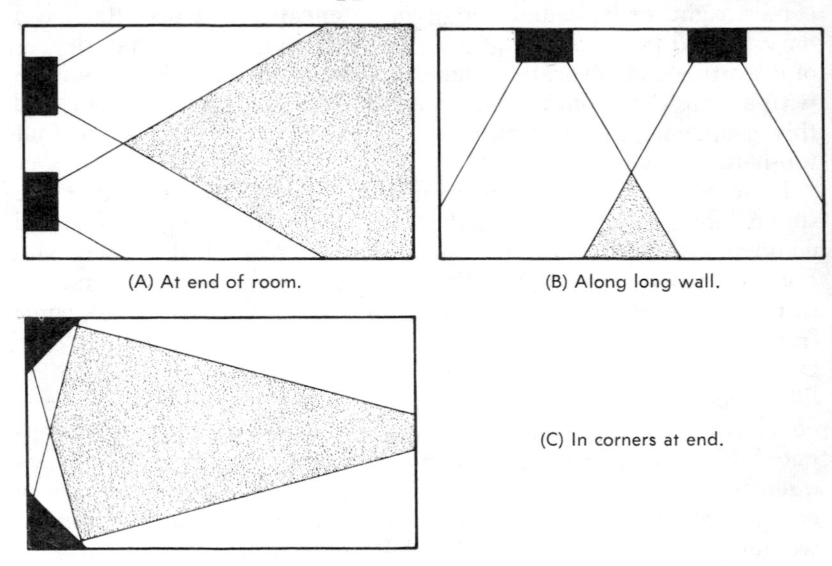

(A) At end of room.

(B) Along long wall.

(C) In corners at end.

Fig. 8-20. Three ways of placing stereo speakers.

As illustrated in Fig. 8-20, the speakers may be placed along the end wall, along the side wall, or in two corners (for which special "corner speakers" are available). As indicated by the shaded areas, the stereo effect is obtained over a much greater area if the speakers are placed at the end of a room. Only those areas within the coverage of both speakers can provide stereo effect. Notice also that it is not good to plan on being too close to the speakers (between them) because then, too, you will be out of the coverage areas.

At the high frequencies, at which the stereo effect is the greatest, the radiation beam of the speakers is the narrowest. This is something to keep in mind especially when you are using single speakers at the left and right. Most tweeters in dual speakers are now designed to spread the high-frequency energy over most of the beam width of the middle-frequency range.

Where space is limited, and the speakers cannot be spaced far enough apart, various methods are employed to orient the speakers so that they radiate as though they were farther apart. In one such arrangement, the speakers are mounted at the ends of a rectangular cabinet, and doors in the end are opened part way to deflect the sound outward to the listener.

The Middle Speaker—As has been mentioned, many audiophiles believe in the use of a "middle" speaker, that is, a speaker located between the left and right speakers and reproducing a subdued combination of the right and left signals. The idea of this arrangement is to overcome what is known as the hole-in-the-middle effect and provide a wide curtain of sound. The hole-in-the-middle effect arises when the stereo effect is so pronounced that the listener begins to distinguish two separate sources. This can be partially overcome by diluting the stereo effect with a blending control. However, a middle speaker is best for removing the hole-in-the-middle effect without loss of the stereo effect. Signals from both channels are coupled to the middle speaker, which is operated at a lower level than either the right or left speaker.

The optimum arrangement for providing center-speaker operation is by mixing (blending) the outputs of the preamplifier and feeding the combined signal to a third amplifier channel for the middle speaker, as shown in Figs. 8-8 and 8-16. The preamplifier outputs must be combined in such a way that they do not cross over and produce interaction in the left and right amplifiers. A separate blend volume control will provide optimum results.

A lower-cost way to provide a signal for the middle speaker by combining the signals at the amplifier outputs is given in Fig. 8-21. The L and R signals are fed to the middle speaker through choke-coil filters that are designed to pass low-frequency signal compo-

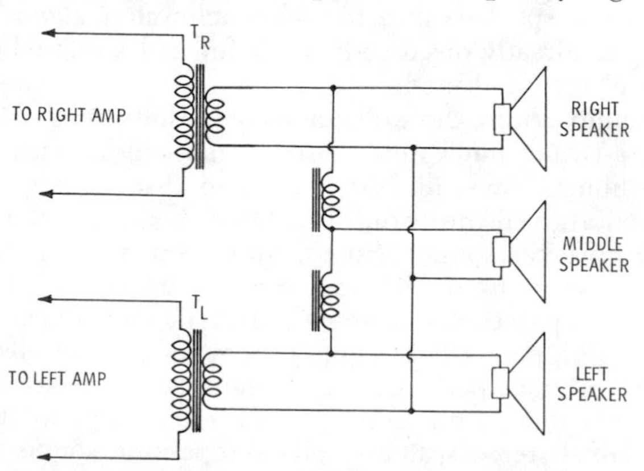

Fig. 8-21. Combining L and R outputs for a middle speaker.

nents and block the middle- and high-frequency components. However, the filter chokes are far from perfect filters, and interaction between the two signals is fed back through them.

A more efficient arrangement is given in Fig. 8-22. The middle speaker is of special design, having a dual voice coil; that is, a separate voice-coil winding is employed for each circuit. Although the filtering is similar to that in Fig. 8-21, the voice coils are so small that the coupling between them is negligible, and separation between the left and right circuits is good. Dual–voice-coil woofers, such as that shown in Fig. 6-35, are now made by nearly all speaker manufacturers.

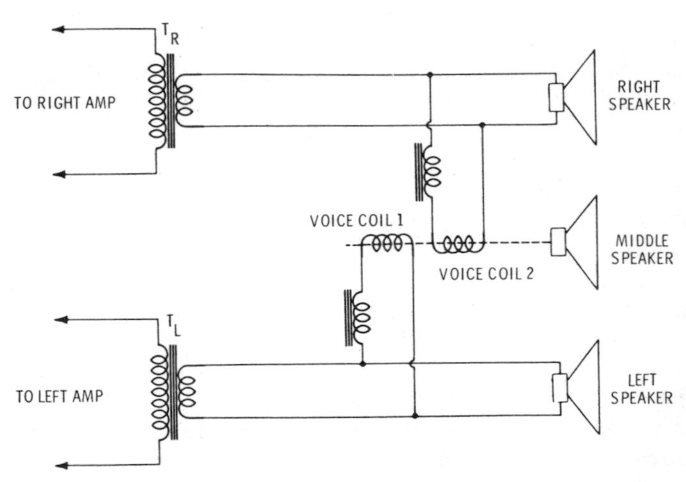

Fig. 8-22. Combining L and R outputs for a two-voice-coil middle speaker.

Four-Channel Speaker Placement

Placement of speakers in a four-channel system should follow all the principles already discussed, plus a few rules related only to the four-channel surround mode.

Many sounds from the sides and rear that were originally recorded on a two-channel stereo disc are masked by louder passages. Since the human ear will "listen to" and discern only the louder sounds from a given direction, the human hearing system will then "mask" the weaker sounds from that direction even though they may be necessary to render the music properly. Therefore, it is desirable to separate these sounds and deliver them from their original bearing points (directions) from the average listener position at the original performance. When these weaker sounds are coming from a different direction, they are more apparent to the ear.

Four-channel stereo systems deliver separated sounds in a mode similar to that in which they were recorded and distributed to the four channels. The speakers in turn should be arranged in the room to deliver the reproduced sounds from directions that simulate the spatial distribution of the original sounds.

For convenience, the speakers are usually placed in the four corners of a room, but a better arrangement is to place the front speakers in the corners and the rear speakers nearer to the listener, somewhat forward of the rear wall and to the side, as shown in Fig. 8-23A. The volume from the rear speakers should be subdued compared with the volume from the front speakers. The left and right speakers should be balanced in volume level.

A method of speaker placement for derived-sound (2-2-4) program reproduction in the absence of a separate specific back signal even with the rear-level control at maximum is shown in Fig. 8-10. This method assures that in the usual listening environment, where the listener sits nearer to the back speaker, proper placement and emphasis of instruments or voices will be retained on the sound stage in front of him (front dominance). Since the back speakers are usually closer, and form a wider listening angle, the fact that the back speakers reproduce some of the front left- or right-channel information, in addition to the reflected sounds from the sides and rear, provides more sharply defined differences in intensity which preserve the maximum effect of directionality. In effect, the ear often senses or perceives greater effective aural separation than the electrical signals actually provide (psychoacoustic response).

Four-directional effects are achieved because the information in the front and rear speakers is different—not because there is some front information appearing with reduced level or different phasing in the rear. As explained previously, the added rear speakers make it possible for the ear to perceive separated signal information from different bearings, which contributes to realism but which has previously been masked when combined with the output from the front.

The derived four-channel sound output to the rear speaker includes a simple method for extracting rear information from two-channel material, and in turn reduces the front information when it is reproduced through the rear speakers. This diversion of front information to the rear enhances the proportion of rear sound to add front and back directionality to the conventional left and right orientation, as shown in Fig. 8-10. Derived four-channel output can also have speaker placements as shown in Fig. 8-23.

Figs. 8-23A through 8-23D show recommended four- and six-speaker placement arrangements. It is suggested that the best arrangement will be found by trial and error, since the shape of the room and furniture placement in the room have a considerable effect on proper speaker placement.

Experiments have shown that it may be desirable to raise the level of the rear speakers above the level of the front speakers to improve the spatial effects, with each rear speaker at a different height.

Figs. 8-23E through 8-23H show suggested arrangements in a rectangular room with furniture layouts taken into consideration. Figs. 8-23I and 8-23J show recommended four-channel arrangements for use in automobiles.

Location of Program Sources, Amplifier, and Control Equipment

The units involved here may include everything in the system, may include all units but the speaker system, or may include all

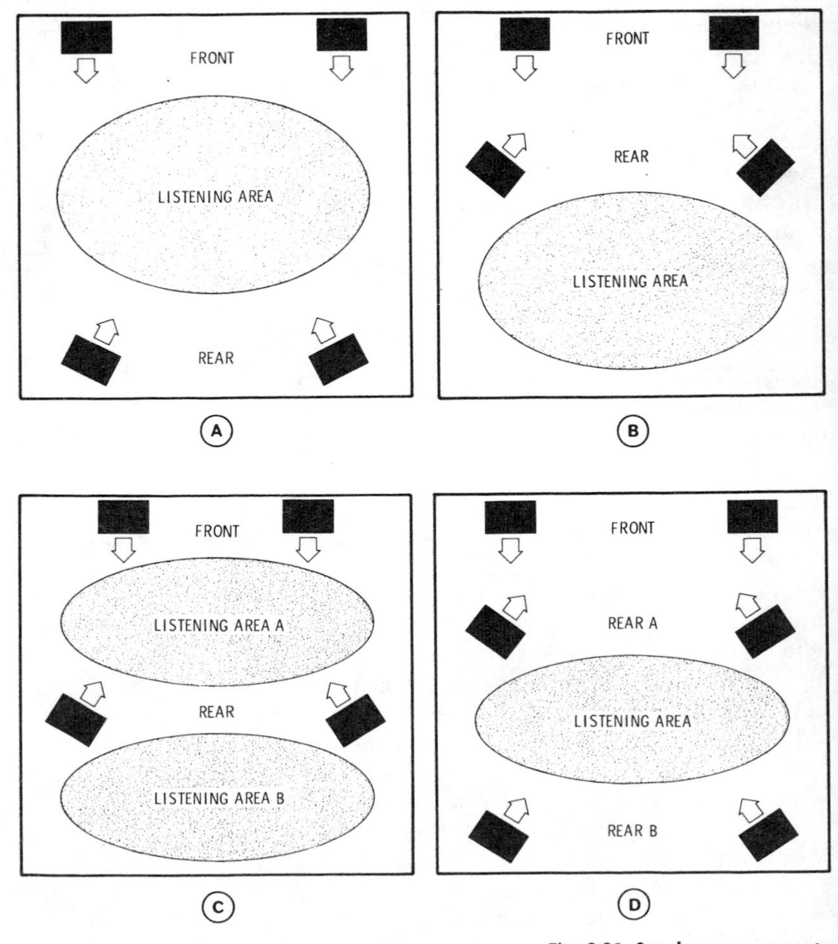

Fig. 8-23. Speaker arrangements

units but the speaker system and the basic amplifier. A basic amplifier completely controlled from a preamplifier could be kept anywhere in the building as long as it has sufficient ventilation.

Wall Cabinets—Sectionalized wall cabinets of all types are available. These cabinets are usually of modern design with spaces to fit any conceivable arrangement, including television and speaker with a full-sized bass-reflex enclosure. One such arrangement is shown in Fig. 8-24. Many others are available. Provision can also be made for extras such as books and ornamentation if desired.

Record player, tuner, controls, and amplifier are all located near each other, and the point of operation is at the nearest point to the listener that it can be without remote control. Speaker systems

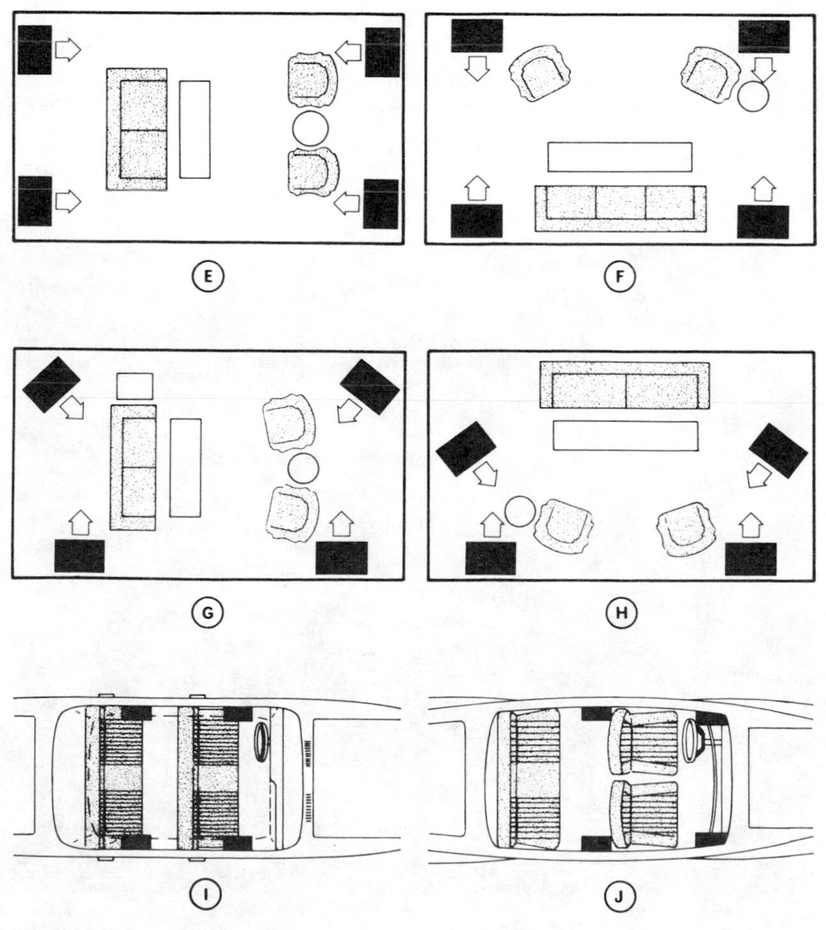

for four-channel systems.

should be mounted at ear level or above and slightly off center of the room. The best position for such a system mounting is on the narrow wall nearer one corner. Make all connections short, shock mount the pickup units and preamplifier, and ground everything metallic.

The same ideas can be used in novel home-modified break fronts and other odd furniture pieces for matching period decorated or early American homes. Usually, the speaker is best mounted separately to gain the advantage of a properly designed enclosure.

Packaged units are commercially available at prices from $160 to $1800. The advantage of these is that one gets a complete set of matched components with no further engineering or construction required—just plug it in. The limitation is that one has no choice in variations of arrangement to suit his individual taste.

Units in Separate Furniture Pieces—Several pieces of furniture may be separately set up to house the various components that must be operated. Table equipment and consoles are available for individually holding record players, tuners, and amplifiers. It is often convenient to have the operating position of the player, tuner, and controls located at an accessible spot convenient to the listening area,

Courtesy Jensen Mfg. Co.

Fig. 8-24. A typical home stereo layout.

to save steps. Here again, custom installation in separate furniture pieces to match the decor can be beautiful and practical.

Wall Units—If mounting the pieces in furniture is not desired, it is modern to mount operation control units in the wall. Several units are made to suit this application with modern and antique panels. The type of unit most suitable for this kind of installation is a tuner with all controls built-in to provide a control center—otherwise several holes in the wall may be necessary. This may make a very convenient and spacesaving operation, but it has disadvantages.

Courtesy NuTone, Inc.

Fig. 8-25. Example of wall-mounted stereo components.

The equipment requires a good-sized aperture in the wall, possibly with equipment protruding through the other side and internal wiring. The equipment is generally required to be within 4 feet of the phono pickup. These problems may be easily worked out when constructing new houses, but in old houses they may present many difficulties.

A variety of equipment for wall mounting is available. One example is shown in Fig. 8-25. The advantage is that valuable room space is conserved because a substantial portion of the equipment is located in the wall. The record-player unit swings outward during use and remains closed into the wall like a door when not in use. Stereo speakers are mounted on another wall.

The equipment may in time become obsolete, and if it is removed, the new equipment may not fit in the same place. Holes in the wall are inconvenient to patch, and new plaster causes decorating problems. With these cautions in mind, it is suggested that one get the best equipment available and be sure the installation is the one that will be desired permanently before proceeding.

Closet Arrangements—Only shelves and enough room to move around in are required in a closet. If an available closet is located conveniently near the principal listening area, one may save considerable money in furnishing and matching, and the equipment will be protected from mechanical disturbance, prying fingers, and other problems. Certain types of home-built equipment or less expensive kits that "don't go" in the living area may be kept out of sight here. The wiring will be short and convenient, and all units will be located in one spot for convenience of operation and control. While speakers may be mounted on the same closet doors, this partly defeats the purpose of unit isolation, and the speaker system thus located does not compete with a good folded corner horn arrangement.

Combinations of all of the foregoing are permissible for any arrangement desired, except that one should keep in mind that the phono lead to the preamplifier should not be over 4 feet long.

SYSTEMS INSTALLATION

After the components of a sound system have been selected, and the positions of all the components and speakers have been determined, the actual job of making the installation begins. The work of installation consists mainly of mounting the equipment, connecting the control center, the amplifier, and the record player to the source of power, and connecting the speakers and program-source equipment to the amplifying equipment through suitable lines and cables. There are a few problems peculiar to each hi-fi installation which must be solved individually at the time the installation is being made.

Installing Program-Source Equipment

When leads from record players, tuners, tape equipment, and microphones are installed, care must be taken to avoid hum pickup, cross talk, and losses in frequency response and level.

High-impedance circuits are very susceptible to hum pickup. Correct grounding of the turntable and pickup arm is exceedingly important in preventing pickup of hum. Connections carelessly made can negate all other precautions in design of equipment. The turntable frame and pickup should be connected by a flexible wire from the turntable frame to the pickup-arm base to the shield of the lead to the preamplifier. Be careful to avoid ground loops between all equipment. Interunit cables provide a complete ground system. Alternate ground wires create ground loops which will usually increase the hum level.

Keep the magnetic phono pickup or microphones more than 2 feet away from power lines and transformers to prevent induction of hum.

When long lines are used with high-impedance sources, with the exception of the crystal type, loss of high frequencies results. When low-impedance microphone circuits are used, there is very little frequency discrimination or loss of level in cabling. These facts should be kept in mind when one is designing any system in which the input circuits must be run a considerable distance.

The cables used with high-impedance equipment usually consist of a single conductor with an overall shield and rubber covering. The shield acts as one of the conductors in the microphone circuit. The shield is grounded and is connected to the microphone case and microphone stand.

Either two- or three-wire shielded cable is often used. When two-wire shielded cable is used, the shield is connected to ground at the amplifier, and to the chassis of the program-source equipment at its other end. The shield does not act as one of the microphone-circuit conductors, and the likelihood of hum pickup with two-wire cable is less than with single-wire shielded cable.

Where microphone or other low-level circuits are carried for long distances in flexible cables, three-wire shielded cable is most effective. Two of the wires serve in the input cable circuit; the third is used as a ground lead. The shield is connected to ground at the amplifier end of the cable only. The source-equipment case is grounded through the third conductor.

When microphone or other input cables are installed permanently in floors or walls, lead-covered twisted or parallel pair may be used. The capacitance of lead-covered cable is quite high; therefore, it can only be installed in short lengths when high-impedance pickups are used. The recommended overall cable length for low-level phono pickups is 4 feet or less. The capacitance of the long cable also impairs the high-frequency transmission. Since crystal pickups are not affected by capacitance, longer lengths of lead-covered cable may be used with them.

Permanently installed input cables can be plug-in connected to the control center and switched so that only those cables in use at any time are fed to the amplifier.

Control-Center and Amplifier Installation

When installing the controls and amplifier, be sure that they will receive proper ventilation; otherwise, components may be damaged by excessive heat.

When an amplifier must be located where vibration from a speaker or other source may cause microphonic noises, it should be mounted on shock-absorbing mountings or rubber pads. Such mountings are available in various sizes and thicknesses, based on the number of pounds the mounting will support while operating normally. To determine the size of the mountings required for an amplifier, the corners of the amplifier should be weighted separately. This is necessary because many of the heavy components are usually located near one end of the amplifier chassis with the result that the weights at the different corners are not the same. When the approximate weights have been determined, mountings of the proper sizes can be chosen. When the amplifier is mounted in this fashion, it should be grounded through a length of heavy braid.

The amplifier should be suitably fused. If the amplifier is not equipped with fuses, a fused power receptacle should be provided. To protect the amplifier components, a 500-mA fuse can be connected between the center tap of the high-voltage winding of the power transformer and ground, as shown in Fig. 8-26. The entire hi-fi equipment is best powered from a single line from the electric-power distribution board, with a 5-ampere fuse in the circuit.

Speaker Installation

When the speakers for a system are installed, the problems of impedance matching and power distribution must be solved. If the speakers are not properly matched, considerable loss in output will result. When a system uses more than one speaker, it is often necessary to have each speaker in the system radiate a different amount of power. Output impedances are usually 4, 8, 16, and 32 ohms to match directly a speaker or combination of speakers. Sometimes 500- or 600-ohm outputs are provided to match long lines to speakers at distances of several hundred feet or more.

Generally, hi-fi speakers are rated at 8 or 16 ohms. Speaker-matching networks and crossovers are generally rated at 16 to 32 ohms.

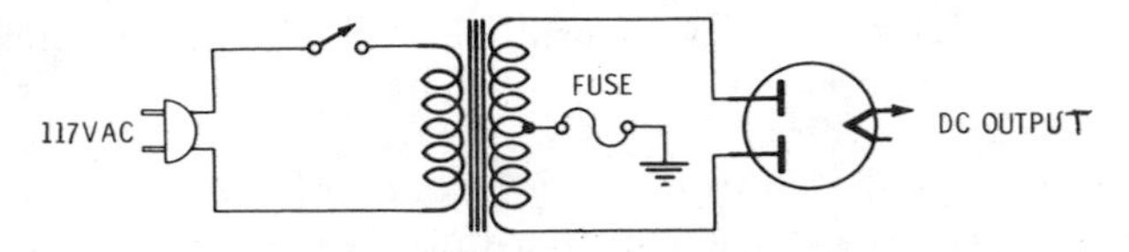

Fig. 8-26. Method of fusing a power amplifier.

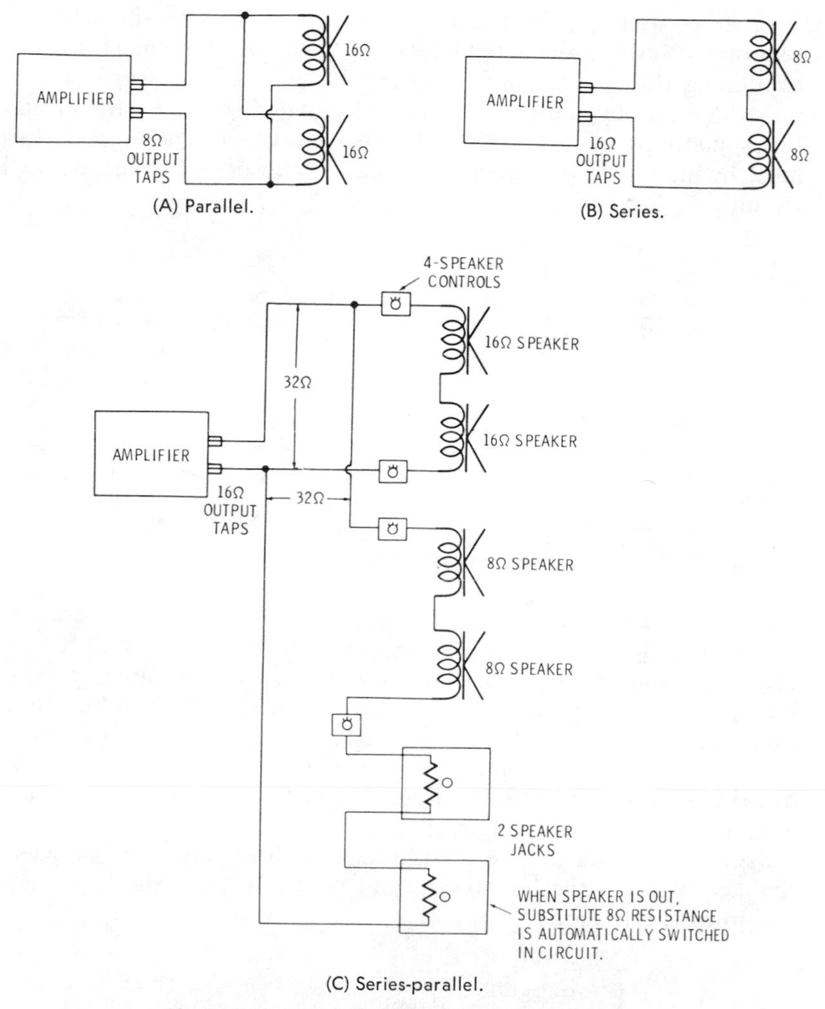

(A) Parallel.

(B) Series.

(C) Series-parallel.

Fig. 8-27. Typical multiple-speaker connection circuits.

It is important to determine the specified impedance of the speaker or speaker system (usually marked on the speaker, the speaker box, or the speaker literature) and to connect the terminals of the speaker or system to the matching terminals on the output of the amplifier.

Two 8-ohm speakers may be connected in series to match a 16-ohm output. Two 16-ohm speakers may be connected in parallel to match an 8-ohm output. Any combination of arrangements of this nature may be used. When connecting two or more speakers, check speaker phasing as described in following paragraphs. Typical multiple-speaker connections are shown in Fig. 8-27.

In stereo systems, the amplifiers usually come with speaker connections clearly indicated. However, sometimes connections for combining the two channels into one for monophonic performance must be made by the owner. Fig. 8-28 gives both the parallel and series methods of connecting the two channels. The impedances must be matched by following the basic rules of series and parallel circuits.

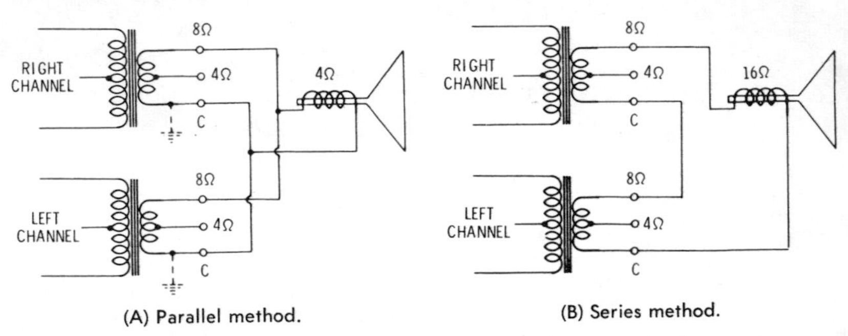

(A) Parallel method.

(B) Series method.

Fig. 8-28. Connection of stereo outputs for single-speaker monophonic operation.

Another situation in which the connections of speaker leads are involved is the addition of a third, or "middle," speaker to a two-channel stereo system. One simple way to do this is by connecting the middle speaker to taps on the two transformers, as indicated in Fig. 8-29. This method is not ideal, however, since there is likely to be considerable interaction between the left and right channels, thereby diluting the stereo effect. Other, more efficient, methods were discussed earlier in this chapter.

Speaker Lines—The wire used in speaker lines must be of sufficient size according to the impedance and power to keep the line losses within tolerable limits.

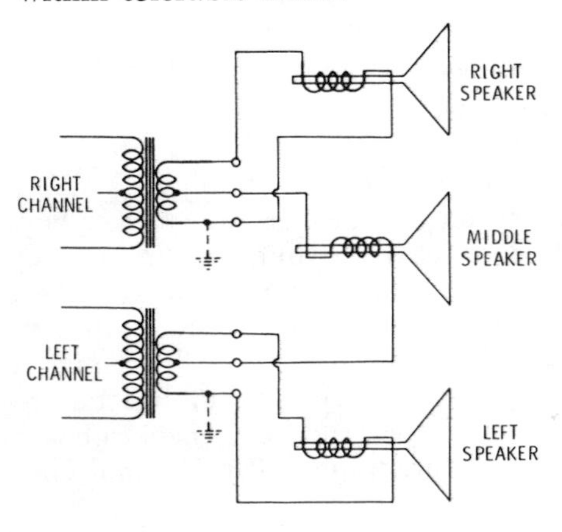

Fig. 8-29. One method of connecting a middle speaker in a stereo system.

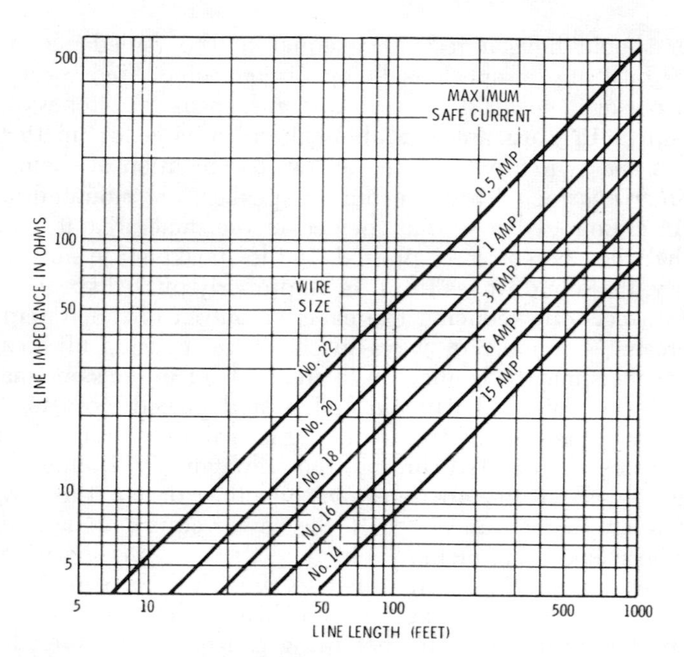

Fig. 8-30. Wire sizes for speaker lines.

When the speaker-line impedance is under 60 ohms, only short lines, not to exceed 100 feet, should be used. When a line impedance of 60 ohms or more is used, lines may be run for a considerable distance, providing wire of sufficient size is used. The curves of Fig. 8-30 give recommended wire sizes for lines of various impedances and lengths.

Speaker Switching—In many installations, it is necessary to have facilities for controlling the volume of auxiliary speakers and for switching speakers in and out of the system. When more than one speaker is used and one or more speakers must be separately controlled, a variable T-pad may be placed in the circuit to control the speaker volume and to provide a substitute resistance in order to maintain constant load impedance to the system when the speaker is partially or completely turned off. Fig. 8-31 shows a circuit of an available unit which can be used to accomplish this. The substitute

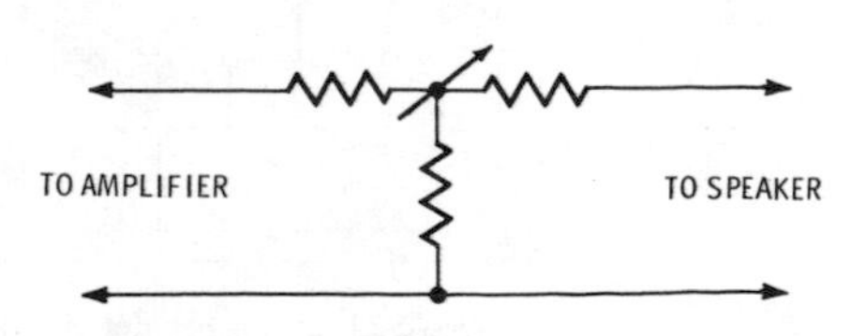

Fig. 8-31. A variable T-pad speaker control.

resistor should have a resistance equal to the impedance of the speaker or group of speakers being disconnected from the circuit. To cut out or in speakers in various rooms, speaker switches able to carry up to 10 watts are available. These provide for up to three speakers and switch on any one or any combination of them.

Speaker Phasing—When a group of speakers are mounted at one point or close together so that they cover the same area, it is necessary that they be correctly phased. If the speakers are not phased properly, the sound waves from the improperly phased speakers will tend to cancel out, reducing the effective output of the system.

There are a number of ways to check the phasing of speakers. Two speakers may be connected to an amplifier in the same manner they will be connected when installed, and placed close together, facing each other. A low-frequency signal from an audio generator or a record should be fed through the amplifier. By listening to the speakers, it is possible to determine whether or not the low frequencies are being cancelled. If the low frequencies are absent when the speakers are facing each other, then the phasing is correct if the speakers are to be mounted so that they face in the same direction. If the low frequencies can be heard when the test is made, then the low frequencies are not being cancelled, and the phasing is correct if the speakers are to be mounted facing away from each other. If the phasing is incorrect in either of the two cases just described, all that is necessary is that the connections to one of the voice coils be reversed.

The phasing of speakers may also be checked by using the circuit shown in Fig. 8-32. The apparatus consists of a pair of headphones connected to the input of an amplifier through two long cords and a double-pole, double-throw switch. The amplifier is equipped with an output indicator. The sound system whose speakers are to be checked for correct phasing should be turned on and a constant tone fed into its input. The double-pole, double-throw switch should

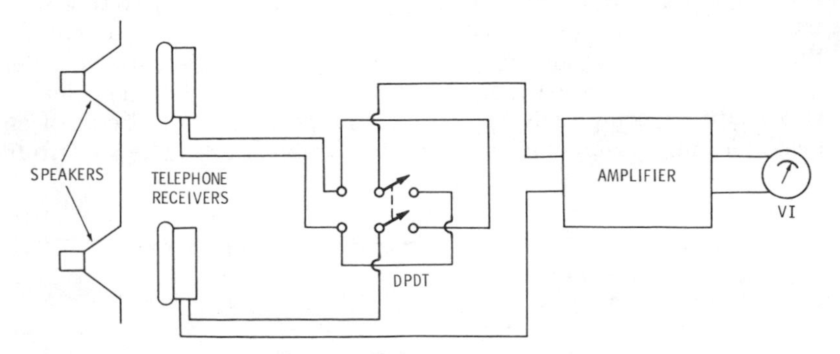

Fig. 8-32. Setup for determining speaker phasing.

be marked to show "in phase" and "out of phase." This may be done by holding both phones in front of one speaker and noting the position of the switch which gives the greatest indication on the output meter. This position should be marked "in phase," and the other position should be marked "out of phase." The equipment may now be used to check the phasing of two speakers. One phone is held in front of each speaker. If the greatest output is indicated when the switch is in the "in phase" position, then the speakers are phased properly.

In stereo systems, phase checking and correcting are made relatively simple. In most stereo systems, a phase-reversing control is included, connected either in the speaker output circuits or earlier in the amplifier chain. However, this control does not provide for phasing of the center speaker. If a single tone or other signal is applied to the inputs of both channels at the same time, the relative outputs from the right and left speakers will indicate the phasing. If the phasing is wrong, the two signals will differ by 180 degrees and tend to cancel each other. If the phases are correct, the signals reinforce each other. Thus, the correct phasing should produce much greater volume.

It should be remembered that there is a difference between speaker phasing and system phasing. The final output phase depends on the windings of the output transformers, connections between preamplifiers and amplifiers, connections of cartridges and input devices, and other things besides speaker connections. Therefore, checking the relative phasing of the speakers themselves is not always enough.

FEEDBACK

Careful attention must be given to the factors which create feedback; otherwise, considerable trouble will be experienced. There are two forms of feedback. They are *acoustic feedback,* which is caused by coupling between the speakers and the program-source units, and *electrical feedback,* due to capacitive coupling between the output circuits and the input circuits.

Electrical feedback can be avoided by keeping the input and output wiring and components well separated and shielded.

Mechanical or acoustic feedback is often difficult to eliminate and, in severe cases, may limit the usable output of a system. Mechanical feedback effects and microphonics may be caused by direct or resonant coupling between the speaker system and the program-source equipment, or through mechanical pickup in the first-stage elements of the preamplifier or the mechanical elements of phono and microphone pickups. This may be corrected by mounting the units concerned on rubber, as has been described, or by moving the units concerned out of the mechanical-contact area or resonant-condition area. Moving the speaker should have the same effect.

Feedback manifests itself as a continuous tone or tones and spurious noises radiated by the speakers of a sound system. Feedback can be avoided by careful attention to the location of speakers and microphonic elements of the system.

There are a few general principles of microphonics and speaker location which, if followed, give reasonable assurance that little or no trouble will be experienced with feedback other than feedback due to direct microphone pickup.

A microphonic element may be described as any element located in the hot circuits of the front end of the system. The most typical examples are tubes which are highly sensitive to any vibrations of their elements. A vibrating element in a tube in the input stage may induce the same effect as a microphone in the circuit by variation of plate-to-cathode resistance, thereby introducing undesirable interference.

A microphonic element should never be located in front of a speaker. If the speaker or speakers used in a system are located in front of the microphonic elements, the possibilities of feedback will be greatly reduced.

Speakers should not be mounted close to reflecting surfaces. When there is a flat surface close to and in front of the speakers, sound will usually be reflected by the surface.

If the system is installed in a location and is intended for use in a room filled with people, feedback may occur when the room is empty, but may not occur when the room is full of people.

Cardioid and other directional microphones are often very effective in eliminating feedback. The side of the microphone which is not sensitive to sound should be pointed in the direction from which the reflected sound is coming.

Attenuation of high-frequency response when equipment is used in a room with hard walls is desirable. Under these conditions, feedback usually occurs due to reflected sound, and, since the high frequencies reflect more efficiently than do the low frequencies, attenuation of the high frequencies usually reduces feedback effects. The high-frequency tone control is usually useful when background music is being reproduced as at dinner, etc. Under these conditions, the high frequencies should be attenuated, since they tend to make conversation difficult.

Control of low-frequency response when speech is being reproduced is desirable. Under these conditions, attenuation of the lower frequencies helps remove the "tubbiness" which occurs at times in certain types of installations.

Index

A

Acoustic
  energy, 266
  feedback, 391
  impedance, 274-276
  inductance, 273
  labyrinth, 316-317
  phase inverter, 316
  watt, 346
Acoustical suspension, 322
Adapters
  four-channel, 233-243
  stereo, two-channel, 232-233
Afc; *see* automatic frequency control
Agc; *see* automatic gain control
A-m
  section, 131-132
  superheterodyne receivers, 82-97
  tuners, 81-82
Amplifier(s), 243-246
  bridge, 194
  circuits, tetrode, 182-194
  class-
    A, 179-180
    AB, 179, 180
    B, 179, 180
    C, 179
  i-f, 85, 109-111
  installation, 386
  location of, 380-384
  power, 166-178
    circuits for, 178-206
    stereo, 196-199
    tube versus transistor, 176-178
  push-pull, 180
  rf, 87-90, 101-103
  tape-recorder, 136-137
  voltage, 202
Amplitude
  distortion, 12-14
  modulation, 81
Audio
  power, 345-349
  preamplifier section, 132
  watt, 346
Automatic
  frequency control, 99-100
  gain control, 93
  volume control, 93-97
    forward, 95-96
    reverse, 95-96
Avc; *see* automatic volume control
Axes, crystal, 60

B

Back loading, 319
Baffles, 305-308
  infinite, 307-308
Balance
  channel, 67-68
  control, 230-231
  instruments, 257-258
Bandpass, hi-fi, 85-86
Bar, 149
Bass-reflex enclosures, 312-316
Bender element, 61
Bias
  control, 50
  recording, 133-134
Binaural, 17
  system, 24-25
Biortho system, 196-197
Bipolar transistor, 79, 103, 104, 106-107
Blend control, 232
Boffle, 311

Brakes, 136
Broadcasting, stereo, 51-53

C

Cabinets, wall, 381-382
Cables, input, 385
Capacitor
  drivers, 265-266
  microphones, 156
Carbon microphones, 151-152
Cardioid pattern, 156
Cartridge(s), 47-48
  dynamic, 63-64
  paralleling, 232
  stereo, 64-68
Cassette, 48-50
Cathodyne phase inverter, 200, 201
Ceramic pickup(s), 42-43, 59-62
Channel(s), 23, 80
Chromium dioxide tape, 147
Closet arrangements, 384
Coaxial arrangement, 284-285
Coboloy, 50
Combination units, 247-257
Compacts, stereo, 247
Compatibility, 30-31, 53
Compensation, 207-208
Complementary-symmetry circuit, 189
  quasi, 192-194
Compliance, 40, 66-67, 273
Condenser microphone, 156
Cone
  diameter, 268-269
  -type radiators, 266-272
Control
  balance, 230-231
  blend, 232
  -center
    installation, 386
    preamplifier circuit, 215-221
  equipment, location of, 380-384
  functions, 230-232
  gain, master, 231-232
  phono, 212-215
Conversion, frequency, 82-84
Core losses, 169
Crolyn, 50
Crossover
  distortion, 188
  electronic, 303-304
  networks, electrical, 294-304
Crystal
  drivers, 265
  microphones, 152-153
Crystal—cont
  pickups, 59-62
  piezoelectric, 59

D

Damping, 275-276
  variable, 205-206
dB; *see* decibel
dBm, 352-353
Decibel, 9, 351-352
De-emphasis, 22, 116
Delay stereophony, 21
Derived four-channel system, 360-364
Detector(s)
  fm, 114-115
  operation of, 91-93
  ratio, 114
  second, 90-93
  section, fm, 130
Diameter, cone, 268-269
Diamond needles, 58-59
Difference signal, 120-121
Diode, 79
  tuning, 107
Directivity
  microphone, 150-151
  speaker units, of, 280-283
Disc(s)
  four-channel, 43
  recording, 35-45
  systems, stereo, special factors in, 40-43
Discrete four-channel tape system, 359-360
Discriminators, 112-114
  Foster-Seeley, 130
Distortion, 351
  amplitude, 12-14
  frequency, 11-12
  harmonic, 14
  intermodulation, 14
  listener tests regarding, 342
  nonlinear, 12
  phase, 15
  sound, 10-15
  spatial, 15
  trackability, 43
  transient, 15
Divider networks, electrical, 294-304
Dolby noise-reduction system, 116-117, 144-148
  A, 145
  B, 145
Driver(s)
  -amplifier stages, 199-200

Driver(s)—cont
capacitor, 265-266
crystal, 265
dynamic, moving-coil, 260-264
electrodynamic, 260-263
moving-coil dynamic, 260-264
permanent-magnet, 263-264
speaker, 260-266
Driving methods, turntable, 73-76
Dynamic
cartridges, 63-64
microphones, 153-154
pickup, 42

E

Eddy currents, 169
Editing, 23
Efficiency, 267
speaker, 346
Electrical
axis, 60
feedback, 391
Electrodynamic driver, 260-263
Emphasis, 22
Enclosures
bass reflex, 312-316
horn-type, 318-321
infinite-baffle, 310, 311-312
simple, 308-312
Encoded matrix four-channel system, 364-370
Equalization
input, 232
record, 163-164
Equalizers, speaker, electrical, 329-333
Erase, 134-135
Expander plates, 60

F

Fantasia, 24
Feedback, 391-392
dc, 189
eliminating, 392
level control, 222-223
negative, 190, 202-205
positive, 190, 205
FET; *see* field-effect transistor
Fidelity, 349-350
Field-effect transistor, 79, 103
Fm
detectors, 114-115
receivers, 100-115
tuners, 100
Four-channel
adapters, 233-243

Four-channel—cont
record discs, 43
sound equipment,
derived, 234
matrixed, 234-243
speaker placement, 378-380
stereo, 28-30
systems, 358-370
derived, 360-364
encoded matrix, 364-370
tape system, discrete, 359-360
Frequency(ies)
audible, 8
conversion, 82-84
distortion, 11-12
division, low-level, 302-304
modulation, 100
requirements, 350-351
response, 65
microphone, 149-150
Front end, 103
Furniture pieces, units in separate, 382-383
Fusing, 191

G

Gain, 351-354
control, master, 231-232
Gate, 80

H

Hangover, 15
Harmonic distortion, 14
Headphones, stereo, 337
Heads
in-line, 51
magnetic-recording, 135-136
staggered, 51
Hearing, human, 8
Helmholtz resonator equation, 314
Hi-fi
arrangements, elaborate, 370
speaker systems, compact, 322-329
High fidelity, 7-8
distortion in, 10-15
stereo versus, 54
system, building up, 354-358
"Hill-and-dale" recording, 36
Horn(s), 276-280
drivers, 276-280
exponential, 277
multicellular, 282-283
-type enclosures, 318-321
Hum pickup, 45
Hysteresis, 133, 169

I

I-f
amplifier(s), 85, 109-111
integrated circuits in, 86-87
section, fm, 130
Image
frequency, 86
rejection, 85-86
Impedance
bridge, transformer tests with, 173-174
output, 176
feedback effect on, 205
speaker, 272-274
Inductance
acoustic, 273
leakage, 169
checking, 174-175
primary, 170
Inertance, 272-273
Infinite baffle, 307-308, 310
enclosure, 310, 311-312
Injection, oscillator signal, 84-85
Insert, 22
Installation
amplifier, 386
control-center, 386
program-source equipment, 384-385
speaker, 386-391
systems, 384-391
Integrated circuit(s), 80
application of, 79-80
avc, 96-97
detector, 90-91
i-f, 109-111
amplifier application of, 86-87
preamplifier, 221-225
stereo, 223-225
Intensity-difference system, 34
Intermediate frequency, 83
Intermodulation distortion, 14

J

JFET; *see* junction field-effect transistors
Junction field-effect transistors, 128-130

L

Labyrinth, acoustic, 316-317
Layout, systems, 370-384
Leakage inductances, 169
checking, 174-175
Level control, feedback, 222-223
Lights, stereo indicator, 124-125
Limiters, 111-112
Limiting, 22
Listening tests, speaker, 343-344
L—R signal, demodulating, 122
Load, cartridge, 65
Loudness controls, 165-166

M

Magnetic
-armature speaker, 264-265
pickups, 62-63
Masking, 116
Matching speaker systems, 375
Matrix four-channel
sound equipment, 234-243
system, encoded, 364-370
Mechanical
axis, 60
feedback, 391
Metal-oxide-semiconductor field-effect transistor, 79-80, 103, 104, 106, 107, 131
Meters, signal-strength, 98-99
Microphone(s), 55, 148-159
capacitor, 156
carbon, 151-152
combination, 156-157
condenser, 156
crystal, 152-153
directivity, 150-151
dynamic, 153-154
frequency response, 149-150
output
impedance, 150
level, 149
placement, 25-28
ribbon, 154
selection, 157-159
stereo, 35
velocity, 154-156
Microphonic element, 392
Middle speaker, 377-378
Mixer-oscillator, 104-108
Modulation
envelope, 81
frequency, 100
Monaural, 17
Monophonic, 17
sound, 18
MOSFET; *see* metal-oxide-semiconductor field-effect transistor
Motors, tape-drive, 135

Mouth, 276
Moving-coil pickup, 42
Multicellular horn, 282-283
Multichannel systems, 23-24
Multipath reception, 258
Multiple-room installations, 348
Multiplex
  adapters, stereo, 118-119
  modulation signal, stereo, 121-122
  operation, stereo, 119-124
  stereo, 52-53

N

Needle(s)
  elliptical, 58
  force, 65
  materials for, 58-59
  reproducing, 56-59
  -tip size, 65
Noise
  converter, 87
  figure, 126-128
    tuner, 126-128
  -reduction system(s)
    Dolby, 144-148
    other, 148
Nonlinear distortion, 12

O

Octave, 296
Operational transconductance amplifier, 96-97
Optical axis, 60
Oscillator signal injection, 84-85
Osmium needles, 58
OTA; *see* operational transconductance amplifier
Output
  circuits, 179-182
  impedance, 176
    microphone, 150
  level, microphone, 149
  transformers
    choosing, 171-175
    performance factors in, 170-171
    testing, 171-175
  voltage, cartridge, 65
Overdub, 22

P

Pentafilar winding, 186
Performance goals, 15-16
Permanent-magnet speaker, 263-264
Phantom circuit, 196
Phase
  distortion, 15
  inversion, 200-202
  inverter(s)
    acoustic, 316
    cathodyne, 200, 201
    stages, 199-200
    transistor, split-load, 202
    tube-type, 200-202
  -reversing switch, 232
Phasing, speaker, 375, 390-391
  -system, 343
Phono
  control, 212-215
  preamp, 212-215
Pickup(s)
  arm, 69-72
  capacitance, 69
  cartridges, 59-69
  ceramic, 42-43, 59-62
  crystal, 59-62
  dynamic, 42
  magnetic, 62-63
  moving-coil, 42
Piezoelectric crystal, 59
Pinch effect, 40-41
  distortion, 58
Placement, speaker, 375-377
  and microphone, 25-28
Plates, crystal, 60
Playback, 137-144
Polyvinyl chloride, 322
Port, 312-316
Power
  amplifier(s), 166-178
    circuits, 178-206
    tube versus transistor, 176-178
  audio, 345-349
  required, determining, 347-349
  supply, 132
Preamp, phono, 212-215
Preamplifier(s), 206-208, 225
  circuit(s), 209-225
    control-center, 215-221
  integrated circuits, 221-225
Pre-emphasis, 115-116
Presence, 8
Pressure, tip, 41
Primary inductance, 170
Program source(s)
  equipment, installing, 384-385
  location of, 380-384
Psychoacoustic response, 379

Q

Quasi-
  complimentary-symmetry circuit, 192-194
  stereo switch, 198, 232

R

Radiators
  cone-type, 266-272
  single-cone, 284
Radio-
  controlled station selector, 257
  frequency section, fm, 128-130
Radius, tip, 41
Ratio detector, 114
Receivers
  a-m superheterodyne, 82-97
  fm, 100-115
Reception, stereo, 51-53
Record
  changers, turntables versus, 76-78
  discs, four-channel, 43
  Industry Association of America, 207
  players, 55, 56-78
Recorders, mechanical features of, 136
Recording, 137
  disc, 35-45
  stereo, 33-35
  tape, 46-51
  techniques, 20-23
Reproducing needles, 56-59
Requirements, disc-recording, 35-37
Resistances, winding, 168
Resistors, low-noise, 216
Resonance, 274-276
  rim, 284
Resonant frequency, 275
Response, frequency limits of, 350-351
Reverberation, 22
Rf amplifiers, 87-90, 101-103
RIAA; *see* Record Industry Association of America
Ribbon microphone, 154
Rochelle salt, 59
Rumble, 45, 72

S

Sapphire needles, 58
SCA signal; *see* Subsidiary Communications Authorization signal
Second detector, 90-93
Semiconductors, application of, 79-80
Sensitivity, tuner, 126-128
Separation, channel, 65, 365
Shear plates, 60
Sidebands, 81-82
Signal-strength meters, 98-99
Simplex system, 196-197
Single-
  channel system, speaker placement in, 372-374
  ended push-pull stage, 188
Sound, 8-9
  chamber, 279
  monophonic, 18
  stereophonic, 18-20
Spatial distortion, 15
Speaker(s)
  acoustic suspension, 330
  center, 304
  coaxial type, 286-288
  construction features of, 285-294
  drivers, 260-266
  dual, 283-285
    -voice-coil, 304, 378
  efficiency of, 346
  equalizers, electrical, 329-333
  extended-range, single-cone type, 285-286
  impedance, 272-274
  installation, 386-391
    multiple-room, 348
  lines, 388-389
  listening tests, 343-344
  magnetic armature, 264-265
  middle, 377-378
  multiple, 283-285
    combinations of, 293-294
    connections of, 387
  phasing, 375, 390-391
  placement, 25-28, 375-377
    four-channel, 378-380
    single-channel system, 372-374
    stereo systems, 374-378
  room effects on, 331
  single-cone, extended range type, 285-286
  switching, 389-390
  system(s)
    choosing, 340-343
    dual, 283-285
    hi-fi, compact, 322-329
    matching, 375
    multiple, 283-285
    parts of, 259-260
    phasing in, 343
    stereo, 333-339

Speaker(s)—cont
  units, directivity of, 280-283
Speeds, turntable, 72
Spider, 271
Squawker, 293
Stereo
  adapter(s)
    circuits, 122-124
    two-channel, 232-233
  amplification, 225-258
    physical arrangements for, 226-230
  amplifier units, 246-247
  broadcasting and reception, 51-53
    two-station, 51-52
  cartridges, 64-68
  center speaker for, 304
  control centers, transistor, 246-247
  converter, 118
  disc systems, special factors in, 40-43
  four-channel, 28-30
  headphones, 337
  high fidelity versus, 54
  history of, 18-20
  indicator lights, 124-125
  integrated circuits, 223-225
  multiple-channel, 25-28
  multiplex, 52-53
    adapters, 118-119
    modulation signal, 121-122
    operation, 119-124
    section, 130-131
  power-amplifier circuits, 196-199
  recording, 33-35
  speaker systems, 333-339
  switching, automatic, 125
  systems
    components of, 32-33
    speaker placement in, 374-378
  translator, 118
  tuners, 117
Stereophonic, 17
  sound, 18-20
  systems, 20-30
Stereophony, delay, 21
Subcarrier, 53
Subsidiary Communications Authorization signal, 124
Sum signal, 120-121
Superheterodyne receivers, a-m, 82-97
Suspensions, 271
Systems
  installation, 384-391
  layout, 370-384
  requirements for, 345

T

T-pad, 389
Tab, record-defeat, 49
Take, 22
Tape, 55
  decks, 136
  formulations, 50
  modes of operation, characteristics of, 138
  recorders, 132-148
  recording, 46-51
  systems, four-channel, discrete, 359-360
  transports, 138
Television, 55, 160
Temperature, crystal affected by, 61-62
Tetrode amplifier circuits, 182-194
Throat, 276
Time-intensity pickup, 33
Tip
  pressure, 41
  radius, 41
  size, needle, 59, 65
Titanium, 50
Tone arm, 69-72
Trackability distortion, 43
Tracking
  error, 69, 70
  force, 72
  problems, 43-44
Tracks, tape, 46
Transducers, 10
Transduction, 10
Transformer(s)
  impedance bridge, testing with, 173-174
  output, 167-168
    choosing, 171-175
    performance factors in, 170-171
    testing, 171-175
  weight of, 171-172
Transient distortion, 15
Transistor
  bipolar, 79, 103, 104, 106-107
  field-effect, 79
    metal-oxide-semiconductor, 79-80
  stereo
    amplifier units, 246-247
    control centers, 246-247
  tuned amplifiers, 86
Triaxial speaker, 293
Tubes, electron, application of, 79-80
Tuned amplifiers, transistor, 86

Tuner(s), 55, 80-132
  a-m, 81-82
  compensation, 115-117
  fm, 100
  high-performance, 128-132
  noise figure, 126-128
  sensitivity, 126-128
  stereo, 117
Tuning
  diodes, 107
  instruments, 257-258
Turntables, 44-45, 72-76
  driving methods for, 73-76
  record changers versus, 76-78
  speeds of, 72
Tweeter, separate, 285,288-293
Twister element, 61
Two-channel stereo adapters, 232-233

U

Ultralinear operation, 182-183
Unity coupling, 183-188

V

Velocity microphones, 154-156
Voltage amplifiers, 202
Volume unit, 353
VU; *see* volume unit

W

Wabble, 72
Wall
  cabinets, 381-382
  units, 383-384
Watt
  acoustic, 346
  audio, 346
Westrex system, 37-40
Winding resistances, 168
Woofer
  separate, 285, 288-293
  -tweeter-squawker, 293
Wow, 73